Science and Building

Science and Building

STRUCTURAL AND

ENVIRONMENTAL DESIGN

IN THE NINETEENTH

AND TWENTIETH CENTURIES

HENRY J. COWAN

Professor of Architectural Science
University of Sydney

A WILEY-INTERSCIENCE PUBLICATION

JOHN WILEY & SONS, New York • London • Sydney • Toronto

For

Judith

Because she likes history

Copyright © 1978 by John Wiley & Sons, Inc.

All rights reserved. Published simultaneously in Canada.

Library of Congress Cataloging in Publication Data

Cowan, Henry J
  Science and building.

  "A Wiley-Interscience publication."
  Includes bibliographical references and index.
    1. Building—History. 2. Structural engineering—
History. I. Title.

TH15.C63   721'.09'034   77-7297
ISBN 0-471-02738-3

Printed in the United States of America

10 9 8 7 6 5 4 3 2 1

PREFACE

This is the second of two volumes on the history of building science from ancient times to the present day; it deals with the nineteenth and twentieth centuries, when science and technology transformed the traditional craft-based practice of architecture and building. The preceding volume, entitled *The Master Builders*, relates the story of the earlier centuries. The two books are, however, separate entities, and this one can be read without having read the other.

I am indebted to Dr. Valerie Havyatt for a great deal of bibliographical research; to Professor J. M. Bennett, Mr. T. Callaghan, Mr. B. Cassidy, Mr. L. A. Challis, Professor R. W. Clough, Professor J. S. Gero, Dr. Valerie Havyatt, Mr. T. F. Heath, Mr. E. B. Huddleston, Mr. J. J. Keough, Mr. R. J. Mallett, Dr. M. Newman, Professor R. K. Macpherson, Mr. W. H. Pilkington, Mr. D. Saunders, Professor P. R. Smith, Mr. G. C. Thompson, Mr. A. Wargon, Mr. B. Webb, and Mr. J. Whittemore for helpful comments and corrections; to Mr. John Dixon for all the photographic work; and to Mrs. Rita Arthurson, Mrs. Hilda Mioche, and Miss Ann Novitski for typing the script.

HENRY J. COWAN

Sydney, Australia
June 1977

CONTENTS

CHAPTER ONE

Introduction

*The one duty we owe to history
is to rewrite it.*

OSCAR WILDE

This chapter surveys briefly the history of architecture and the history of science. It then considers the effect on the design of buildings of the social changes that followed the end of the Napoleonic wars.

New building types, such as factories, offices, and mass housing, required in the newly industrialized countries, changed the character of construction and led to the formally organized professions of architecture and engineering.

1.1 THE HISTORY OF ARCHITECTURE AND THE HISTORY OF SCIENCE

Science is taught, traditionally, as a logical development from experimental data, without regard to the historical sequence of the solutions or the difficulties encountered in deriving them. Architecture, on the other hand, is taught largely by studying the work of great masters. The sense of history is never entirely absent from this treatment even when the work is recent.

We regard it as relevant to our discussion of a certain building that it was designed by Filippo Brunelleschi, Christopher Wren, or Kenzo Tange. It is, at least partly, the result of the architect's personality and of the time in which it was conceived. We see a significant difference between the architecture or the art of the fifteenth and sixteenth centuries in Italy, and with only a small margin of error an expert can generally identify the time at which a building was designed or a painting executed.

It is also possible, to some extent, to time the results of scientific research. There are data that could not have been obtained before certain techniques become available. We do not, however, normally consider a theory in relation to the personality of its discoverer or the time in which it was derived; for example, the lever principle (which describes the relation between parallel forces) was discovered by Archimedes in Sicily in the third century B.C. The parallelogram of forces (which describes the relation between converging forces) was discovered by Simon Stevin in Holland in the sixteenth century. These two discoveries, nineteen centuries apart, are usually taught together as two complementary parts of a single problem.

Thus science is progressive. Although results are occasionally proved wrong and techniques are superseded by better methods, the body of scientific knowledge is constantly growing and ancient results may be equally useful today. The ancient Egyptians discovered and cataloged new stars. We have continued to do the same thing, with better instruments, but it does not really matter whether a certain star was discovered in the thirteenth or the twentieth century.

By contrast, the erection of a church in the Gothic style would in the twentieth be an exercise in revivalism, whether the architect used traditional methods or prestressed concrete. The same applies to painting and music.

This difference reflects a wider distinction in the outlook of the scientist and the creative artist. Attempts to bridge the gap have not been lacking. Many scientists have shown a great interest in the history of their subjects, and the discipline imposed on modern architecture by engineered structures and new materials has been explored by several eminent architects, engineers, and art historians.

In spite of these efforts, science, architecture, and general history remain in essentially separate, watertight compartments, and it is interesting to note events that occurred at the same time in the different fields.

Galileo, an early contributor to structural mechanics, was a contemporary of Carlo Maderna, a Baroque architect. Urban VIII was pope, the inquisition had recently been introduced into Rome, and the Venetians had defeated the Turks at Lepanto.

The late eighteenth century, which saw the development of iron structures in England and the basic structural theory in France, was also the age of the Classic Revival. The ironmaster Abraham Darby III, the civil engineer Thomas Telford, and the French engineering theoretician Charles Augustin de Coulomb were contemporaries of the architects Claude Nicolas Ledoux in France and Robert Adam in England. France had the Revolution. England was ruled by George III; it lost the American colonies and built a new empire in India.

In the year 1900 the steel frame was well established, and the Americans were building "skyscrapers"; reinforced concrete was not yet a respectable material (the Institution of Civil Engineers still expressed reservations about its safety in 1913). The theory of statically determinate structures* was well developed and theories had been produced for the design of statically indeterminate structures.* Most major cities had recently acquired a piped water supply. Artificial lighting by oil or gas was common and electric lighting was a novelty. Wallace Clement Sabine had written the first publications on theater acoustics. Norman Shaw was at the height of his architectural career in England, and Frank Lloyd Wright was just becoming known in the United States. Most European countries were monarchies, and most of Asia and Africa had been colonized by the Europeans.

1.2 THE NINETEENTH-CENTURY DICHOTOMY BETWEEN ARCHITECTURE AND ENGINEERING

The present meaning of the words architect and engineer dates only from the late nineteenth century. The Greek word *architekton,* which meant master builder, described a builder of bridges and aqueducts as well as theaters and temples.

The word *ingeniator* encountered in medieval texts denoted a person who devised ingenious contrivances, mainly for military purposes. In 1716 the French army founded the *Corps des Ingénieurs des Ponts et Chaussées* and, in 1747, the *École des Ponts et Chaussées* which became the prototype of modern university engineering schools. Shortly after John Smeaton was the first person to call himself a civil engineer (Ref. 1.1, Section 8.7), by which he meant an engineer not employed by the military.

The distinction was institutionalized by the foundation of the Institution of Civil Engineers in London in 1818 and of the Royal Institute of British Architects in 1834. The difference between the professions lay in the purpose of the construction (roads, bridges, or canals as opposed to monumental buildings) rather than in the methods employed, for both architects and engineers in the early nineteenth century used traditional materials in a traditional way.

Engineers customarily designed the new factory buildings along with the engines. Iron was a natural material to use in a factory since the engineer in any case was using it for the engines. From there it spread gradually to architect-designed build-

* See the glossary for a definition.

ings, and some architects called in engineering consultants to help them with an unfamiliar material. As the design of iron and concrete structures became more mathematical, consulting engineers were more regularly employed, a development that dates only from the later nineteenth century. This cooperation has not always been friendly, and at the turn of the century the gap between the two professions had widened. By 1900 it was regarded as a particular honor for an English architect to be elected to the Royal Academy and for an engineer to be made a Fellow of the Royal Society. The architect became a creative artist who relied for his technical advice on the engineer and the engineer concentrated more and more on the mathematical aspects of engineering.

In the late nineteenth century there was merit in a complete division of labor from both points of view. The steel frame did not need to be, and generally was not, apparent when the building was complete. The architect could therefore design his Classic or Gothic facade and then hand the drawings over to the engineer who would thread through the architect's building a steel skeleton to support its weight.

The physical destruction and the social changes brought about by World War I (1914–1918) created an unprecedented need for new buildings. Because of the rise in wages in relation to salaries and profits, the absolute cost of building had increased, and facades laboriously carved from natural stone became too expensive even for monumental buildings. Many architects, particularly in defeated Germany and Austria, reexamined the traditional values and a new style of architecture emerged. It is debatable whether this owed more to economic necessity or to a new theory of architectural design; however, the architectural revolution was completed by World War II (1939–1945).

1.3 WHY THIS BOOK DEALS MAINLY WITH STRUCTURES

The new architecture emphasized the structure as an integral, to some theorists even a dominating, part of the design. Hence it was not something that could be left to be inserted by the consulting engineer. The architect had to learn about structural systems, and a way had to be found to make this subject intelligible to him in spite of the highly technical nature of the calculations.

This is the most important aspect of the work done by scientists and engineers and this is why it occupies the greater part of this book.

The inclusion of environmental considerations in the basic design is of more recent date. Traditional construction had generally provided satisfactory sound and heat insulation and, in the humid tropics, natural ventilation; the mechanical and electrical services were relatively simple in the 1920s and even in the 1950s air conditioning was uncommon outside America.

This is one reason why the environmental aspects of building science are of more recent date than structural design. Another important consideration is the fact that a building that falls down ceases to exist. A building that is intolerably hot or cold or noisy may survive, and indeed there are many acknowledged architectural masterpieces that are quite unsatisfactory environmentally but continue in use. A building that collap
s
es. . . .

1.4 THE EFFECT OF SOCIAL CHANGES DURING THE NINETEENTH CENTURY ON THE ARCHITECTURAL DESIGN OF BUILDINGS

The history of building technology can be divided into two distinctive parts. The earlier, during which it was mainly empirical (Ref. 1.1), and the later, during which it was based more and more on science.

I have chosen 1815, the date of the Battle of Waterloo, as the beginning of the second period, which is the subject of this book. The causes of industrialization, which changed the character of the cities of Europe, belong to the preceding century, as do the first iron structures and the birth of the mathematical theory of structures (Ref. 1.1), but their effect on the design and construction of buildings before the end of the Napoleonic wars was only slight.

This book, therefore, deals with the relatively short time between 1815 and the present, a century and a half that contributed more to building science than all the years that preceded it.

The Napoleonic wars were followed by an era of social change. The attempt to restore Europe to conditions existing before the French Revolution failed, and disturbances were common in most European countries, particularly in 1830 and 1848. Britain was the only European country not affected by the revolutionary movement of 1848, possibly because in 1832 some measure of parliamentary reform had been introduced. Denmark achieved male suffrage for the lower house of the *rigsdag* in 1849 (not introduced in Britain until the Reform Act of 1884), but in most countries the revolution was succeeded by a period of "law and order." Many who were in danger of persecution or did not like the new order migrated to North and South America and Australia. This was at least one reason for the rapid development of these continents, and in due course they produced a new architecture less concerned with tradition than builders of the Old World had been. Since 1907 the commercial buildings of America have exceeded the cathedrals of Europe in height.

The second half of the nineteenth century and the first half of the twentieth showed fast economic progress. The rise in prosperity was general and the working and middle classes shared in it, mainly because of advanced productivity and the greater use of mechanical power; for example, in the United States, where mechanization was being pressed hardest because of a shortage of labor, the amount of power consumed by industry rose from 1.6 million hp in 1859 to 9.6 million (1190 to 7160 MW) in 1899. During the same time the workforce climbed from 1.3 to 5.1 million, thus increasing the average horsepower per worker by 50%. The value added to the product per worker almost doubled during the same period (Ref. 1.2, p. 681), a trend that accelerated in the twentieth century.

The relation between water supply, sewage disposal, and disease was established in the 1840s, and in northern Europe and North America drastic changes were made for the better in the short space of about twenty-five years (Section 7.5). Toward the end of the nineteenth century new medical procedures, such as sterilization and vaccination, greatly improved public health. As a result the increase in population was rapid. Although the birthrate declined slowly, the deathrate dropped much faster. What seemed to have been a distinct social advance in the last century had by 1950 become the world's greatest problem: how to cope with population growth resulting from a failure to match death control with birth control.

This population growth was accompanied by added urbanization. The capital cities of Europe were rebuilt during the late nineteenth century and most of the "old" buildings are, in fact, Victorian. This demand for new construction completely altered the problems of architecture. The emphasis changed from churches and palaces to administrative and office buildings, factories, hospitals, schools, and homes for the middle and working classes.

The period from 1850 to 1914 was a great age for science and technology in general and an age of notable advances in building construction and building science, but it was not a good age for the art of architecture.

By 1900 architectural traditions were already disintegrating and the social changes introduced by the two world wars completed the process.

The era from 1850 to 1939 is thus difficult to evaluate. New problems were being solved by procedures developed for entirely different building types. How does one adapt the Greek orders or Perpendicular Gothic to a tall building (Section 3.1)? What is the correct historical style for a bank, a university laboratory, or a hospital with new specialized services? How is a steel frame fixed into a Classic facade? How does one provide modern lighting, heating, and acoustics in a Neo-Gothic town hall (Section 8.4)? The answer given today is don't try; yet the problems seem so much clearer in hindsight.

Modern architecture began looking for a new basis. One school sought the answer in the engineered structures of the late eighteenth and nineteenth centuries, which in their own day had not been considered architecture at all (Section 10.1). Another sought it in contemporary painting and sculpture.

The difference between the old and new in the twentieth century was much greater than the change from Romanesque to Gothic or from Romanesque and Gothic to the Renaissance. Sufficient time was not always allowed for testing new ideas and for examining the technical side effects. Some of the early buildings by acknowledged masters of modern architecture, such as Le Corbusier and Mies van der Rohe, proved to be technical failures. Plain white surfaces, which looked splendid on the drawing board and in early photographs, were spoiled by cracked concrete or rendering or by unexpected deterioration of the paint. Flat roofs leaked, and interiors were intolerably hot because of an excess of unprotected glass (Section 8.2). These problems also seem so much clearer in hindsight.

Today we accept the need for proper consideration of the technical aspects and for the break in architectural tradition produced by the industrial revolution, but whether the new architecture is better than the old is open to question.

Most people would agree that modern factories, schools, and hospitals function better than those of the last or any preceding century. We have produced few residential buildings comparable in quality to a good sixteenth-century Italian palazzo or eighteenth-century English country house, but the *average* standard of residential buildings is higher. In particular, working-class housing has improved. On the technical and social levels modern architecture has been successful, even though there is room for further improvement; but it is a matter of opinion whether the best modern architecture is as good artistically as the Parthenon, Chartres Cathedral, or the Duomo of Florence.

The problems look different in the New and Old Worlds. It is arguable that modern buildings have ruined the capital cities of Europe and Asia, conceived on an entirely different scale. It is equally arguable that tall buildings have made New York, Chicago, Rio de Janeiro, and Sydney what they are.

Inevitably the social, political, and artistic aspects of this subject are controversial. This book, however, is concerned with them only incidentally. We are considering the science and technology of buildings, and in this field there is more to report on the period between 1815–1975 than for the preceding 2500 years.

This book is therefore divided by subject matter, and not by chronology, although some of the material follows naturally in chronological order. Thus most of the theory of statically determinate structures* (Chapter 2) predates most of the theory of statically indeterminate structures* (Chapter 3). The mechanization of structural design (Chapter 5), the theory of tall buildings (Chapter 4), and the renewed interest in curved structures (Chapter 6) are all recent developments.

In the twentieth century environmental aspects became an important and, ultimately, a controlling factor. Today the cost of building services is often higher than the cost of the structure, and we are devoting Chapters 7 and 8 to them. Chapter 9 deals with new building materials, and the last chapter considers the problems associated with industrialized building.

* See the glossary for a definition.

CHAPTER TWO

Statically Determinate

Structures

A formula is something that worked once, and keeps trying to do it again.

HENRY S. HASKINS

The second half of the nineteenth century was productive for the theory of statically determinate structures.* Its first practically useful solutions were obtained for the calculation of timber and iron trusses. The design of beams and columns, discussed for more than a century, was solved shortly after.

The steel frame originated in Chicago in the 1880s, but the development of reinforced concrete was slower and geographically more dispersed.

2.1 THE DESIGN OF TIMBER, IRON, AND STEEL TRUSSES

The question of equilibrium had been considered since the days of Archimedes (Ref. 1.1, Section 2.5), solver of the lever problem. The complementary problem of the parallelogram of forces was solved only in the late sixteenth century by Stevin (Ref. 1.1, Section 7.4). Also in the sixteenth century Palladio revived the Roman method of constructing triangulated timber trusses (Ref. 1.1, Fig. 7.18). It should therefore have been possible in the seventeenth century to derive a numerical method for determining the sizes of these trusses, particularly because timber trusses were being used in bridge construction and architectural design.

In fact, theoretically based calculations were derived only in the midnineteenth century when the rapid advance in railway construction made them essential. The first railway was built in England in 1825. The first railway in the United States was built in 1830, the first Russian railway, in 1837. Railways were based on the concept that an engine with smooth iron wheels pulled a train over smooth iron rails so that the loss of traction effort due to friction could be kept to a minimum. This implied that the gradients would have to be low because the friction would be insufficient to allow the engine to pull the train up a steep incline. On uneven terrain it would be necessary therefore, to build, sometimes in combination, tunnels or cuttings, bridges or embankments. Bridges were generally cheaper than tunnels. Until all bridges had been built and the line was continuous, the entire railway would be useless.

The early railways in England were built with masonry bridges, including long viaducts resembling the Roman aqueducts. Opposition to railway construction was considerable in England and other countries of western Europe from people who today would be called conservationists. Pressure of public opinion, and sometimes legislation, enforced the use of traditional bridges built of traditional materials that blended into the landscape. In the United States and Russia the railways passed through long stretches of wild country, the distances were much greater, and consequently the criterion for bridge design was the cheapest construction, namely, triangulated trusses. Moreover, timber was often available near the railway line. It was in America and Russia therefore that the theory of the design of triangulated trusses was first formulated.

The first American railway bridges were made of timber, the best known type being Howe's truss, which consisted of upper and lower timber chords, crossed diagonals of timber considered to be in compression, and vertical wrought iron ties; it was in use before 1840. Because of the danger of fire, these trusses were later made entirely of iron.

* See the glossary for a definition.

2.1

Pratt iron truss over the Delaware River near Trenton, N.J.

The first all-metal truss was built by Squire Whipple in the United States in 1840. It had an upper chord of cast iron, vertical members of cast iron which were held to be in compression, a wrought iron bottom chord, and diagonal wrought iron bars. In 1844 Thomas Willis Pratt patented a more economical iron truss in which the verticals were also held to be in compression and the diagonals in tension (Fig. 2.1); the Pratt truss is still popular, (Fig. 2.5a). In England in 1846 James Warren built the first iron truss without verticals (Fig. 2.4c); Warren trusses are also still in use.

Whipple and D. I. Jourawski, both railway engineers, one in the United States the other in Russia, and working independently of one another, obtained the first practical design solutions. Whipple published his in *An Essay on Bridge Building*, printed in Utica, New York, in 1847. Jourawski published his essentially similar method of analysis in a Russian journal in 1850.

Today we call this the *method of resolution at the joints*. Because the truss is in equilibrium, each joint of the truss is in equilibrium. We can therefore resolve horizontally and vertically at each joint in turn, and thus obtain $2j$ equations, where j is the number of joints in the truss. As we shall see in Section 2.2, there are $2j - 3$ members in the truss, and therefore three more equations than are needed; these serve as check equations. The method relies entirely on calculations.

The next advance in the design of trusses was the *method of sections,* published in 1862 by A. Ritter, a German engineer, in *Elementare Theorie und Berechnung eiserner Dach und Brücken-Constructionen* (Elementary Theory and Calculation of Iron Roof and Iron Bridge Construction). In this method an imaginary cut was made through the truss and the cut members were replaced by internal forces (Fig. 2.2). Assuming that we wish to find the force in the bottom chord, we take moments of these internal forces and of the external forces on one side of the cut about some

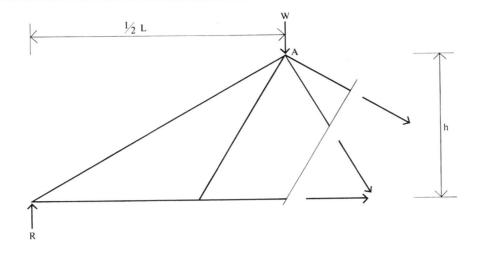

2.2

Determination of the force in the bottom chord of a roof truss by the method of sections. Taking moments about A, force in bottom chord × h = left-hand reaction × ½ L. The other three forces pass through A and therefore have no moment about A.

point; the most suitable point is A. The forces in two of the cut members pass through A as does the load at A. If, instead, we take moments about an unsuitable point we obtain the same answer but the calculation is longer. The method of sections is particularly convenient for a preliminary design when we merely wish to know the size of one or two critical members. It is laborious if the forces in all the members are required.

The third method became the most popular. In the eighteenth century and even in the nineteenth graphical methods were faster than those based on arithmetic. The slide rule that utilized John Napier's logarithms was invented by Edmund Gunter in 1620, but a slide rule with a cursor was not produced until 1859 (Section 5.1). Hence Monge, the founder of the *École Polytechnique* in Paris, encouraged graphic methods where they could be used. Graphic statics was developed independently by Karl Culmann, Professor of Engineering Science at the newly founded Federal Institute of Technology in Zurich, and by Clerk Maxwell, then Professor of Natural Philosophy at Kings College, London; the latter is best known for his work on electromagnetism. Maxwell published his method earlier, in 1864, in the *Philosophical Magazine* (Vol. 27, p. 250), but it was not accepted until Robert H. Bow explained it in his book *Economics of Construction in Relation to Framed Structures* (Spon, London, 1873). Bow introduced the notation that still bears his name, whereby the spaces between the members and forces are lettered instead of the joints (Fig. 2.3).

In all three methods we obtain the (tensile or compressive) internal forces in the members of the truss. We then calculate the cross-sectional areas required for the

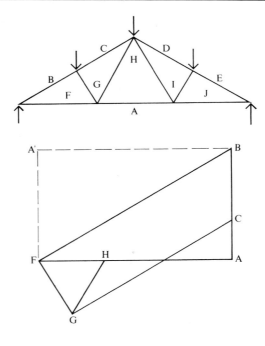

2.3

Determination of the forces in a roof truss by the reciprocal stress diagram. The upper diagram is marked according to Bow's notation, in which the spaces between the internal and external forces are lettered. The joints are not marked. Each force is denoted by letters on either side. Consider the equilibrium at the left-hand support. The joint is held in equilibrium by the left-hand reaction *AB,* the internal force in the rafter *BF,* and the internal force in the horizontal chord *AF.* We know the magnitude and direction of the reaction *AB.* The (tensile or compressive) force in the rafter BF must be parallel to *BF* and the (tensile or compressive) force in the horizontal chord *AF* must be horizontal. Consequently we draw *AB* to scale, *BF,* parallel to the rafter, and *AF,* horizontal. The intersection gives the point *F.* This is the triangle of forces for the joint. If desired, we can complete the parallelogram of forces by drawing the dotted lines *BA′* and *A′F.* Consider the equilibrium at the next joint above. We already know the external load *BC* and the internal force in the member *BF.* We draw lines parallel to *CG* and *FG* to complete the quadrilateral of forces, and so on until we obtain the complete reciprocal diagram. The unknown forces can then be scaled off; we determine whether they are tensile or compressive from the direction of the force. If this is drawn on a sufficiently large scale with a finely sharpened pencil on a good drawing board, the method is accurate, but since the development of mechanical aids (Section 5.4) it has fallen into disfavor.

members by dividing each force by the maximum permissible (tensile or compressive) stress for the material (see Section 2.4).

The three principal methods for the analysis of plane trusses, after centuries of empirical design, were developed in less than twenty years. It demonstrates how an urgent need can act as a stimulus.

2.2 THE CONCEPT OF STATICAL DETERMINATENESS

It is implicit in the methods described in the preceding section that the structure can be calculated by the laws of statics alone.

This problem was considered quite early. In 1837 August Ferdinand Möbius, Professor of Astronomy at the University of Leipzig, published in *Lehrbuch der Statik* (Textbook of Statics) a theorem that now bears his name.

Let us assume that all members of a truss are joined by pins, which was an early and convenient method of joining timber or cast iron compression members and wrought iron tie bars before screw-cutting machines were invented. The term pin joint is now used for any joint that allows a reasonable amount of rotation so that the members are in compression or tension but not in bending. Thus a pin joint is defined as a joint that does not transmit bending.

The simplest truss then consists of three members and three joints (Fig. 2.4a). If we remove one member we are left with a mechanism. For each additional joint we

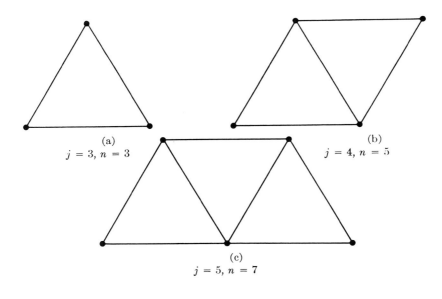

(a)
$j = 3, n = 3$

(b)
$j = 4, n = 5$

(c)
$j = 5, n = 7$

2.4

The simplest statically determinate truss has three members and three pin joints. Each additional pin joint requires two additional members. The truss (c) is a short Warren truss.

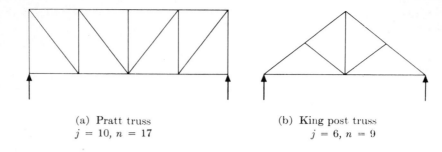

(a) Pratt truss
$j = 10, n = 17$

(b) King post truss
$j = 6, n = 9$

2.5

Statically determinate trusses with j pin joints require $n = 2j - 3$ members. If there are more, the truss is statically indeterminate (Chapter 3). If there are fewer, the truss becomes a mechanism and collapses.

must add two members. The number of members required is therefore

$$n = 2j - 3 \qquad (2.1)$$

where n is the number of members and j, the number of joints.

If the truss has fewer members, it collapses. If it has more members, we cannot work it out by statics. A truss with $2j - 3$ members is therefore *statically determinate* (Fig. 2.5). We shall look at statically indeterminate structures in Chapter 3.

Möbius was ahead of his time. His law does not appear to have been known to any of the people who devised the methods for solving trusses explained in Section 2.1, but when Otto Mohr, Professor at the Dresden Polytechnikum, rediscovered this law and published it in 1874 it became generally accepted.

Möbius' law can be written in a different form by including the reactions required. A truss supported at its ends needs two vertical reactions (Fig. 2.5). In addition it needs one horizontal reaction at either end to ensure that it will not be pushed off its supports by a horizontal force. Only one horizontal reaction is appropriate because two horizontal reactions would restrain the elastic expansion or contraction of the bottom chord. By including these three reactions in (2.1) we obtain

$$n + m = 2j \qquad (2.2)$$

where m is the number of reactions required.

2.3 THE DESIGN OF BEAMS

Galileo provided a solution for the bending of a cantilever carrying a single load at its end (Ref. 1.1, Section 7.4). The solution was incorrect but the bending moment for the cantilever was correctly stated. Most subsequent solutions of the bending problem were discussed in terms of Galileo's cantilever. Mariotte (Ref. 1.1, Section 8.3) at the end of the seventeenth century determined by experiment the relation between the strength of a cantilever and a simply supported beam. Thomas Tredgold (Ref. 2.6) derived bending moments for both cantilevers and simply supported beams by using a rather complicated geometric method based on Euclid's *Elements*. The problem was first stated in mechanical terms in 1826 by Louis Marie Henri Navier in his *Résumé de leçons données à l'Ecole des Ponts et Chaussées, sur l'application de la mécanique à l'etablissement des constructions et des machines* (Summary of Lectures given at the School for Bridges and Roads on the application of mechanics to the design of structures and machines) in the manner shown in Fig. 2.6.

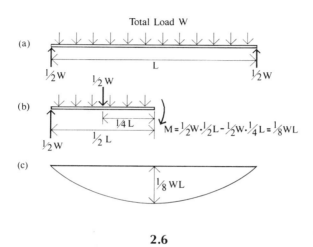

2.6

The bending moment in a simply supported beam carrying a uniformly distributed load is statically determinate; that is, it can be obtained by using the conditions of equilibrium. By symmetry each end reaction is $\frac{1}{2}W$, where W is the total load carried by the beam (a). The maximum bending moment occurs at midspan. Let us make an imaginary cut at midspan. The forces acting on the left-hand half of the beam are the end reaction $\frac{1}{2}W$ and half the load W carried by the beam, which acts midway along the half span; that is, at quarter span. Taking moments about the cut at midspan (b), the bending moment $M = \frac{1}{2}W \cdot \frac{1}{2}L - \frac{1}{2}W \cdot \frac{1}{4}L = \frac{1}{8}WL$.

The bending moment varies parabolically from zero at the ends to the maximum value (c).

(a) Forces acting on beam.
(b) Imaginary cut at the center of the beam, balanced by the maximum bending moment.
(c) Bending moment diagram.

Having determined the bending moment M, we can then calculate the section modulus Z from the maximum permissible stress f by using Navier's theory of bending (Ref. 1.1, Section 8.3):

$$M = fZ$$

The advantage of using an I-section, which concentrates most of the material as far as possible from the neutral axis, was known empirically even before the theory of bending had been derived. Tredgold (Ref. 2.6) recommended the use of cast-iron I-sections in 1824; Henry Cort started rolling flanged wrought-iron sections in the early nineteenth century, and I-sections for structures were rolled before 1850.

There are evident advantages in having tables with the geometric properties I (second moment of area) and Z (section modulus) of the various sections, and also safe-load tables which show the load-carrying capacity of the various sections over different spans. Tredgold supplied simple safe-load tables in his book on cast iron. Soon manufacturers found it useful to issue these tables for their products as part of their trade catalogues; some survive. With appropriate tables the design of simple beams is easy.

Although wrought iron was frequently used in railway construction because of its greater ductility and better resistance to shock and vibrations, the cheaper cast iron was more common in buildings in the first half of the nineteenth century. Cast iron had a much lower strength in tension than in compression, and toward the middle of the century it became the practice to use a tension flange that was larger than the compression flange.

In 1847 Henry Fielder obtained a patent for making compound beams by riveting cast- and wrought-iron plates and angles together to make use of cast iron in compression and wrought iron in tension. At the time the price of wrought iron was about double that of cast iron, and in spite of the cost of labor these compound sections proved economical.

In 1856 Henry Bessemer invented the process named after him for producing steel by blowing air through fluid pig iron instead of reducing the carbon content by the traditional laborious process. The price of steel fell drastically as a result. Bessemer's steel combined ductility with a strength much higher than wrought iron.

In 1885 Dorman Long and Co. in England and Carnegie, Phipps and Co. in the United States began rolling steel joists, and cast-iron and wrought-iron beams soon became uneconomical. Cast-iron columns continued to be made until 1914.

2.4 THE BUCKLING PROBLEM

Leonard Euler provided the solution to the elastic buckling problem in 1759 (Ref. 1.1, Section 8.4), but it remained an academic exercise because few practical columns buckle that way.

There are two entirely different types of column failure. If a short column of wrought iron or steel is overloaded, the material is squashed and never recovers its shape. The load at which failure occurs depends only on the cross-sectional area of the column and on the stength of the material but not on its elastic modulus. A short cast-iron column cracks but the result is otherwise the same.

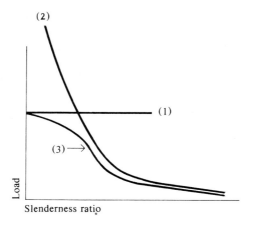

Practical design of columns. The straight line (1) represents the short-column load P_s, which is independent of the slenderness ratio. The hyperbolic line (2) represents the Euler load P_e, which is inversely proportional to the (slenderness ratio)^2. Evidently the sharp corner formed by curves (1) and (2) is not followed by a practical column and line (3) represents the empiral load P obtained from the Rankine formula (2.3). Although it has a different theoretical basis, the Perry-Robertson formula also gives a curve similar to (3).

The slenderness ratio $= L/\sqrt{I/A}$ (L is the length of the column, A is its cross-sectional area, and I is its second moment of area); it follows from the Euler formula and is proportional to the ratio of length to thickness.

A column with a high ratio of length to thickness, generally referred to as a long column, shows an entirely different sort of failure. On being overloaded it buckles sideways and thus ceases to be a useful column but the material remains elastic; that is, the original shape is recovered on unloading and the column is not damaged by the failure. This is the case solved by Euler. The Euler load is independent of the strength of the material but it depends on its elastic modulus and on the ratio of length to thickness.

The great majority of practical iron and steel columns are neither so short as a "short" column nor so long as a "long" column. In this intermediate range both the short-column formula and the Euler formula, but specifically the latter, are unsafe (Fig. 2.7). The Euler formula was therefore one of the main reasons why many practically trained British engineers were suspicious of the French mathematical theory of structures.

In 1824 Thomas Tredgold published an empirical column formula of the type

$$P = \frac{fA}{1 + aL^2/d^2}$$

where P is the safe load for a circular column, f is the safe column stress, A is the cross-sectional area of the column, L is its length, d is the diameter, and a is a constant.

Referring to this formula, Tredgold noted:

It may be useful to remark, that the most refined methods of analysis have been applied to the same subjects by Euler, Lagrange and other continental mathematicians, without arriving at results more accurate, more simple, or more convenient in practice (Ref. 2.6, p. xiii).

Tredgold's empirical formula, as rephrased by Lewis Gordon, was still widely used in the twentieth century.

The first theoretically based solution was given by J. W. McQuorn Rankine in his *Manual of Applied Mechanics*, published in 1858. Rankine had succeeded Gordon three years before in the chair of engineering at Glasgow University. Rankine argued that the short-column load P_s was correct for a slenderness ratio of zero and that the Euler load P_e was correct for a slenderness ratio of infinity. For intermediate values the column load P could be obtained as a reasonable approximation by

$$\frac{1}{P} = \frac{1}{P_s} + \frac{1}{P_e} \tag{2.3}$$

This is shown in Fig. 2.7. The formula is similar to Tredgold's and Gordon's and the constants required were still determined by experiment.

A formula of the Rankine-Gordon-Tredgold type still appears in the 1969 Specification of the American Institute of Steel Construction (the building code for steel).

The problem to which Rankine addressed himself lay in the structure of the short-column and Euler formulas. One does not contain a term for the bending of columns and the other does not contain a term for the strength of the material. This can be overcome by considering a column whose load has a slight eccentricity or one that is slightly bent out of shape. It is then possible to set up a formula that will produce a theoretically accurate result not merely for slender and short columns but for all intermediate values of slenderness. The problem was solved for an eccentrically loaded column by Professor R. H. Smith (*Engineering*, Vol. 44, 1887, p. 303) and for an initially curved column by Professor John Perry (*Engineering* Vol. 42, 1886, p. 464).

It can be argued that it is as difficult to balance a needle on its point as it is to load a column perfectly concentrically or to produce a column that is perfectly straight. In order to match the behavior of a real column with the theoretical results, it is therefore appropriate to assume a slight minimal eccentricity of loading, which is used even when the column is, in theory, loaded concentrically, or to assume that the column is initially slightly curved; the effect of eccentricity and curvature on load-carrying capacity is similar.

The first approach was recommended by the (American) Joint Committee Report on Ultimate Strength Design (Ref. 2.1) in 1955, and has since been introduced into the American and several other national concrete codes. Concrete columns are, on the whole, much less slender than those of iron and steel and buckling need not always be considered; however, the use of a minimal eccentricity, which is determined empirically, is appropriate even for "short" columns.

The (British) Steel Structures Research Committee (Ref. 2.7) recommended in its first report in 1931 that a formula based on the assumption of a slight initial curvature be adopted. Known as the Perry-Robertson formula, it has since been introduced into the British and several other national steel codes. The constants used are based on experiments carried out by Professor Andrew Robertson (Ref. 2.2). Steel columns are now designed on the assumption that buckling must be considered.

Evidently the compression member in a truss (Section 2.1) behaves like a column, and permissible compressive stresses have, since the late nineteenth century, been decreased to allow for the slenderness of the member, either by an empirical or the Perry-Robertson formula. This was not done in the early iron bridges, nor was adequate bracing always provided to reduce the slenderness ratio. Consequently the compression flanges at the top of some unbraced bridges buckled thus causing some disastrous failures (e.g., Ref. 2.8, Fig. 181, p. 296).

2.5   THE BUCKLING OF BEAMS

The compression flanges of beams also buckle if the slenderness ratio is too high, but this type of failure is not so easily identified.

It was first reported by Sir William Fairbairn who had started work as an apprentice at the age of 15. Fairbairn taught himself some mathematics in the evenings, but he was essentially a practical engineer who built many factories and other structures, particularly in the industrial north of England. Although he was born only one year after Tredgold, he had a greater respect for French theory. After visiting the Paris Exhibition of 1855 he made a report that commented on education:

I firmly believe, from what I have seen that the French and the Germans are in advance of us in theoretical knowledge of the principles of the higher branches of industrial art; and I think this arises from the greater facilities afforded by the institutions of those countries for instruction in chemical and mechanical science. . . .

Under the powerful stimulus of self-aggrandisement we have perseveringly advanced the quantity, whilst other nations, less favoured and less bountifully supplied, have been studying with much more care than ourselves the numerous uses to which the material may be applied, and are in many cases in advance of us in quality. (Ref. 2.8, p. 126).

In his more complex investigations Fairbairn therefore cooperated with Eaton Hodgkinson, who was born in the same year, and also received only an elementary education. At the age of 22 Hodgkinson met John Dalton, formulator of the atomic theory of chemistry. Dalton helped Hodgkinson with mathematics and gave him the works of the classical French writers on mechanics, such as Bernoulli, Euler, and Lagrange. Hodgkinson later presented many of his investigations in papers read before the Manchester Literary and Philosophical Society, of which Dalton was a leading member. In 1847 Hodgkinson became Professor of the Mechanical Principles of Engineering at University College, London.

Robert Stephenson (son of the man who built the first passenger railway) obtained in 1845 the parliamentary act that authorized a railway from London to Holyhead, from which passengers proceeded by boat to Dublin. This railway involved the crossing of the Menai Strait between the mainland and the Isle of Anglesey. The

2.8

The tubular railway bridge over the River Conway has a span of 420 ft
(126 m) and was completed in 1849. Fairbairn investigated not only the
effect of the vertical loads on the tube but also the effect of lateral wind
pressure and of nonuniform heating by sunlight.

British Admiralty imposed restrictions on the size and location of the piers in the
interest of shipping, and it was eventually decided to use iron tubes through which
the train passed as if through a tunnel. There were to be two central spans of 460 ft
(140 m) each and two shorter spans. In addition to this longer bridge, called the
Britannia Bridge, there was to be a 420 ft (128 m) tubular bridge over the River
Conway (Fig. 2.8).

Fairbairn was retained as a consultant. In his fabricating yard he tested several
tubular sections simply supported at their ends. It was well-known that cast-iron
beams supported at their ends failed at the bottom face when the tensile strength of
the material was reached. The wrought-iron tubes, however, failed on top. In his
Account of the Construction of the Britannia and Conway Tubular Bridges, published
in 1849, Fairbairn gave the following explanation:

Some curious and interesting phenomena presented themselves in the experiments—many of
them anomalous to our preconceived notions of the strength of materials, and totally different
to anything yet exhibited in any previous research. It has invariably been observed, that in
almost every experiment the tubes gave evidence of weakness in their powers of resistance at
the top side, to the forces tending to crush them (Ref. 2.9, p. 291).

Statically Determinate Structures

Fairbairn called on Hodgkinson to give a theoretical explanation, but Hodgkinson wrote that

any conclusions deduced from received principles, with respect to the strength of thin tubes, could only be approximations: for these tubes usually give way by the top or compressed side becoming wrinkled, and unable to offer resistance long before the parts subjected to tension are strained to the utmost they could bear.

Hodgkinson therefore proposed further tests

to ascertain how far this defect, which had not been contemplated in theory, would affect the truth of computations on the strength of tubes (Ref. 2.8, p. 158).

Fairbairn then built a model tube of the longest span to a scale of 1:6. The upper side was strengthed until failure occurred simultaneously on top and bottom. Before this condition could be achieved, however, waves occurred in the plates forming the sides, a phenomenon now called shear buckling. It was therefore decided to rivet stiffening angles to the webs. The two concepts proposed by Hopkinson, namely, reduced permissible stresses for the compression flanges of beams and stiffeners for the webs, have been used ever since.

2.6 THE CHICAGO SCHOOL AND THE SKELETON FRAME

It is symptomatic of the future development of high-rise architecture that the first iron-framed building was erected in 1885 not in Europe but in the American Middle West.

No particular structural problem delayed this development. Factory buildings five and six stories high, with iron columns and iron beams, had been erected in England in large numbers since 1803. The theory for the design of a building frame presented no difficulties (Fig. 2.9). It could be regarded as a series of columns that carried simply supported beams. Navier had published the solution to the bending problem in 1826 (Section 2.3). Tredgold had solved the problem of column design and published it in 1858, together with the bending theory, in a lucidly written book (Section 2.4).

Wind loading was not considered in the design of the early iron and steel frames, for the buildings had substantial masonry walls that provided adequate shear panels to resist wind loads.

The invention of the skeleton frame was therefore not held back by technical problems but by social and economic considerations. Two factors encouraged the construction of the first iron-framed building in the 1880s and specifically in Chicago. One was the invention of a safety catch in 1854 that made elevators acceptable for transporting passengers (Section 8.7). Elevators had been used for centuries for hoisting goods and for transporting miners who presumably accepted an occupational risk, but the first passenger elevator was built in New York only in 1857; the first in Chicago was an Otis Patent Steam Passenger Elevator, installed in the Honore Block I in 1870 and destroyed by the fire of 1871.

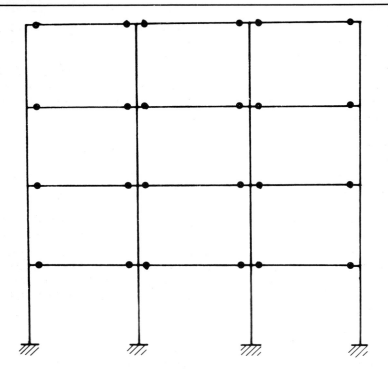

2.9

The Chicago iron frame consisted of cast-iron or steel columns on which wrought-iron or steel beams were considered to be simply supported. Wind load, if it was considered at all, was provided by the masonry walls or by crossed diagonals of wrought-iron or steel bars, not by the flexural stiffness of the frame.

The other obstacle to be overcome was the inertia of the architectural profession. Iron had been used successfully for a century in bridges and factories, but these structures had been designed by engineers. The 1851 International Exhibition in London (Section 10.1) had made iron a respectable and, indeed, a much admired material, but the Crystal Palace was built by a designer of greenhouses with the help of an engineering firm.

The architect-designed buildings that used iron before 1880 were mostly market halls or libraries and there were not many of them; they included the reading room of the British Museum by Sydney Smirke, the Bibliothèque Nationale in Paris by Henri Labrouste, the Grandes Halles of Paris by Victor Baltard, and the Hungerford Fish Market in London (see also Section 3.3).

Among the few iron-framed buildings not in this category were the Théâtre Francais and the Brighton Pavilion (Ref. 1.1, Section 8.6).

Chicago, however, was not impeded by tradition. It grew quickly from a small settlement. Fort Dearborn, a two-story log house, was built in 1804 and burned to the ground by Indians in 1812. In 1830 the population was still less than 100. In 1840 it was 4000, in 1850, 30,000, in 1860, 109,000, and in 1870, 299,000. The Great Fire started on Sunday, October 8, 1871. By the following morning 18,000 buildings, including the entire city center had been destroyed (Ref. 2.10, p. 4).

Chicago had already become the birthplace of the balloon frame which revolutionized the construction of timber houses. It was invented by George Washington Snow and first used in a Chicago church in 1833. The weakness of the traditional timber structure and one of its most expensive parts was the joint. In the balloon frame the ancient mortised and tenoned joints were omitted and the pieces of timber joined with mass produced iron nails which were stronger and cost only a fraction of the price of the handmade nails of the early eighteenth century. In addition, the timber studs were smaller and more closely spaced. The name "balloon frame" was originally a term of ridicule, but the new technique was quickly accepted in the Middle and Far West because it saved a great deal of time and money and required less skilled labor. It soon spread to the rest of the United States and Australia. In Europe, however, the traditional methods of jointing were more firmly established.

After the fire Chicago was rebuilt with great energy. By 1880 the population had increased to half a million, to a million by 1890, and by 1910 had exceeded two millions. The power-operated passenger elevator had made its appearance just before the fire, and many of the buildings in the city were classified as "elevator buildings." In due course this produced an increase in height.

Most of the post-fire buildings had interior cast-iron columns which supported cast-iron or wrought-iron girders; they, in turn, supported iron or timber joists. The column spacing was relatively small, rarely exceeding 10 ft (3 m), and usually the same in both directions. Wind bracing was never used in the 1870s and 80s. The exterior walls were generally of solid masonry divided into piers of brick or dressed stone and designed to support themselves and the floor and roof loads of the half-bays immediately adjacent to the walls. Presumably they also provided wind resistance. Construction was fast; most buildings were completed in less than a year. Their facades had a simplicity of treatment and lack of reliance on historically derived ornamentation that commended the "Chicago School" to the pioneers of modern architecture. In this respect Chicago architecture of the late nineteenth century differed not merely from that of Europe but also from that of the eastern states.

Below ground also there were notable advances. The limestone bedrock, approximately 125 ft (38 m) below street level in Chicago's city center, is overlain by sand and clay interspersed with water pockets. This caused appreciable settlement of the buildings which consequently were designed with spread foundations consisting at first of stepped masonry and later of grillages of iron rails. Normally each column and wall was built with a separate isolated foundation. In some buildings an attempt was made to predict the settlement and the buildings were built above grade by the predicted amount.

There were many talented and original architects in Chicago in the 1880s, but the evolution of the skeleton frame owes most to William Le Baron Jenney, who was born in Massachusetts in 1832. In 1853 he went to France and in 1856 graduated in engineering from the *École Centrale des Arts et Manufactures*. Jenney became a major in the Union army during the Civil War. After his discharge in 1867 he moved to Chicago, where, except for a brief appointment in 1876, as Professor of Architecture at the University of Michigan, he practiced until 1905.

Jenney's first building, the seven-story Portland Block in 1872, had loadbearing walls, interior cast-iron columns, and cast-iron beams. There was no adornment on the flat external walls except for shallow rustication and narrow quoins at the corners, but every office had an outside exposure. In the five-story Leiter Building, erected in 1879, 8 × 12 in. (200 × 300 mm) cast-iron columns were set flat against the east and west walls to support the timber girders. On the south wall continuous cast-iron mullions which partly supported the floor structure were used from the foundations to the roof. The outer walls were partly loadbearing and no iron was set into the north wall.

The first skeleton frame was used by Jenney in the nine-story Home Insurance Building, erected in 1885. The structural design was credited to Jenney's engineering assistant George B. Whitney, a graduate in civil engineering at the University of Michigan. Cast-iron columns were used for the interior and inside the exterior walls. The beams were designed to be wrought iron, but "permission was granted to substitute the first shipment of Bessemer beams from the Carnegie-Phipps Company in the upper stories for the wrought-iron beams."

The first skeleton frame went almost unnoticed at the time. In 1896 F. T. Gates, president of the Bessemer Steamship Company, wrote to the editor of the *Engineering Record* to ask "to whom the honour is due of discovering or practically working out the idea of lofty steel construction of buildings." Among the replies the most significant came from Daniel H. Burnham, who had by that time become one of Chicago's leading architects:

The principle of carrying the entire structure on a carefully balanced and braced metal frame, protected from fire, is precisely what Mr. William LeB. Jenney worked out. No one anticipated him in it, and he deserves the entire credit belonging to the engineering feat which he was first to accomplish (*Engineering Record*, July 25, 1896, quoted by Randall, Ref. 2.10, p. 106).

The skeleton was the second ingredient required for the successful design of tall buildings, the other being the elevators. Before the fire the height of buildings in the Chicago business district varied from four to six stories; elevators were used only in the last year before the fire. After the fire elevators became common (Section 8.7), and buildings rose to eight stories. In 1882 the Montauk Block became the first ten-story building. Designed by Burnham and Root, it had heavy loadbearing walls, interior cast-iron columns, and wrought iron beams. It was also the first building in which iron-rail grillages were used to reduce the depth of the foundations.

In the early 1890s more than a dozen buildings in Chicago exceeded ten stories in height, and in 1891 the term "skyscraper" was coined. The Montauk building was thus the first skyscraper.

As the buildings with loadbearing walls became taller, the walls became thicker. The limit was reached in the sixteen-story Monadnock Block, built in 1891 (Fig.

2.10a). Its solid external walls, which measured 72 in. (1.83 m) at the base, were the thickest built in Chicago, and possibly anywhere in the world, for loadbearing purposes in modern times. Again, the architects were Burnham and Root; however, the decision to use loadbearing walls was the client's, Shepherd Brooks, who in 1892, when the additions to the Monadnock Block were being planned, wrote to a business associate:

As to erecting a tall building entirely of steel, for a permanent investment, I should not think of such a thing. It would no doubt pay well for many years, but there is so much risk and uncertainty in regard to its lasting strength." (Ref. 2.3, p. 160).

Burnham and Root had built the walls 8 in. (200 mm) above grade. By 1905 the walls had settled that "and several inches more." By 1940 they had settled 20 in. (500 mm). The settlement, due to the heavy weight pressing on the Chicago clay, presented an insoluble problem, and the Monadnock Block was the last of the skyscrapers with loadbearing walls until the 1960s (Section 9.2).

The Capital Building, later known as the Masonic Temple, set a new record in 1892. Another Burnham and Root design, it had twenty-one stories and was 302 ft (92 m) high. It had a steel frame and its massive columns were boxshaped, built up from plates and channels. Crossed wrought-iron bars were used to brace the frame against the wind. The foundations were deep raft footings of reinforced concrete.

Thereafter the leadership passed to New York. The St. Paul Building, erected in 1896, had twenty-six stories; the Singer Building, erected in 1907, had forty-seven stories and a height of 675 ft (206 m). It was the first building to exceed the height of the Gothic and Renaissance cathedrals and the Great Pyramid of Gizeh in Ancient Egypt.

The Home Insurance Building, discussed above, was the first to use steel in place of wrought-iron beams (in the upper floors only). Thereafter steel became more common for beams but it was considered too expensive to be used in the construction of columns.

In 1890 the four-story Reliance Building (Burnham and Root), the ten-story Rand McNally Building (Burnham and Root), and the twelve-story Caxton Building (Holabird and Roche) became the first to use steel columns. Cast-iron columns were still used in the sixteen-story Manhattan Building in 1890, and the sixteen-story Unity Building in 1892. By the turn of the century, however, the riveted-skeleton steel frame had become standard construction for all American skyscrapers (see Fig. 2.10b).

Rigid-frame design was first used by Corydon T. Purdy in the Old Colony Building, erected in 1893–1894. This narrow sixteen-story building was exposed to the wind on all four sides and had large windows so that the walls could not be depended on for lateral resistance. Purdy connected the steel girders in the outer bays to the wrought-iron columns with a deep rounded fillet (Fig. 2.11), which produced a series of rigid portal frames capable of resisting horizontal forces. In the early twentieth-century American skyscrapers were normally designed as a series of stacked portal frames to resist the wind, the system still used in New York's Empire State Building (Section 4.1).

Statically Determinate Structures

2.10a

The sixteen-story Monadnock Building, erected in 1891. Designed by Burnham and Root, it had load-bearing walls 72 in. (1.83 m) thick at the base because the client lacked confidence in the performance of steel-framed construction. (See also section 9.2). (Photograph by Ralph Ruebner.)

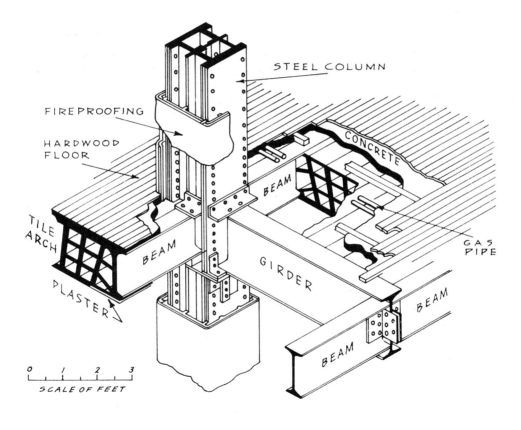

2.10b

Typical detail of steel-framed building in Chicago. (Fair Store, a nine-story structure built by W. LeB. Jenney in 1892. From *Industrial Chicago*, Goodspeed, Chicago, 1891, reproduced in Ref. 2.19, Plate 51.)

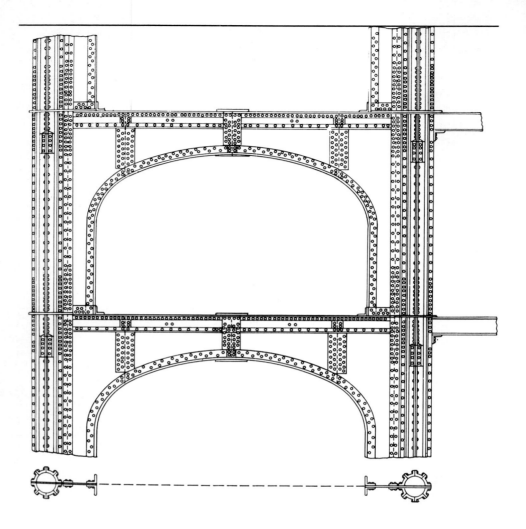

2.11

Frame built up of portals. The example shown is the Old Colony Building, Chicago, erected in 1893 and 1894. The simply supported beam spanning the central bay is visible on the right. The architects were Holabird and Roche and the structural consultant was Corydon T. Purdy (from J.K. Freitag, *Architectural Engineering*, Second Edition, Wiley, New York, 1901.)

2.7 THE INVENTION OF REINFORCED CONCRETE

Plain concrete had by the mid-nineteeth century proved itself an excellent material for engineering structures, particularly those in contact with water (Ref. 1.1, Section 8.7). Some quite substantial arch bridges were eventually erected in plain concrete. The longest was a railway bridge with a span of 64.5 m (215 ft), built by Dyckerhoff and Widman (1904–1906) in Kempten, Germany, over the river Iller.

The acceptance of concrete in building was much slower, although concrete with waterproof (natural or portland) cement replaced lime concrete in foundations and cement mortar was substituted for lime mortar in external renderings.

There were a few isolated examples of the use of concrete in walls. In 1837 John Bazley White, who three years before had bought a cement works in Kent, built a large two-story Tudor-style house for his own use (Ref. 2.11, p. 20). In 1834 the French architect Jean-Auguste Lebrun built a one-story schoolhouse at St. Agnan in Tarn-et-Garonne (southern France); his offer to roof it with a concrete barrel roof was declined.

In 1835–1836 he built a concrete church at nearby Corbarieu with walls that were 700 mm (2 ft 4 in.) thick. The building was to have been roofed with plain concrete barrel vaults spanning 8 m (26 ft 8 in.). When the formwork was removed, a small fissure in the end wall spread to the crown and eventually ran the length of the barrel vault. Lebrun had provided no ties to absorb the horizontal thrust of the concrete vault, and in 1837 the roof was replaced with timber (Ref. 2.12, p. 26).

The attempts to use concrete in the walls of multistory buildings inevitably led to its use in floor construction. This required reinforcement. Brick arches topped with sand and tiles had been used since the late eighteenth century for fireproof factory floors (Ref. 1.1, Fig. 8.11, Section 8.6). In the early nineteenth century the sand was sometimes mixed with lime or cement to form a weak concrete.

In 1844 Dr. Fox patented a form of floor construction, claimed to have been used in the 1830s, which consisted of cast-iron girders set 18 in. (0.5 m) apart with lime concrete supported on wooden laths and subsequently plastered. This structure was essentially a concrete floor with cast-iron beams. It was used, for example, in Balmoral Castle, Queen Victoria's Scottish residence (Ref. 2.12, p. 29). A later version with wrought-iron beams is shown in Fig. 7.2.

In the 1840s William Fairbairn, who had designed many iron-framed factories, used cast-iron beams at 10 ft (3 m) centers, then a large spacing, with concrete arches spanning between the beams (Fig. 2.12). The concrete was supported on wrought-iron plates and the beams were tied in the customary manner with wrought-iron bars. Both the iron plates and bars may have acted as reinforcement for the concrete (see also Section 7.2).

The first serious attempt to use concrete in building was made by François Coignet who in 1855 patented *béton agglomeré*. This was mixed with as little water as possible in a continuous mixer (turned by a horse). It was rammed into the forms, not cast, and the forms were regarded as reusable. Coignet stated in an addendum that the technique could be used for the more complex parts of buildings, such as balconies and ornaments, if the forms were made with sufficient care.

Later in the same year Coignet took out a patent in Great Britain in which he referred to iron bars crossing one another like the squares on a chess board; the bars

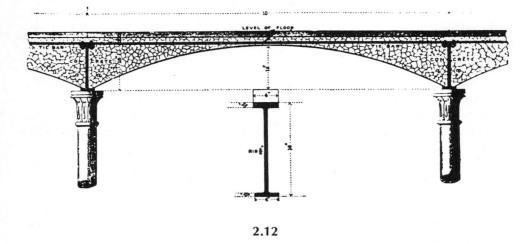

2.12

Fireproof floor built by Sir William Fairbairn in the 1840s. The cast-iron beams were placed at 10-ft (3-m) centers. This produced a comparatively long span for the concrete arches. The concrete was cast on wrought-iron plates as permanent framework. The customary wrought-iron ties for the arches were cast within the concrete (from William Fairbairn, *The Application of Cast and Wrought Iron to Building Purposes*).

were to penetrate into each of the four walls supporting the floor slab. In the following year Coignet took out a new patent in France which used the word *tirans* (tension members) for the iron bars. Evidently Coignet had at that time quite a good appreciation of their function.

Coignet submitted a proposal to erect a concrete house at the Universal Exhibition to be held in Paris in 1855; in it he referred to the new material both as *béton agglomeré* and *béton pisé*. His explanatory comments showed remarkable vision:

Our customs no longer require these enormous and solid constructions which impose on future generations the tastes and architecture of the past. Is not the best solution for one house to shelter one generation? . . . This solution is certainly opportune today, for the reign of stone in building construction seems to have come to an end. Cement, concrete and iron are destined to replace it. Stone will only be used for monuments (Ref. 2.12, p. 28).

Coignet's proposal to erect a complete house was refused, but he was allowed to exhibit some concrete components and received a bronze medal.

At the same exhibition Joseph Louis Lambot exhibited a boat, patented earlier that year, made from cement mortar and iron bars. It excited much interest and some historians have credited Lambot with being the inventor of reinforced concrete. One of the boats was still in use in 1902, and in 1955, on the anniversary of the Exhibition, two of Lambot's concrete boats were raised from the bottom of a lake in southern France.

Statically Determinate Structures

In the early 1850s Napoleon III and his prefect of the Seine Department, Baron Haussmann, were rebuilding Paris, and Coignet received a number of contracts. A French magazine reported in 1859 how whole quarters of Paris had been constructed in an "incredibly short time" by using Coignet's "pastry-construction" which employed lime concrete made of 1½ parts of river gravel, 2½ parts of sharp sand, 1½ parts of lime, and 1½ parts of boiling water (Ref. 2.11, p. 18*).

In 1862 Coignet built himself a three-story private residence in St. Denis, a suburb of Paris, using concrete throughout; it is still standing. It had moldings in the style of Palladio and a concrete roof reinforced with 120-mm (4 ¾-in.) deep I-girders at 1 m (3 ft 4 in.) centers (Ref. 2.11, p. 14*). Coignet ran into financial difficulties in 1867 and was ruined by the economic collapse that followed the fall of Paris to the Germans in 1871.

In the meantime William Boutland Wilkinson had developed reinforced concrete independently in Newcastle-upon-Tyne in northern England. A number of patents of a similar kind were registered in Britain in the midnineteenth century, but Wilkinson's patent of 1854 is the only one known to have been used in practice:

The invention of improvements in the construction of fireproof buildings, warehouse and other buildings or parts of the same, including staircase, pavings, etc. also with Paris Plaster hollow tubular fireproof and soundproof partitions, concrete floors to be reinforced with wire rope and small iron bars embedded below the central axis of the concrete (Ref. 2.11, p. 2*).

Although Wilkinson lived to 83 and his successful building firm survived him, his effect on reinforced concrete design was slight. Indeed, his patent had been forgotten when, in 1955, the contractor demolishing a building erected by Wilkinson in 1865 noticed the unusual nature of the construction and called in Professor Fisher Cassie of Newcastle University, who subsequently located several more Wilkinson buildings (Ref. 2.13). All were small and all used wire rope discarded from colliery hoists as the main reinforcement (Fig. 2.13). This reinforcement was fixed at the top of the

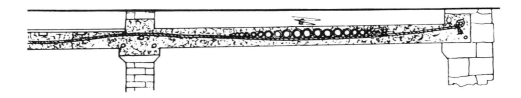

2.13

W.B. Wilkinson's patent reinforced concrete floor. The main reinforcement consisted of discarded colliery cables fixed at the top of the beams and draped towards midspan so that the shape of the reinforcement conformed to that appropriate to a continuous beam. Drainpipes were inserted near midspan to reduce the weight of the concrete.

beams over the support and draped toward midspan; that is, the shape of the reinforcement conformed to that appropriate to a continuous beam (see Section 3.2). In that respect Wilkinson was either ahead of his contemporaries in technical insight or had produced the correct answer by accident. Some small iron bars provided additional reinforcement. Cassie tested some concrete and obtained a crushing strength of 3800 to 4500 psi (3.8 to 4.5 ksi or 26 to 31 MPa); this is comparable to the strength of present-day concrete (Ref. 2.11, p. 28*).

The most daring of the early concrete buildings was erected in England by Andrew Petersen, an admirer of Sir Christopher Wren, who had made a fortune in India. In his retirement he built a thirteen-story concrete tower 218 ft (66.5 m) high at Sway in Hampshire (1879–1885). Petersen used climbing formwork and a hoist operated by one horse to lift the concrete. The thickness of the walls varied from 24 in. (600 mm) at the base to 12 in. (300 mm) at the top. The floors had iron sections as reinforcement. It is not clear why Petersen built this remarkable structure, apparently without expert advice, but it is still standing.

Although telegraphic communication between the United States and Europe had been established, by submarine cable, in 1866, after several unsuccessful attempts, there was still a communication gap, and the American development of reinforced concrete was essentially separate from that of Europe.

An early example was a house that William E. Ward built for himself in 1873 at Port Chester, a suburb of New York. Ward was a mechanical engineer who owned a factory for making bolts. Fire was a common occurrence in the timber-floored (and frequently all timber) houses of the New York region, which has hot summers and cold winters that make heating essential. Ward was determined to build a fireproof residence. After he saw a copy of *A practical treatise on Coignet-Beton and other artificial stone,* published in New York in 1871 by Q. A. Gillmore, he performed a series of experiments on concrete beams reinforced with iron bars. He first ascertained that the reinforcement should be at the bottom and then conducted tests on shear resistance, deflection, and fire resistance.

In 1872 he commissioned Robert Mook, a New York architect, to design a house that would contain virtually no timber. It was built in a style modeled on the French Renaissance, and common for villas in the Hudson Valley, with a mansard roof, two crenelated towers, and an open verandah with Tuscan columns. The only significant amount of wood was in the window frames, doors, and rails for the stairs. The mansard roof was concrete reinforced with iron rods, and the carpets were fixed to nailing strips cast in the concrete floors.

Ward did his own structural design and personally supervised the work. The floors were mostly 3½ in. (90 mm) thick, reinforced with rods 5/16 in. (8 mm) in diameter and with spans as long as 6 ft (1.8 m). The hollow verandah columns acted as downpipes for the rainwater. All the moldings and structural coffered ceilings were cast in place with concrete made of portland cement, sand, and a crushed bluestone. The house is still standing. In 1877 The *American Architect* described it and published several articles on the use of concrete. Ward himself wrote a paper *Beton in combination with iron as a building material,* which appeared in Volume 4 (1882–1883) of the *Transactions of the American Society of Mechanical Engineers.*

Of more significance was the experimental work undertaken by Thaddeus Hyatt, starting in 1855. In 1877 he published for private circulation *An account of some experiments with Portland-Cement-Concrete combined with iron as a building material, with reference to economy of metal in construction, and for security against fire in the making of floors, roofs and walking surfaces.* This paper contained data from tests of about fifty beams for strength and fire resistance and a method of calculation. Hyatt took out several patents in the United States and England.

Ernest Leslie Ransome was born in Ipswich, England, in 1844. In the same year his father started a factory for making precast concrete blocks. In 1866 the son was sent to a branch established by the senior Ransome in Baltimore, but in 1870 he was already working independently. It is not known how much Ransome had heard of Ward's and Hyatt's experiments or any European work; in 1884, however, he patented a spiral-twisted square bar to improve the bond between iron and concrete. From 1889 to 1891 he built the Leland Stanford Junior Museum near San Francisco at the newly founded Stanford University, a Classic three-story fireproof building in which no timber was used. The floors were reinforced concrete, the concrete tiles on the roof were supported on interlocking iron trusses, the windows were metal-framed, and the hall was marble-surfaced. Ransome was probably the first to remove the cement film on the surface of concrete by tooling and thus expose the aggregate. For a nineteenth-century concrete building the finish was exceptionally good.

Ransome developed his ideas on reinforced-concrete frame construction on several multistory buildings before erecting in 1902 the first reinforced concrete skyscraper, the sixteen-story Ingalls Building in Cincinnati which was 177 ft (54 m) high. This building was cast with a frame in which the walls acted merely as curtain walls. It was a functional building in the tradition of the Chicago School (Section 2.6). In the same year Ransome patented his system of concrete-iron frame construction for which he claimed, as the main advantage, that the frame structure obviated the need for thick concrete walls, a method that made large windows possible wherever required.

Joseph Monier, the man most commonly named as the inventor of concrete, was, like Paxton (Section 10.1), first interested in the layout of gardens. In 1863 he patented a cage of iron bars plastered with cement mortar to form a large flower pot. It has been said that Monier needed large pots for growing trees in greenhouses, which were just becoming popular (Section 9.3), in winter and moving them outdoors in summer and that clay pots proved too fragile. On the other hand, Monier may have had other applications in mind and was prevented from taking out a more general patent because Lambot's patent for a *fercement* boat and Coignet's patent for *béton aggloméré* were still valid. In 1873 he took out another patent for constructing bridges, beams, and vaults in reinforced concrete. Monier's main claim to fame was the skill with which he established his foreign patent rights. In 1879 Gustav Adolf Wayss, who had had some experience in the use of plain concrete construction with Roman cement, acquired the German rights to the Monier system. He later formed a partnership with Conrad Freytag, and the firm of Wayss and Freytag established reinforced concrete as a scientifically designed material. We return to this subject in the next section.

In the twentieth century reinforced concrete became the most important structural material, and the numerous professional and trade associations were naturally concerned with its history. The men most commonly named as its inventors are, first, Monier and, second, Lambot. However, one invented a flowerpot, and the other, a boat. Coignet, at an earlier date, devised a method of building construction, but the concrete, although reinforced, was different from the modern form. At the same time Wilkinson in England invented a method of construction comparable to modern reinforced concrete, but it was forgotten, and modern reinforced concrete design in England is based on French inventions. There is, therefore, no simple answer to the question who invented reinforced concrete in Europe.

American development was essentially separate. Depending on whether more importance is attached to theory or practice, reinforced concrete in America was invented, a little later than in Europe, by Hyatt or Ransome.

2.8   THE THEORY AND DESIGN OF REINFORCED CONCRETE

Most of the early work on the theory of reinforced concrete design took place in France and Germany. There were three important problems to be solved. One was the behavior of the concrete structure as a rigid frame. Ransome established the principle at the turn of the century in America, and François Hennebique did so at the same time in Europe.

Hennebique, born in 1842, became a master carpenter and worked for many years on the restoration of medieval cathedrals. He first used concrete reinforced with iron bars to construct fireproof floors in 1879 and conceived the idea of bending up the reinforcement to resist the tension developed in the concrete over the supports. He patented this process in 1892 (Fig. 2.14). Hennebique's building firm gradually did more and more work in reinforced concrete. In 1892 he closed down his construction firm and set up a consulting engineering office; the construction was done by contractors licensed and trained by the Hennebique organization.

Hennebique emphasized that reinforced concrete was not a cheap substitute material. He stressed the need for constant supervision to achieve good workmanship and was the first to substitute steel for wrought iron as reinforcement.

Like Ransome, he regarded the concrete structure as a frame consisting of columns and floors; the external walls carried no load and did not need to be thick. Lightness was a quality evident in all of Hennibique's structures and may be one reason why they compared favorably in cost with others built at the turn of the century.

Among Hennebique's designs was a curved freestanding staircase for the Petit Palais de Champs-Elysées, built in 1898 for the Paris Exhibition of 1900 (the apparently supporting walls were built subsequently to provide storage space). The design was largely intuitive; twenty years later when the torsional strength of reinforced concrete was better understood, few engineers would have dared to build this light structure (Ref. 2.5).

Louis Gustave Mouchel, Hennebique's representative in England, designed the eleven-story skeleton-frame of the Royal Liver Building in Liverpool and set a European record for a tall building in reinforced concrete. Built in 1908 and 1909, its height is 167 ft (51 m) to the main roof, but this is topped by a clock tower and a giant effigy of the legendary liver bird, which takes the total height to 310 ft (95 m). It was the first British skyscraper.

Hennebique introduced the rigid-frame concept for reinforced concrete structures in Europe but did not design his floor structures as continuous beams and slabs,

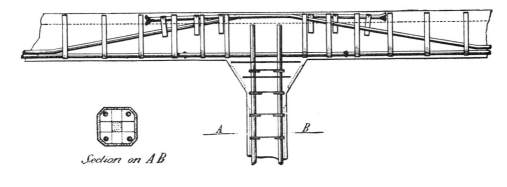

Section on A B

2.14

Arrangement of the reinforcement in the Hennebique system, registered in the United Kingdom and its colonies in 1897 (Patent No. 30143). The opening clause of the specification states the advantages claimed:

The use of strengthened beton in buildings has within recent years greatly developed. It has been thought possible by mixing beton and iron or steel to replace the purely metallic elements of building construction by parts equally incombustible but lighter and more simply and rapidly made. In any case the mixture of cement or hydraulic lime which resists perfectly compression, with iron or steel which most particularly resists tension and flexion, has not hitherto been capable of being carried out in a judicious and rational manner.

although the theory had been published forty years earlier (Sections 3.2 and 4.2). Instead, he used coefficients for the bending moment at the supports and midspan ranging from $1/8$ to $1/30$; that is, the bending moment was taken as $WL/8$, etc., where W is the total load carried by the beam, and L is its span (see Section 2.6). These coefficients appear in most early textbooks on reinforced concrete (Refs. 2.14 and 2.15) and in early building regulations for concrete construction (Ref. 2.16). Reinforced concrete floors have been designed as continuous structures only since 1910.

The cross-sectional dimensions were calculated from these bending moments. It was generally assumed that reinforced concrete beams behaved in accordance with Navier's hypothesis (Section 2.3), namely, that sections originally plane and parallel remained plane and converged on to a common center of curvature. On the other hand, there was disagreement on whether concrete obeyed Hooke's law.* Marsh (Ref. 2.15, pp. 268–275) described a number of hypotheses based on different nonlinear relations between stress and strain, and some exceedingly complicated formulas resulted from these assumptions.

In the end a method proposed by Mattias Koenen in a book *Das System Monier*, published by Wayss and Freytag (Section 2.7) in 1886, became the basis of reinforced concrete design. Koenen proposed that Hooke's law be assumed to apply to concrete as well as iron, that the iron be assumed to resist the whole of the tension, and that

* See the glossary for a definition.

perfect adhesion be assumed between the iron and the concrete. He also noted that the coefficient of thermal expansion of iron and concrete was nearly the same and consequently no stresses would be due to changes of temperature set up between them; this was essential to a fireproof material. Koenen, however, placed the neutral axis at half depth, which ignored the difference in the moduli of elasticity of iron and concrete.

This mistake was corrected in a paper presented to the *Société des Ingénieurs Civils de France* in 1894 by Edmond Coignet (son of François Coignet, Section 2.7) and Napoleon de Tedesco. In 1897 Charles Rabut gave the first course of reinforced concrete design at the *École des Ponts et Chaussées* in Paris, and in 1899 in Liège Paul Cristophe published *Le Béton Armé et ses Applications* which was based on the straight-line theory of Coignet and Tedesco. Charles F. Marsh, in the first English-language book on reinforced concrete design (Ref. 2.15), in turn used Cristophe's method.

The theory of reinforced concrete design was greatly simplified by Emil Mörsch, chief engineer at Wayss and Freytag, in a book *Der Betoneisenbau* (later renamed *Der Eisenbetonbau,* Ref. 2.17), published by the company in 1902. (Mörsch became professor at the Federal Institute of Technology in Zurich in 1904). This remained the standard method until the 1970s when ultimate strength design was introduced in most countries (Section 4.10).

Wayss and Freytag financed the first major experimental investigation of reinforced concrete at the Technical University of Stuttgart by Professor Carl von Bach and Dr. Otto Graf, who succeeded to Bach's chair. The thoroughness of this investigation is impressive even today. Bach and Graf established four ways in which a reinforced concrete beam can fail. One is by crushing of the concrete on the compression face of the beam while the steel is still elastic (Fig. 2.15a). This occurs suddenly, with little warning, and is therefore a dangerous type of failure to be avoided (Section 4.10). The second is by yielding of the steel (Fig. 2.15b) which causes a gradual rise of the neutral axis as the steel continues to deform plastically; eventually the concrete is crushed and the beam collapses. This, however, is a gradual process, and there is warning of impending failure. Beams are normally designed to fail in this manner. The third type of failure, caused by shear which produces a diagonal tension crack (Fig. 2.15c), can be delayed but not entirely avoided by shear reinforcement. Finally, a beam can fail because its bars pull out when inadequately anchored to the concrete. This is prevented by a deformed bar surface or a mechanical anchorage. By 1914 the basic principles were established.

At the turn of the century Hennebique had become "le Napoleon du Béton Armé," but in 1901 he suffered a serious setback when the five-story structure of the *Hotel zum Goldenen Bären* in Basel collapsed during construction with loss of life (Ref. 2.11, p. 124, and Ref. 2.4, p. 62). The commission of inquiry, under the chairmanship of Professor W. Ritter of the Zurich Technical University, blamed faults in the design by the Hennebique organization in Paris and poor workmanship by the local contractor. In particular, it criticized the use of unwashed sand and gravel, including some taken directly from the building site, and the failure to test the quality of the cement and the compressive strength of the concrete.

The collapse led in 1903 to the introduction in Switzerland of the first building regulations for reinforced concrete, and other countries soon followed. In the United

2.15a

Failure of reinforced concrete beams by crushing of the concrete on the compression face.

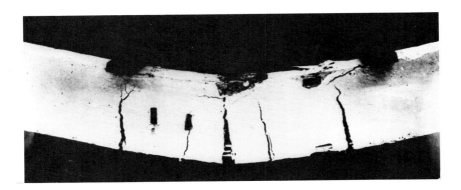

2.15b

Failure of reinforced concrete beams by yielding of the steel on the tension face, followed by a rise of the neutral axis which induces crushing of the concrete.

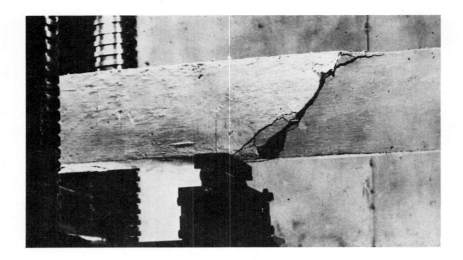

2.15c

Failure of reinforced concrete beams by diagonal tension cracks caused by shear.

States a Joint Committee set up by the American Society of Civil Engineers produced its first code in 1908.

The collapse also highlighted the third important problem in reinforced concrete design, namely, the need for quality control of a material as variable as concrete (see Section 4.6). In the first two decades of this century concrete-mix proportioning was put on a scientific basis. In 1918 Professor Duff Abrams of the Lewis Institute in Chicago enunciated the rule that the strength of concrete is inversely proportional to the water/cement ratio, that is, the ratio that the amount of water bears to the amount of cement in the mix. The ancient Romans knew that good concrete needed as little water as possible, but this rule supplied a quantitative basis.

Durability continued to be a cause for concern. In 1908 the (British) Institution of Civil Engineers appointed a Special Committee on Reinforced Concrete which published an interim report in 1910 and a second report in 1913 (Ref. 2.18). These reports expressed doubts about the safety of reinforced concrete and are typical of the change in British engineering from the adventurous spirit of the eighteenth and nineteenth centuries to the conservatism of the twentieth. They also drew attention to the lack of evidence of the durability of reinforced concrete structures. Some early structures provided insufficient cover over the reinforcement or used porous or inadequately compacted concrete so that the steel rusted and the entire structure deteriorated. Some specifications of the early twentieth century called for only ½ in. (12 mm) of cover and this was sometimes reduced by poor workmanship; today ¾ in. (20 mm) is considered a minimum. The excessive use of hooks to anchor the

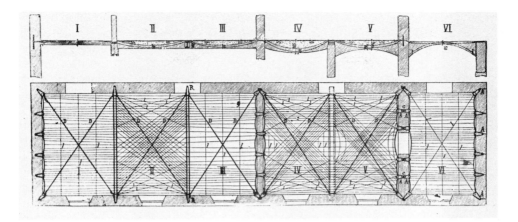

2.16

The Matrai system, originally patented in Hungary. Professor Matrai likened his system to a spider's web firmly attached to the points of support. The system required "three or four times less metal" than ordinary reinforced concrete beams.

In bays marked I and III the wires are placed so as to distribute the load equally over the beams, which only support half the load, being assisted by diagonal cables which are calculated to take the other half. In bays II, IV, and V the diagonal cables are not sufficient to transmit half the load to the extremities of the beams, and they are supplemented by diagonal wires. The transverse wires in bay VI are sufficient to distribute the load equally over the walls, and the diagonals and longitudinals are only employed to intertie the transverse wires.

The upper surface of the concrete followed the curve of the wires, and the hollow was filled with coke breeze concrete.

reinforcement, which caused stress concentrations that lead to cracking, and the lack of shear reinforcement were causes of deterioration in early structures.

Until 1919 reinforced concrete was largely dominated by the various patent systems. Marsh (Ref. 2.16) in 1904 listed forty-three different systems of which fifteen were patented in France, fourteen in Germany or Austria-Hungary, eight in the United States, three in Britain, and three elsewhere. These systems varied from the exceedingly complex (Fig. 2.16) to simple arrangements similar to those employed today (Fig. 2.14).

The latter triumphed because even in the early twentieth century the extra cost of labor (including extra supervision because of the complexity) did not compensate for the material saved.

Because reinforced concrete is composed of two materials, the mathematics of its theory is more complex than for single materials. The principle, however, is essentially simple, and the complexity of its equations in the 1900s was due to factors that are not important in practice. By 1930 the theory had been greatly simplified by the deletion of insignificant variables.

Design can also be made easier by charts and tables; by 1910 they were already making their appearance in textbooks. Although the designer's task is not so simple as in structural steel work, it is mainly because of the lack of standardization of concrete sections. The ability to vary reinforced concrete freely to suit the architectural requirements of a building is one of its great advantages. Reinforced concrete, unlike steel, can be used to form not only the skeleton but also the body of the building; floors and roof are usually included in the structural frame (Section 4.2).

The assessment of the bending moment that can be resisted by a reinforced concrete section is therefore not the main problem posed by reinforced concrete design, even though it occupied the greatest amount of space in the textbooks of the early 1900s. The real difficulty lies in the continuity of the structure. Because the floors and the columns are normally cast in one piece, bending moments are transmitted from each floor slab to the adjacent beams and from each beam to the adjacent columns.

Hennebique solved this problem by using empirical bending-moment coefficients, a method that is still included in building codes and employed for small and simple structural members. It can, however, give entirely erroneous answers, although this was largely ignored before 1910. We examine the problem of continuity in the next chapter.

CHAPTER THREE

Statically Indeterminate

Structures

*Certain truths regarding the material
(concrete) are clear enough. First, it
is a mass material; second, an impressionable
one as to surface; third, it is a material
which may be made continuous or monolithic
within very wide limits; fourth, it is a
material which can be chemicalized,
colored or rendered impervious to water;
fifth, it is a willing material when fresh,
fragile when still young, stubborn when old,
lacking always in tensile strength.*

FRANK LLOYD WRIGHT

Architectural Record, 1928

Our discussion is first related to aspects of the use of reinforced concrete as a structural material, notably the problems created by its surface finish, and the approximate methods of structural design current in the earlier years of this century.

Statically indeterminate structures* are not necessarily superior to statically determinate ones. In the late nineteenth and early twentieth centuries long-span structures were frequently made deliberately structurally determinate. We shall look at the reasons for doing so and at the pin-jointed cantilever girders, arches, and portals that resulted.

We shall then consider the effect of prestressing on both steel and concrete. Finally we shall review briefly the accurate design methods for statically indeterminate structures during the nineteenth century. The more practical design methods of the twentieth are discussed in Chapter 4.

3.1 ARCHITECTURAL CONCRETE

The triumphant progress of reinforced concrete from its hesitant beginnings in the midnineteenth century to its present prominence in long-span and, to a lesser extent, multistory architecture was held back by two problems. One was the lack of satisfactory design theory for statically indeterminate structures and the other was the need for a better method of finishing concrete surfaces.

New concrete can be made to look similar to stucco, highly esteemed in the nineteenth century, particularly if it is troweled to a smooth surface. The gray color can be masked by paint. However, the smooth finish produced by lengthy troweling is due to an excess of cement being worked up to the surface. When the cement dries, it shrinks and produces surface crazing, and sometimes large cracks, if the contraction of the concrete is restrained. Paint can cover surface crazing but not deep cracks. Moreover, the paints available in the early years of this century were not durable on concrete and required frequent renewal.

The late nineteenth and the early twentieth centuries were notable for eclectic revivals, some of which were unsuitable for concrete. Concrete structures in the style of the early Renaissance or Greek Revival performed reasonably satisfactorily, but buildings with sculptural decorations, such as the Neo-Gothic, were troubled by cracking due to the stress concentrations that inevitably occur at the reentrant corners. Casting twenty concrete gargoyles from a single mold instead of sculpting them individually from natural stone proved a great economy, but unlike natural stone the concrete did not age gracefully.

Although concrete was a material much admired by some of the pioneers of the modern school, their use of it often showed a lack of appreciation of its limitations. Le Corbusier in *Towards a New Architecture* illustrated some of his concrete houses and underneath wrote this caption:

The concrete was poured from above as you would fill a bottle. A house can be completed in three days. It comes out from the shuttering like a casting. But this shocks our contemporary architects, who cannot believe in a house that is made in three days; we must take a year to build it, and we must have pointed roofs, dormers, and mansards (Ref. 3.11, p. 212).

* See the glossary for a definition.

Le Corbusier's client, Henry Frugès, described the construction of the concrete house at Pessac in 1926–1927:

A further difference of opinion between Le Corbusier and myself: with his inveterate hatred of all forms of decoration (which stemmed from his Protestant background and the general austerity of his personality) he wanted to leave the walls completely unfinished so that they still showed the marks of the shuttering. I was flabbergasted. He told me that if we wished to offer the houses at the lowest possible price, we could not afford to spend money on unnecessary luxuries. He then launched into a diatribe against ornamentation exclaiming "We are tired of décor, what we need is a good visual laxative! Bare walls, total simplicity, that is how we restore our visual sense"! I understood him only too well, because we both wanted to build economically, but he did not understand me! (Ref. 3.1, p. 9).

The garden suburb of Pessac was not a success. It was an interesting experiment because it provided an early exercise in standardization but the concrete weathered badly. Similar problems were encountered in buildings designed by other modern architects in the 1920s and 30s.

Auguste Perret, who was thirteen years older than Le Corbusier, was the first architect to produce a high quality finish on exposed concrete. He achieved it by acting both as designer and builder and by paying careful attention to detail. Perret's formwork carpentry was always immaculate and his concreting operation was properly controlled. His buildings are still in excellent condition (Ref. 3.10). Perret made extensive use of precast concrete blocks and used them for pierced screen walls in conjunction with glass. The earliest large-scale use of this technique was in the Church of Notre Dame at Le Raincy, a suburb of Paris, constructed in 1922; its walls consisted largely of glass set in concrete.

Perret also appreciated the ease with which arches, vaults, and staircases could be built in reinforced concrete. Many of his buildings have elegantly curved stairs. In the Esders Clothing Factory, built in Paris in 1919, Perret used semicircular arches to support the roof at midspan, thus doubling the clear span of the beams carrying the roof without a central line of columns. Apart from the functional advantage of the long span, the arches created visual interest in an otherwise plain interior.

Most of Perret's reinforced concrete frames emphasized the horizontal and the vertical members, and the columns were usually shaped in accordance with Vitruvius' rules for entasis. Perret was a great innovator but not a revolutionary like Le Corbusier, and he retained his loyalty to classicism even in his late buildings in the 1930s and 40s. The great increase in the cost of labor in the years following World War II made Perret's methods of construction obsolete. It was no longer economically feasible to build precision formwork for site-cast concrete and then fit precisely precast concrete blocks between the site-cast columns.

Among the great architects of the early twentieth century the one most percipient of the qualities of concrete was Frank Lloyd Wright, although the material was never so important in his work as in Perret's or Le Corbusier's. Wright used concrete for the first time in 1905 in the E-Z factory in Chicago, but the reinforced concrete frame was faced with brick. In 1906 in the Unitarian Church in Oak Park, Illinois, the entire facade was made of site-cast concrete with an exposed pebble-aggregate finish. In the 1920s Wright used precast concrete blocks in a number of buildings, in some as permanent formwork for the reinforced concrete frame, and thus solved the problem that had defeated Le Corbusier, namely, how to produce an acceptable concrete finish on vertical surfaces. At the same time he avoided the high labor cost of Perret's precision-made formwork.

In the 1930s concrete was still an engineer's rather than an architect's material. Even cement manufacturers in their publicity tended to stress "the ease with which concrete can be veneered with natural stone, producing a building which is indistinguishable from one built from natural stone, but so much cheaper." The problem of the surface finish of concrete was solved only in the 1950s and 60s (Section 9.5).

3.2 THE THEORY OF BUILT-IN AND CONTINUOUS BEAMS

About 1900 reinforced concrete became the new material's accepted name; it gradually displaced the terms ferroconcrete, armored concrete, iron concrete, metal concrete, and other less-used terms like calceolithe-metallico-neurophore.

Rules for the design of reinforced concrete frames were included in the first textbooks on the subject, published in the first decade of the twentieth century (Refs. 2.14, 2.15 and 2.17). Before 1910 the frame was generally split up into slabs restrained at the ends and supported on beams, which in turn were considered restrained at the ends and supported on columns.

The appropriate theory had in 1826 been included by Navier in his *Résumé des leçons* (Section 2.3). Navier noted that a beam fully restrained at its ends (or built-in) had a zero slope at each end. He derived the relation between the deflection y and the bending moment M in the form (Ref. 1.1, Section 8.4).

$$EI = \frac{d^2y}{dx^2} = M \qquad (3.1)$$

where E is Young's modulus* and I is the second moment of area. Furthermore, the slope of any curve is its differential coefficient so that the slope

$$\theta = \frac{dy}{dx} \qquad (3.2)$$

In a built-in beam we have $y = 0$ and $dy/dx = 0$ at both ends because the beam is supported at both ends and prevented from any rotation by the end restraints. These four statements about the geometry of the beam are sufficient to solve the problem, and Navier did so by twice integrating the bending moment and then determining the constants of integration from $y = 0$ and $dy/dx = 0$ at each end of the beam.

This is rather laborious and a method devised by Otto Mohr, Professor of Engineering Mechanics at the Stuttgart Technical University, later at the Dresden Technical University, is generally preferred today. In 1868 Mohr published a paper in the *Zeitschrift des Architekten und Ingenieurvereins, Hannover* (Journal of the Society of Architects and Engineers, Hanover) in which he pointed out that Navier's equation

---

* See the glossary for a definition.

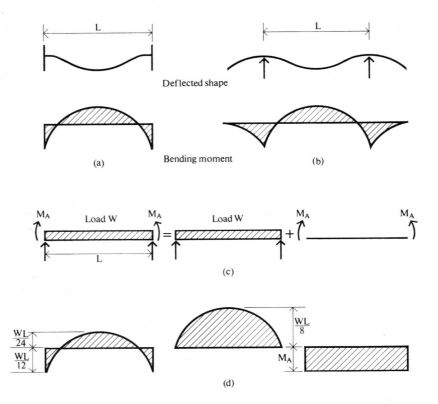

Deflected shape

Bending moment

(a)   (b)

(c)

(d)

(3.1) was mathematically similar to the equation that relates the bending moment to the load; this could therefore be used to determine slope and deflection by a mathematical analogy now known as *Mohr's theorem*.

We can divide the bending moment diagram for a built-in beam (Fig. 3.1) into two parts: that due to a simply supported beam carrying the same loads over the same span (see Fig. 2.6b) and that due to the end restraints, which is a constant bending moment. The first causes the beam to flex downward and the other causes it to flex upward. Because the beam is in fact rigidly restrained at the ends, the slope at the ends is zero. From Mohr's theorem it can be shown that the areas of the two parts of the bending-moment diagram must be equal, and this gives the magnitude of the restraining moments at the supports. It follows that the maximum bending moment occurs at the supports, not at midspan. This, however, is less than for a simply supported beam. Moreover, it is a negative moment which causes tension on top of the beam and compression on the bottom face. Thus we require the reinforcement for the concrete on the top face at the supports. Another smaller maximum bending moment occurs at midspan, and there we require the reinforcement at the bottom.

This approach was used by Hennebique in his early structures (Section 2.8), although with modified numerical coefficients to allow for the fact that reinforced concrete beams are restrained but not fully built in at their ends.

In the 1910s the theory of continuous beams was applied to the design of reinforced concrete frames. It had been derived by B. P. E. Clapeyron in connection

3.1

Bending moments in built-in beams.

(a) Because the beam is rigidly restrained at the supports, its deflected shape must remain horizontal at those points. The supports impose restraints on the beam which are equivalent to vertical reactions and to restraining moments M_A.

(b) The beam behaves therefore in the same way as the central portion of a simply supported beam with cantilever overhangs that are just long enough to keep the beam horizontal at the supports, and the magnitude of the restraining moments M_A is therefore equal to that of the cantilever moments.

(c) The bending moment diagram is composed of two parts: the usual bending moment diagram for a simply supported beam (Fig. 2.6) and a superimposed diagram due to the constant restraining moment M_A.

(d) It can be shown by Mohr's theorem that the areas of these two component parts must be equal. Because a parabola has two-thirds of the area of a rectangle enclosing it,

$$M_A L = \tfrac{2}{3} \, \tfrac{1}{8} \, WL \cdot L$$

which gives $M_A = WL/12$. This is also the maximum negative bending moment, which requires reinforcement on *top* of a concrete beam. By subtraction the maximum positive bending moment is

$$\frac{WL}{8} - \frac{WL}{12} = \frac{WL}{24}$$

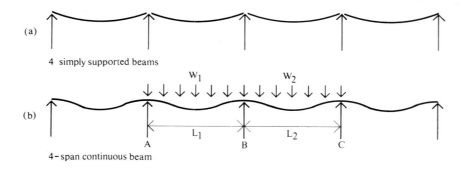

(a)

4 simply supported beams

W_1 W_2

(b)

L_1 L_2

A   B   C

4-span continuous beam

3.2

Simply supported beams deform independently of one another, and there is no continuity of curvature (a). In a continuous beam the spans are joined together and interact with one another (b). The deflected shape of the beam shown at (b) is thus a continuous curve. At point B the slope (or angle) at the right-hand end of the span L_1 is thus the same as the slope at the left-hand end of the span L_2. This equality is caused by the interaction between the two spans, which in turn is due to a restraining moment M_B. We can derive the slope at B in the span L_1 in terms of the restraining moments M_A (at A) and M_B (at B), and the slope at B in the span L_2 in terms of M_B and M_C, using Mohr's theorem. Equating these two slopes we obtain the theorem of three moments:

$$M_A L_1 + 2 M_B (L_1 + L_2) + M_C L_2 = -\tfrac{1}{4} (W_1 L_1^2 + W_2 L_2^2)$$

with the reconstruction of the d'Asnieres Bridge near Paris in 1849. It was later modified by H. Bertot in a paper published in 1855 in the *Mémoires* of the *Société Ingénieurs Civils de France* (Vol. 28, p. 278). In this more convenient form it is known as the *theorem of three moments* (Fig. 3.2). For several decades it was the principal method used in the design of reinforced concrete frames. It was assumed that the columns were sufficiently flexible to exercise no appreciable restraint on the much stiffer continuous floor structure; this is reasonable for a beam-stiffened floor on relatively slender columns.

By the middle of the twentieth century reinforced concrete was rivaling steel as the principal structural material for tall buildings. Because of the increase in height, columns (which had to carry all the floor loads above them) greatly increased in size, and the wind pressure produced bending stresses in the columns. The use of a rigid-frame design thus became a necessity (Section 4.2).

3.3   ADVANTAGES AND DISADVANTAGES OF CONTINUITY AND STATICALLY INDETERMINATE STRUCTURES

Although the maximum bending moment, and therefore the amount of material required, is lower for continuous beams than for the simply supported, the advantages are not all on the side of continuity. Foundation settlement, if not excessive, is relatively harmless in beams that are statically determinate but dangerous in the continuous beam (Fig. 3.3).

In 1866 W. Gerber patented a type of continuous beam that now bears his name in which statical determinateness was restored by inserting a sufficient number of pin joints (Section 2.2) and which was therefore unaffected by the slight settlement of one or more of the supports. The Gerber beam can be worked out by statics alone (Fig. 3.4).

It also has a constructional advantage, for the cantilevered and short-span beams can be constructed separately. This principle was used by Sir Benjamin Baker in the Firth-of-Forth Railway Bridge, built between 1883 and 1889 (Fig. 3.5). In this bridge the girders cantilever from *rigid* supports. Consequently two pin joints per span are required for statical determinateness. The Forth Bridge had the longest span of the

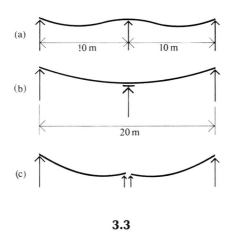

3.3

Effect of differential settlement on simply supported and on continuous beams.

(a) Deformed shape of a continuous beam.
(b) Deformed shape of the beam after sufficient settlement has occurred at the central support to turn it into a single simply supported span. The bending moment is thus quadrupled and its sign changes from negative to positive.
(c) Two simply supported beams are not affected by differential settlement at the central support.

Advantages and Disadvantages of Continuity and Statically Indeterminate Structures

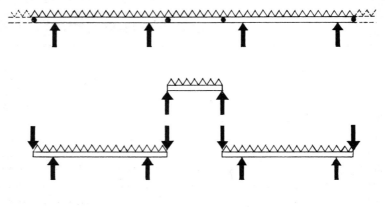

3.4

(a) In a Gerber beam two pin joints are introduced into alternate spans.
(b) The continuous beam is thus broken up into short inset spans and simply supported beams with cantilever overhangs.

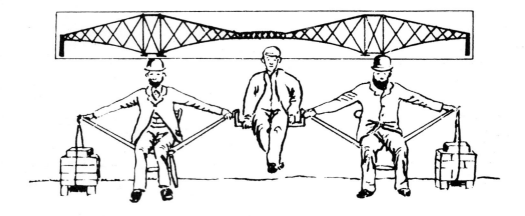

3.5

The structural principle of the Forth Bridge.

During one of his lectures Sir Benjamin Baker demonstrated the principle of the cantilever bridge with a living model consisting of two men supporting a boy. The outstretched arms and the sticks of the men represented the cantilever spans, the boy represented the inset span and the piles of bricks, the anchorages (Ref. 3.2, p. 38).

3.6

The Firth-of-Forth Railway Bridge, built 1883–1889. Overall length 5350 ft
(1630 m); maximum span 1710 ft (513 m).

nineteenth century, and even today it is still one of the world's most impressive
structures. It has two clear spans of 1710 ft (521 m), a length of 5350 ft (1630 m), and
three intermediate piers (Fig. 3.6); pinned inset spans make it statically determinate.

Statical determinateness was also introduced into long-span buildings in the late
nineteenth century. Long spans were a convenience in covered market halls, more
and more of which were built of iron in the nineteenth century; they were a necessity
in railway stations in which intermediate columns got in the way of trains switching
from one track to another. Iron arches were capable of providing the necessary clear
span. One of the first of the great iron railway station arches, built in 1848 by
Isambard Kingdom Brunel for Paddington Station, London, had a span of 104.5 ft
(31.9 m). In 1866, W. H. Barlow built a parabolic arch for London's St. Pancras
Station with a span of 244 ft (74.4 m), still the greatest in any European terminal (Fig.
3.7). These structures could not, however, be accurately designed by the mechanics
then known, and in 1865 J. W. Schwedler used the principle of the three-pin arch
(Fig. 3.8) for the first time in two German railway stations which rendered the arches
statically determinate.

3.7

The parabolic arch of St. Pancras Railway Station, London, designed by W. H. Barlow in 1866. The span is 244 ft (73.4 m). The great arch is in curious contrast with the Neo-Gothic building in front of it which housed the offices and the station hotel. The latter was designed by Sir Gilbert Scott whose remark "that it was possibly too good for its purpose" was characteristic of the widening split that developed during the late nineteenth century between architecture and engineering.

In 1889 the *Galerie des Machines* was erected for the Paris Exhibition to be held later that year. It was designed by the architect Ferdinand Dutert and the engineer Cottançin and set a new record for a buiLding span (113 m or 370 ft). It was built from three-pinned portal frames with curved sides and a gabled roof, and its roof, end walls, and side walls consisted mainly of glass, which emphasized the lightness of the structure. Its length was 420 m (1378 ft). The building was later demolished. Siegfried Giedion commented on the contrast between the appearance of this structure compared with a traditional arch:

Each arched truss is made up of two segments. A pin unites them at a pivotal point high above the centre line of the hall. Moving downward, the trusses become increasingly attenuated until they appear scarcely to touch the ground; moving upwards, they spread and gain weight and power. The usual proportions seem to be exactly reversed. These truly articulated arches disturb, or rather disrupt, traditional static feelings with regard to the rational relations of support and load. Elongated like immensely drawn-out cantilevers, the trusses embody movement in all their parts. Nothing remains of the quiet stone architecture of the barrel vault. A new sort of movement, penetrating space—as new in kind as that achieved in Borromini's cupolas—is created here (Ref. 3.12, p. 271).

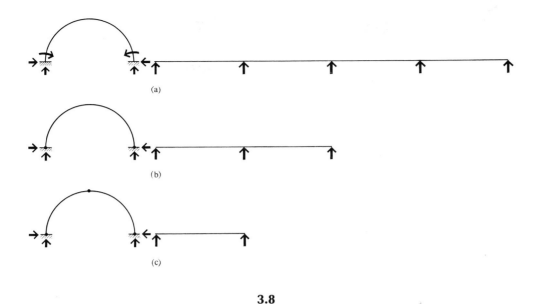

3.8

Comparison of arches and continuous beams. The effect of adding an additional end restraint to an arch is the same as that of adding an additional intermediate reaction to a beam. Statical determinateness can be restored by adding sufficient pin joints.

In the twentieth century this same impression was created in reinforced concrete by Robert Maillart in a number of spectacular bridges in the Swiss mountains, of which the one at Salginatobel, built in 1929–30 with a span of 90 m (295 ft) is the best known (Ref. 3.13). In these bridges the road slab was merged with a three-pin arch to form a massive concrete structure (Fig. 3.9) which reached its greatest thickness near

3.9

The characteristic style of Maillart's reinforced concrete bridges resulted from the junction of a three-pinned arch with the road slab. Because the arch is three-pinned, the bending moments are zero at the crown and supports. The arch is thickened in the region of maximum bending moment.

the point of maximum bending moment and contracted to hinges at both supports and the crown.

The statically determinate long-span structure was particularly popular in the late nineteenth and early twentieth centuries because its theory was clearly understood. No secondary stresses are caused by small foundation settlement, thermal movement, or minor inaccuracies in manufacture. On the other hand, the removal of a single member turns the structure into a mechanism that collapses.

The statically indeterminate structure has therefore, by its very nature, a greater margin of safety. Only when all the redundancies have been removed by the formation of hinges and one further hinge has formed can the structure collapse. This

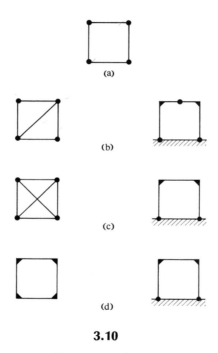

3.10

The square frame.

(a) Mechanism. A square frame with four pin joints is structurally unstable.
(b) Statically determinate frames: panel of a Pratt truss; three-pin portal frame.
(c) Frames with one redundancy: panel of truss with crossed diagonals; portal frame pin-jointed to rigid foundations.
(d) Frames with three redundancies: panel of Vierendeel truss (i.e., a frame with all-rigid joints); portal built into rigid foundation.

3.11a

Gladesville Bridge, Sydney, Australia, completed in 1964. The bridge has a span of 1000 ft (305 m).

collapse mechanism is utilized in the plastic theory (Section 4.10) for structural design. Rigid frames are thus stronger for the same amount of material and provide a greater margin of safety against accidental damage by internal explosion; for example, from domestic gas appliances, earthquakes, and collapse when the structure is weakened by fire.

In a statically determinate structure a slight inaccuracy in the length of any one member causes a slight though generally insignificant distortion in the structure's shape. In a statically indeterminate structure any inaccuracy introduces possibly very high stresses; for example, the diagonal in the Pratt truss in Fig. 3.10b is bound to fit, whatever its length, because if necessary the pin joints will rotate to adjust themselves to the required shape. A second diagonal (Fig. 3.10c) would, however, have to fit accurately or *prestress* the entire frame. Prestressing may seriously damage the truss if it is applied unintentionally. It can be used deliberately to increase the load-carrying capacity of the structure.

3.11b

Gladesville Bridge was formed by four parallel arches consisting of individual precast concrete voussoirs (blocks) assembled on formwork with a floating crane. As each arch was completed it was compressed by hydraulic jacks, which caused it to lift off the formwork. The open joints were then filled with concrete around the hydraulic jacks; after it had hardened, the jacks and the formwork were removed. The arch remained prestressed due to its own great weight.

Thus the bottom chord of many trusses in the late nineteenth and early twentieth centuries was fitted with a turnbuckle that was tightened after the entire truss had been assembled. It flexed the truss upward and induced a deflection opposite to that caused by the load. The turnbuckle could also be used to induce stresses opposite to those caused by the load if the truss was statically indeterminate and this increased the load-carrying capacity. The member containing the turnbuckle, however, needed to be made larger because it had to provide the tensile prestressing force in addition to the tension due to the load carried by the truss.

Toward the end of the nineteenth century hydraulic jacks were occasionally used for the same purpose; for example, they were employed in the Eiffel Tower in 1889 at the base of the legs to correct erection errors and to adjust the reactions. A more recent and much larger jacking operation was undertaken in the construction of the Gladesville Bridge (Fig. 3.11). The same principle is used in *prestressed steel* structures; a high-tensile steel cable provides the prestress. The most useful application, however, is in prestressed concrete.

3.4 PRESTRESSED CONCRETE

The first attempts to use prestressed concrete are almost as old as the early buildings in reinforced concrete (Section 2.7). In 1886 C. W. F. Doehring patented in Germany a method for the manufacture of mortar slabs with steel wire reinforcement. In 1888 P. H. Jackson patented a method for inducing preliminary compressive stresses in concrete arches and floor structures by tightening iron tie rods with turnbuckles (Ref. 2.11).

Doehring stressed his wires before casting the concrete, a method now known as pretensioning. Jackson cast his reinforcement into the concrete and prevented adhesion between the concrete and iron by placing the latter in a tube or covering it with paper and stressing it after the concrete had hardened; this method is now known as posttensioning. Several more patents followed in the United States and in various European countries, but none of the methods was successful because no allowance was made for the loss of prestress due to shrinkage and creep.

The increase of the deformation of concrete with time had been observed by W. K. Hatt (Ref. 3.3) in 1907. In 1915 F. R. McMillan noted an appreciable difference in the long-term deformation of heavily and lightly loaded concrete (Ref. 3.4) and thus separated the influence of shrinkage and creep, but their effect on the loss of prestress was not properly appreciated until Eugène Freyssinet investigated the problem in the 1920s. He filed his first patent application in France in 1928.

Mathias Koenen, for example, who took out a patent in 1888, specified a prestress of 600 kg/cm^2 (60 MPa or 8.5 ksi) (Ref. 2.11). The shrinkage of concrete is approximately 3×10^{-4} and the modulus of elasticity of iron is 200,000 MPa (29,000 ksi). A contraction of 3×10^{-4} causes a change in the stress of the iron of $3 \times 10^{-4} \times 200,000 = 60$ MPa, and therefore the entire prestress was dissipated by shrinkage alone.

The problem was solved empirically by Karl Wettstein (Ref. 2.11) who in 1921 registered a patent in Austria for the use of concrete prestressed with piano wire 0.3 mm (0.01 in.) thick that had a strength of 140 to 200 kg/mm^2 (1400 to 2000 MPa or 200 to 290 ksi). The loss of prestress due to shrinkage was thus only 60/1400 or 4% of the total prestress, and even after loss due to creep (which is caused by the squeezing of water from the pores of the concrete by a sustained high load) there was still sufficient prestress to prevent the concrete from cracking under load. Wettstein's thinnest prestressed concrete planks deflected visibly under load, like planks of timber, and recovered their original shape elastically. The capacity of concrete to sustain a large deformation and recover its original shape came as a surprise to most people, as indeed it still does today to those unfamiliar with the phenomenon.

The results of Freyssinet's research, published in 1933, placed prestressed concrete design on a scientific basis. By 1939 it was an established structural material in France and Germany, and by that time several successful patent systems had been

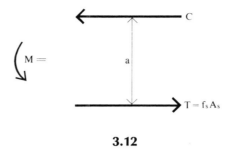

3.12

Resistance moment M of a beam. The moment is formed by the compressive force, C, the tensile force T, and the moment arm a. In a reinforced concrete or a prestressed concrete beam the compressive force is due to the concrete and the tensile force is due to the steel. The tensile force in steel is in either case the product of the stress in the steel f_s and the cross-sectional area of the steel A_s.

granted for its use. In 1940, when France fell, the Germans proceeded to build elaborate defenses for their army and navy along the Atlantic coast.

Steel was one of the materials in short supply during the war, and reinforced concrete required an appreciable quantity of steel reinforcement because it provided the tensile force needed to resist the bending moment (Fig. 3.12) imposed on the concrete beams by the loads. This tensile force is $f_s A_s$, where f_s is the stress in the steel and A_s is the steel's cross-sectional area. The maximum stress for prestressing steel is about ten times that for ordinary reinforcing steel, and therefore the amount of steel required is greatly reduced. It is not possible to use high-tensile steel in ordinary reinforced concrete because its elastic extension (ten times as much as for ordinary steel) would produce excessive extension of the concrete and large cracks would cause excessive deflection and permit corrosion of the steel. This would lead to disintegration of the concrete, for rust has a bigger volume than steel.

Because prestressed concrete saved steel, it developed rapidly in central Europe during the period between 1940 and 1944. After the war the worldwide steel shortage continued for several years, and prestressed concrete made further rapid progress in all the developed countries.

Prestressed concrete is still an important material in the construction of bridges, but when steel became plentiful again the cost of the prestressing operation made the material uneconomical for general use in buildings. It is still used for long spans in which the greater self-weight of reinforced concrete is a disadvantage (see Section 6.7) and has also found a special application in the design of flat plates (Section 4.4).

A new concept of prestressed concrete design was proposed in 1961 by T. Y. Lin, Professor of Civil Engineering at the University of California (Ref. 3.9, p. 339). Lin designed prestressed concrete girders with high-tensile steel cables draped to conform to the bending moment diagram. Thus, if the beam is simply supported and carries a uniformly distributed load, its bending moment diagram is a parabola (Fig.

Statically Indeterminate Structures

2.6). The cable is then also shaped as a parabola, with its eccentricity varying from ϵ at midspan to zero at the ends. The moment set up by the cable therefore varies from $P\epsilon$ at midspan to zero at the supports, where P is the prestressing force in the cable. This moment tends to flex the beam upward opposite to the bending moment due to the imposed loads. If the two moments are made exactly equal and opposite, the beam is free from bending moment and acted on only by the compressive prestressing force P.

We noted earlier in this section that due to creep the deflection of concrete beams increases over a period of time. This is a particular problem with flat plates* (Section 4.4), and those with longer spans are therefore prestressed by the load-balancing method to reduce creep deflection. The load to be balanced is the dead load due to the weight of the building plus about half the live load due to its occupants and contents. Thus the flat plate is flexed slightly upward when the live load is zero (which is rare) and slightly downward when the full live load is acting (also rare). Under the common case of a partial live load the elastic deflection is small because the eccentric prestress almost balances the bending moment due to the total load. The long-term increase of this deflection is therefore also small.

3.5 THE DESIGN OF SMALL STEEL FRAMES FOR LATERAL LOAD

The forces due to earthquakes and wind had been ignored in structures before and during the greater part of the nineteenth century. The importance of wind forces was tragically demonstrated by the collapse of Tay Bridge (Ref. 3.2, p. 23, Ref. 3.5, p. 190). This had been designed by Sir Thomas Bouch and built between 1871 and 1878 over the Firth of Tay, a tidal estuary 35 miles (56 km) north of Edinburgh. The bridge consisted of eighty-five wrought-iron lattice girders supported high above the water on cast-iron columns. At its completion it was the world's longest bridge. On a rainy night in 1879 more than 3000 ft (1000 m) of the bridge and an entire passenger train were blown into the water by a gale; there were no survivors.

The court of inquiry set up in 1880 revealed deficiencies in the workmanship and the quality of the iron castings, but it also started a controversy on the allowance that should be made for wind pressure.

In 1759 John Smeaton presented a paper to the Royal Society in which he suggested a design wind pressure of 6 psf (pounds per square foot) for high winds, 9 psf for very high wind, and 12 psf (575 Pa) for storms or tempests. Because of Smeaton's successful construction of the lighthouse on the stormy Eddystone, these figures found wide acceptance.

Bouch had consulted the Astronomer Royal, Sir George Ayrey, who wrote:

We know that upon very limited surfaces, and for very limited times, the pressure of the wind does amount to sometimes 40 psf or in Scotland probably more. . . . I think we may say that the greatest wind pressure to which a plane surface like that of the bridge will be subjected on its whole extent, is 10 psf.

* See the glossary for a definition.

This was the figure that Bouch had used. Most of the witnesses called at the inquiry gave evidence to prove that it was much too low. George Stokes, a noted authority on fluid mechanics who held the Lucasian Chair of Mathematics at Cambridge, thought it should have been greater than 50 psf (2.4 kPa).

Before Sir Benjamin Baker designed the Forth Bridge (Section 3.3) he made extensive measurements at the site and from them arrived at a wind pressure of 56 psf (2.7 kPa). The (British) Board of Trade subsequently specified this figure as the design wind pressure, and it was used in the Forth Bridge and in the reconstruction of the Tay Bridge by W. H. Barlow.

In retrospect it seems that the Tay Bridge disaster was due less to low design pressure than to the absence of continuous lateral wind bracing.

The need to design iron-framed buildings for wind was not immediately accepted, and wind pressure was not considered in the early iron- and steel-framed buildings of

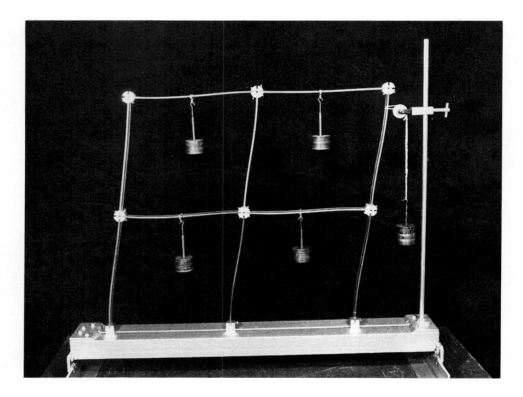

3.13

A rigid frame deformed by a horizontal force forms points of contraflexure at which the curvature changes from convex to concave. At these points the slope (i.e., the angle of the elastically deflected beams and columns) is a maximum. These points occur approximately half-way along the beams and columns.

the 1880s. In the 1890s, however, some buildings more than fifteen stories high were provided with diagonal wind bracing (Section 2.6). The first paper on wind bracing for high buildings appeared in 1892 in the *Transactions of the American Society of Civil Engineers.*

The resistance of building frames to lateral forces cannot be solved by statics alone. If a rigid frame is deformed by a horizontal force (Fig. 3.13), points of contraflexure can be identified where the curvature changes from convex to concave. In the early twentieth century the problem was commonly solved by assuming that these points occurred half-way along the beams and half-way along the columns. At the points of contraflexure the slope has a maximum value, and it can be shown (Ref. 3.6 p. 242) that at these points the bending moment is zero. We have noted (Section 2.2) that at a pin joint the bending moment is, by definition, zero and can therefore assume that a "hypothetical pin joint" exists at each of these halfway points (Fig. 3.14).

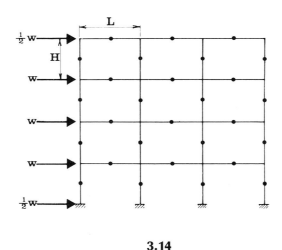

3.14

It can be shown (Ref. 3.6, p. 242) that at the points of contraflexure the bending moment is zero. It is therefore possible to assume a "pin joint" at these points. This assumption renders the structure statically determinate.

The assumption of hypothetical pin joints renders the frame shown in Fig. 3.14 statically determinate; the bending moments can then be easily derived (Fig. 3.15). This simple approximate method proved satisfactory for determining the bending moments due to horizontal loads in small building frames and remained in use until more accurate methods were developed (Chapter 4).

By the late nineteenth century it was known that an earthquake exerted horizontal forces on buildings, although it was not possible to assign a numerical value to this force until much later (Section 4.8).

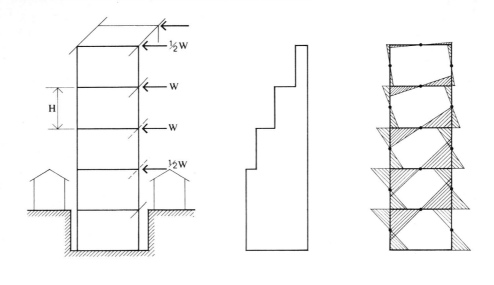

3.15

Bending moments obtained by assuming hypothetical pin joints in accordance with Fig. 3.13. For the frame shown at the left the shear force (at the center) is the cumulative sum of the horizontal forces (i.e., is ½W at the top and increases to 3W at the bottom). The bending moments (at the right) are zero at the hypothetical pin joints. They increase linearly in the columns to a maximum of V. $\frac{1}{2}H$ at the column-beam junctions. These moments are then transmitted to the beams; each beam is subjected to the sum of the bending moments in the adjacent columns at the beam-column junction and this reduces to zero at the hypothetical pin joints.

The San Francisco earthquake of 1906 (Ref. 3.7) provided convincing evidence of the superiority of framed structures over those built with loadbearing masonry or brick walls. Some buildings were designed for horizontal earthquake forces as early as the 1910s (Section 4.8); the Nippon Kogo Bank, a rigid reinforced-concrete frame designed for a maximum ground acceleration of 0.065 g(where g is the acceleration due to gravity) successfully withstood the great Tokyo earthquake of 1923 (Ref. 3.8).

3.6 THE STRAIN ENERGY METHOD

The search for an accurate method for the design of rigid frames started in the 1860s. The first useful solution, based on the principle of the conservation of energy, was

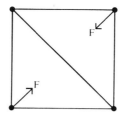

3.16

The Maxwell-Mohr method of solving statically indeterminate structures applied to a frame with crossed diagonals. The second diagonal in Fig. 3.10c is replaced by two equal and opposite external forces F, one acting on the upper and the other on the lower pin joint. The distance between these joints depends on the elongation of the second (redundant) diagonal under the action of the forces F.

obtained by Clerk Maxwell (see Section 2.1) and published in the *Philosophical Magazine* in 1864 (Vol. 27, p. 294). Maxwell started with a statically determinate frame and then considered the additional restraints (or redundant members), which turned it into a statically indeterminate frame (Fig. 3.10), as equivalent to unknown external forces. These restraints are invariably associated with a statement about the geometry of structure; we noted in Section 3.2 that a built-in beam can be solved by utilizing the zero end slopes due to the end restraints. Maxwell replaced the redundant diagonal in Fig. 3.10c (left side) with two equal and opposite external forces F, one acting on the upper pin joint, the other on the lower (Fig. 3.16). The distance between these joints depended on the elastic elongation of the redundant diagonal under the action of the forces F, and this statement supplied the additional equation needed to solve the problem.

Maxwell's paper was written in abstract terms without illustrations and was over-looked until Otto Mohr (see Section 3.2) rediscovered it in 1875. It is now known as the Maxwell-Mohr method.

Alberto Castigliano developed a variation of this method independently in 1873. Castigliano came from a poor family, and after working as a teacher he obtained a scholarship to the faculty of pure mathematics at the University of Turin, where he completed all the examinations for the three-year curriculum in one year. He then presented a dissertation on the *principio del minimo lavoro* (principle of least work) enunciated without proof by General Count Luigi Federigo Menabrea, a military engineer, in a book published in 1867. Castigliano proved Menabrea's theorem and applied it to a number of problems in his dissertation, which he defended in 1873. Following his graduation he worked for the Northern Italian Railway and in 1879 published in French an expanded version of his thesis which contained more applications to practical engineering problems. This book was translated into German in 1886 and into English in 1919 (Ref. 3.14).

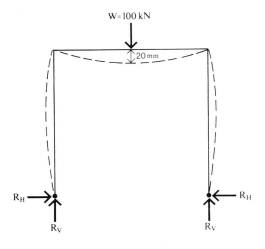

W = 100 kN

20 mm

R_H

R_H

R_V

R_V

3.17

The Castigliano method of solving statically indeterminate structures applied to a portal frame hinged at both supports, carrying a load W. This produces statically determinate reactions R_V and statically indeterminate reactions R_H, which are needed to prevent spreading of the supports. The frame is consequently not displaced along the line of action of R_H, and the differential coefficient of the total strain energy of the frame with respect to R_H is therefore zero.

Castigliano's method is still the most widely used in the design of statically indeterminate curved structures, such as elastic arches [i.e., those built of steel or reinforced concrete, not masonry (see Ref. 1.1, Section 7.6)].

The Castigliano method is based on the strain energy in the structure; for example, let us consider the portal frame in Fig. 3.10.c (right) which carries a central vertical load W (Fig. 3.17) that produces bending moments, direct compressive forces, and shear forces in all three of its members. These, in turn, produce elastic strains that are stored in the structure as strain energy and recovered when the load is removed. The strain energy stored in the structure equals the loss of potential energy of the load; for example, if the load is equivalent to a force of 100 kN and the resulting central deflection is 20mm, the loss of potential energy is 100 kN × 20 mm = 2 kNm and the strain energy must be the same.

Castigliano proved that if the stored strain energy were minimized by partial differentiation with respect to any of the forces acting on the frame the result would be the displacement of this force along its line of action. Let us apply this statement to the specific case of the portal frame in Fig. 3.17. We can determine its vertical reactions R_V by statics, but not its statically indeterminate horizontal reactions R_H. Because the frame is firmly pinned at its foundations, its displacement along the line of action of the horizontal reaction R_H is zero, and consequently the differential coefficient of the strain energy with respect to either of the horizontal reactions R_H is zero. We thus set up equations for the total strain energy stored in the frame; this equals the sum of the products of the forces multiplied by the resulting deformations and the moments multiplied by the resulting rotations. In order to calculate the minimum, we then differentiate this sum, equate to zero, and obtain an equation for the unknown horizontal reactions.

This method is more sophisticated than most of those developed fifty years later (Section 4.2); however, it is laborious for all but the simplest frames. It was not a practical method for multistory rigid frames before the invention of electronic computers because of the amount of calculation required (Section 5.2).

It is possible that Castigliano, who was both an eminent mathematician and a sound practicing engineer, would have developed a more practical design procedure if he had lived longer; he died in 1884 of pneumonia at the age of 37, three weeks after he had received a gold medal from the employees of the railway company for his work on improving the company's pension fund.

In his book (Ref. 3.14) Castigliano applied the method mainly to arched wrought-iron roofs (steel was not yet used in Italy at the time) and two masonry arch bridges. It was the first time that statically indeterminate iron arches had been accurately designed. On the other hand, for the masonry arches the design procedure was a retrograde step because the masonry arch fails by opening of the joints between the blocks of stone. The confusion between elastic and ultimate-strength design was still unresolved (Ref. 1.1, Sections 7.6 and 8.3).

CHAPTER FOUR

Tall Buildings

Multiplication is vexation
Division is as bad
The rule of three does puzzle me
And practice drives me mad.

Elizabethan rhyme

We conclude the story of the multistory building in this chapter. The increasing complexity of the calculations posed many problems which we shall consider in Chapter 5.

The architecture of tall buildings produced many arguments between the traditional and modern schools of architecture eventually resolved in favor of the latter. The structural design also called for a new approach. The moment distribution method satisfied the requirements of the 1930s, 40s, and 50s. Two-way slabs, flat slabs, and flat plates became common during that time.

The search for an ultimate strength theory, abandoned in the nineteenth century, was resumed, and in recent years it has to an increasing extent become embodied in building codes. This necessitated not only a reexamination of the concepts of structural safety and a more accurate assessment of the loads carried by the structure but also a study of the problems caused by deformations.

The construction of tall buildings almost ceased from the beginning of the Great Depression to the end of the era of postwar reconstruction in the 1950s. When it was resumed, new three-dimensional systems that saved an appreciable amount of material and produced more elegant structures were introduced.

4.1 TALL BUILDINGS—THE ARCHITECTURE OF THE NEW WORLD

We noted in Sections 2.6 and 2.7 that the earliest tall iron-framed buildings were erected in the 1880s in Chicago, where this development was not impeded by respect for the existing towers or steeples of an old cathedral or city hall. The first concrete-framed tall building, built in Cincinnati in 1902, was followed in 1908–1909 by the Royal Liver Building in Liverpool, which was a special case of a building being erected on the waterfront where it did not interfere visually with the Victorian city center.

In the European cities building heights were limited by tradition and in some by law. The new skyscrapers therefore developed in the United States and other American countries where there were few restrictions on building heights. In Australia height limitations derived from Britain were in force until the 1950s; when these were lifted, the construction of tall buildings started immediately in Sydney and Melbourne. In my opinion the two cities thereby acquired a distinctive character they had previously lacked.

In Europe the impact of tall buildings was less satisfactory, even though the tallest European buildings are lower than those erected in America or Australia. The medieval city was dominated by the cathedral tower or spire. In the growing industrial cities, such as Manchester or Sheffield in the late nineteenth and early twentieth centuries, the cathedral became a minor center by comparison to the civic buildings, but the dominance of the public over the commercial buildings was still beyond question.

Tall commercial buildings fitted easily into Chicago because no important building in the city center had survived the fire of 1871. In New York the conflict between the existing religious and civic buildings and skyscrapers posed a problem that was ultimately resolved in favor of the new tall buildings. New York lost something but gained more.

Le Corbusier was an early supporter of American skyscrapers, but noted that they were placed much too closely. In 1924 he wrote in *The City of Tomorrow:*

It is 9 a.m.

From its four vomitories, each 250 yards (229 m) wide, the station disgorges travellers from the suburbs. The trains, running in one direction only, follow one another at one-minute intervals. The station square is so enormous that everybody can make straight to his work without crowding or difficulty.

Underground, the tube taps the suburban lines at various points and discharges into the basement of the sky-scrapers, which gradually fill up. Every sky-scraper is a tube station (Ref. 4.1).

The importance of connecting the tall buildings to the public transportation system and to garages for private cars has been belatedly recognized (Section 4.13).

The appearance of the tall building posed another problem that was only gradually solved. The buildings erected in Chicago had been conceived by their designers as utilitarian structures with plain facades. In the closing years of the nineteenth century, however, tall buildings were designed to conform to an appropriate architectural "style." There were two historical precedents.

In ancient Rome some tall public buildings had been erected; for example, the Colosseum, which had a height of 48 m (158 ft), was decorated with four tiers of semiattached columns in the Tuscan, Ionic, and Corinthian orders. In the multistory architecture of the late nineteenth and early twentieth centuries tiers of columns were frequently used and giant columns extended through four or five stories. There is, however, a limit to the height of a building that can be covered with orders of Classic columns.

The second historical precedent, Gothic architecture, did not suffer from this restriction. A Gothic cathedral was as high as its architects could build it and of all historical styles it expressed verticality most perfectly.

In 1913 the Woolworth Building on lower Broadway, New York, set a new record for height (fifty-five storys, 792 ft or 241 m). It was designed to withstand winds of hurricane force, and up to the twenty-eighth floor the frame was built up from riveted portals with rigid, rounded corners (see Fig. 2.11). Above that to the forty-second floor wind resistance was provided by knee braces, and a simple steel frame was designed for the remaining thirteen storys. The technical excellence of this building has never been questioned. Gunvald Aus Company were the structural engineers and Gilbert Cass, the architect. The outside was covered with delicate Gothic ornament which gave the building a strong vertical feeling. Montgomery Schyler, a noted critic, praised the building highly in the Architectural Record in 1913 (Ref. 4.2). In an earlier article he had argued that a change in form was not basically a matter of style. Once the frame was formulated, the exterior details could be borrowed from any one of the "historical styles." On the other hand, in the writings of the protagonists of the modern style of architecture the Woolworth Building became a particular target for attack or ridicule. This conflict became the subject of a public controversy after the competition for a new building for the newspaper *The Chicago Tribune* was won in 1922 by Raymond Hood and John Mead Howells of New York (Fig. 4.1). Their design had some Gothic features derived from Rouen Cathedral, although they were

4.1

The 34-story Chicago Tribune Tower, designed by R. Hood and J. M. Howells and built from 1923 to 1925; its height is 450 ft (137 m).

not so pronounced as in the Woolworth Building. The architects described their design as follows:

We feel that in this design we have produced a unit. It is not a tower or top, placed on a building—it is all one building.

It climbs into the air naturally, carrying up its main structural lines, and binding them together with a high open parapet. Our disposition of the main structural piers on the exterior has been adopted to give full utilization of the corner light in the offices, and the view up and down the Avenue.

Our desire has not so much been an archaeological expression of any particular style as to express in the exterior the essentially American problem of skyscraper construction, with its continued vertical lines and its inserted horizontals (*The International Competition for a New Administration Building for the Chicago Tribune MCMXXII*, Tribune Company, Chicago 1923).

Louis Sullivan, one of the pioneers of the Chicago School, and Frank Lloyd Wright's *Lieber Meister* (teacher), took an opposite view:

Confronted by the limpid eye of analysis, the first prize trembles and falls, self-confessed, crumbling to the ground. Visibly it is not architecture in the sense herein expounded. Its formula is literary: words, words, words. It is an imaginary structure—not imaginative. Starting with false premise, it was doomed to false conclusion, and it is clear enough, moreover, that the conclusion was the real premise, the mental process in reverse of appearance. The predetermination of a huge mass of imaginary masonry at the top very naturally required the presence of huge imaginary masonry piers reaching up from the ground to give imaginary support (the Chicago Tribune Competition, *Architectural Record,* Vol. 53, February 1923, pp. 154–155, 157–158.

Apart from the second prize-winner Eliel Saarinen, who submitted a restrained modern design reminiscent of the Neo-Romanesque style of Henry Hobson Richardson (who had influenced the Chicago School), there were designs in a distinctly modern style by Walter Burley Griffin, Walter Gropius, Adolf Loos, Adolf Meyer, and Bruno and Max Taut (see Section 8.1). Siegfried Giedion was particularly critical of the rejection of Gropius' design:

Both the jury and the public must have considered his scheme quite unstylish and old-fashioned. There is no doubt, however, that it was much closer in spirit to the Chicago School than the Gothic tower which was executed. . . .

By this time, however, the confidence and belief in its own forces which had sustained the Chicago School had completely disappeared. The School might just as well not have existed; its principles were crowded out by the vogue of Woolworth Gothic (a reference to the Woolworth Building of 1913; Ref. 3.12, p. 391-392).

Carl W. Condit in 1973 expressed a different opinion:

The first-prize design will very likely continue to be a matter of controversy on through the years; one can only say that the jurors could have done far worse in selecting a winner, and that their scorn for the modern style was unfortunately born out by the undistinguished work that the modernists submitted (Ref. 4.3, p. 109, © 1973 by the University of Chicago).

It took another quarter-century before a new aesthetic for tall buildings in steel, glass, and concrete was perfected. Even then a number of visually successful buildings in the modern style failed to deal effectively with the problems of durability and interior environment (Sections 3.1 and 8.2).

The thirty-four-story structure of the Tribune Tower (450 ft or 137 m) was complicated by the setbacks of the facade, which required column offsets on plate girders over the lobby and other column offsets at other floors. The structural design by Frank E. Brown and Henry J. Burt followed the conventional column-and-girder system up to the twenty-fifth floor. At that level the octagonal tower was carried by a dense grillage of girders and beams, and two trusses whose depth extended throughout the entire twenty-sixth floor. Some of the columns were continued beyond the twenty-fifth to form the octagonal ring and were connected to the main frame by single horizontal struts encased in masonry to imitate the flying buttresses of the tower of Rouen Cathedral (Figs. 10 and 11 in Ref. 4.3, p. 103). The real wind bracing was provided by crossed diagonals and knee braces within the steel frame. The building was designed for a wind pressure of 30 psf (1.43 kPa) because of its exposed location and its proximity to Lake Michigan.

The Empire State Building in New York was under construction from 1929 to 1931; the architects were Shreve, Lamb, and Harmon, and the engineer, H. G. Balcom. The building has eighty-five stories and a height of 1044 ft (318 m); it is surmounted by a pylon originally intended for mooring airships but has since been found more useful for radio and television transmission. This took the height to 102 stories (1239 ft or 378 m) and set a record that was not surpassed until 1973 by the World Trade Center in New York (Section 4.12) and in 1974 by the Sears Tower in Chicago; however, the new record is only slightly higher (110 stories; 1450 ft or 442 m).

The building was designed with riveted steel frame with portal bracing (see Fig. 2.11); the girders were riveted throughout their depth to the column webs and the beams were similarly riveted to the girders. The New York Zoning Law of 1916 required two setbacks for the facade, and these produced column setbacks at the sixth and the seventy-second floors which were carried on enlarged distributing girders.

4.2 ANALYSIS OF RIGID FRAMES, 1900–1940

The growth in the height of building frames stimulated renewed interest in a method of analysis more accurate than those discussed in Section 3.5 but less laborious than the Maxwell-Mohr and Castigliano methods discussed in Section 3.6.

In 1880 the engineering school of the Technical University of Munich set as a prize subject the determination of the secondary stresses caused in trusses if the pinned joints were replaced by rigid joints. The prize was won by H. Manderla (Ref. 4.4) who showed that these stresses depended on the rotations or end slopes of the joints and derived equations for the bending moments in terms of the stiffness of the members and their rotations. The method was later improved by Mohr (see Section 3.2), and published in *Zivilingenieur* in 1892 (Vol. 38, p. 577). Mohr included the subject in his lectures and subsequently in his book on engineering mechanics (Ref. 4.5, p. 467), from which Axel Bendixen (Ref. 4.6) in 1914 developed the *slope-*

deflection method for analyzing rigid frames. A similar method was developed independently in the United States by G. A. Maney in 1915 (Ref. 4.7).

The slope-deflection method assumes that in a rectangular rigid frame a right-angled rigid joint remains a right angle after the frame has been deformed because the joint is infinitely stiffer in relation to the rest of the frame. The end slopes of the members terminating at the joint must therefore differ by 90° both before and after deformation (Fig. 3.13). This geometric statement yields one equation for each of the redundancies.

Maney demonstrated the potentialities of the slope-deflection analysis by working out a complete multistory frame, a prodigious feat without modern calculating machinery, but it also demonstrated that the method, although simpler than earlier methods, was still too laborious for a tall multistory frame because too many simultaneous equations had to be solved. The only practical approach therefore was to break up the multistory frame into individual stories (Fig. 4.2).

It had been a relatively simple matter to work out the coefficients for the bending moments in built-in and continuous beams (Section 3.2); the information could be given in a few pages.

There are many more variables in the design of rigid frames. The first systematic attempt to tabulate their bending moment coefficients was made in 1913 by Adolf Kleinlogel, Professor at the Technical University of Darmstadt (Ref. 4.8), who later extended his work from single-bay to multibay frames (Ref. 4.9), including solutions for one-story frames of several bays and for single floors of multistory frames of the type shown in Fig. 4.2. A proposed extension of the work to complete multistory frames was not carried out because of Kleinlogel's death in 1958. His and other similar books of tables have, however, obvious limitations because of the complexity of most tall building frames (see Sections 4.1 and 5.4).

In 1930 Hardy Cross, Professor of Structural Engineering at the University of Illinois, published in the *Proceedings of the American Society of Civil Engineers* (Ref. 4.55, pp. 1–28) *the moment distribution method* which greatly reduced the labor of computation. Like the slope-deflection analysis, it is based on the proposition that a right-angled rigid joint remains at a right angle after the structure has deformed. Initially all joints of the frame are assumed firmly clamped to a hypothetical rigid background. The beams and the columns are therefore assumed to be built in at the ends and the moments in them are determined from the appropriate equations (Fig. 3.1). At most joints this clamping procedure produces unbalanced moments; for example, in a rectangular frame carrying vertical loads there are substantial bending moments in the clamped beams and no bending moments in the clamped columns (Fig. 4.3). Each of the joints is then released in turn and the unbalanced moment is distributed to the adjoining members. The process of releasing the joints and distributing the bending moments can be continued indefinitely and the unbalanced moments become smaller and smaller. In practice, however, a sufficient degree of accuracy is frequently obtained with three distributions and two may be satisfactory for a preliminary design. It is therefore possible to adjust the amount of work to the importance of the problem.

Cross's method had two great advantages. By comparison with the older methods of analysis the calculations were less laborious and the method could be learnt by persons with only limited mathematical knowledge. Moment distribution was first

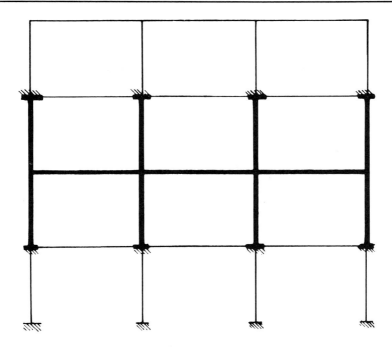

4.2

Simplified method of designing multistory frames by the slope-deflection or moment-distribution methods. The frame is broken up into individual stories, each containing the floor structure and the columns above and below the floor. The columns are considered to be rigidly restrained at their far ends. Each floor can thus be analyzed separately, although with some loss of accuracy.

employed in the design of reinforced concrete rigid-frame bridges used for grade-separated motorways, but it was soon proved to be useful for building frames.

The invention of the flat-plate floor structure in 1940 greatly increased its popularity, for neither the empirical rules for flat-slab design nor the traditional continuous-beam concept for designing reinforced concrete frames could easily be applied to this new form of construction.

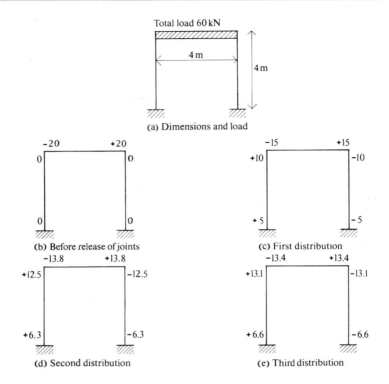

Total load 60 kN

4 m

4 m

(a) Dimensions and load

−20   +20
0   0

0   0

(b) Before release of joints

−15   +15
+10   −10

+5   −5

(c) First distribution

−13.8   +13.8
+12.5   −12.5

+6.3   −6.3

(d) Second distribution

−13.4   +13.4
+13.1   −13.1

+6.6   −6.6

(e) Third distribution

4.3

Solution of a single-bay portal frame built in at the supports by the moment distribution method (a). All the joints are clamped, which gives the bending moments at the ends of the beam (Fig. 3.1) $WL/12 = 60 \times 4/12 = 20$ kNm and zero bending moments in the columns (b). The process of distribution has two parts:

1. The unbalanced bending moment at each joint must be distributed to the other member(s) framing into the joint. If they are equally stiff, the moment is divided equally; if some members are stiffer, they take a larger proportion of the moment to be distributed. Here we must distribute $\frac{1}{2} \times 20 = 10$ kNm.
2. The rotation of a member at the joint released induces bending moments at its far end. This "carry-over" moment is due to the elastic deformation of the member. Cross showed that the "carry-over factor" for this frame is $\frac{1}{2}$; therefore the carry-over moment is $\frac{1}{2} \times 10 = 5$ kNm. After the first distribution the bending moments are as shown at (c), and the difference between the bending moments at the top joints is $10 - 5 = 5$, or 67% of the average moment. The second distribution (d) reduces this to 10%; the third distribution (e), to 2%. An infinite number of distributions is needed to eliminate the difference, but 2% is sufficiently accurate for most purposes.

4.3 TWO-WAY SLABS

So far we have discussed bending in only a single direction. The traditional timber-framed factory building consisted of timber columns that carried primary timber beams or girders which supported secondary timber beams or joists at right angles; these, in turn, carried the timber planks that ran at right angles to the joists (and parallel to the girders).

When this structure was translated into iron, and later into steel, the same system was used, except that the floor surface was normally of a different material, such as timber planks, brick jack arches, or concrete. Each structural member thus spanned and bent in one direction only.

The earliest reinforced concrete structures were designed on the same principle. The reinforced concrete columns supported reinforced concrete girders which supported reinforced concrete joists. The reinforced concrete floor slab spanned across the joists (Fig. 4.4), provided the ratio of the span of the joists to that of the slabs was at least 1.5, a reasonable assumption.

If, however, the two spans were approximately equal, the concrete slab would, in fact, bend in two directions (Fig. 4.5). We can consider that the floor slab is divided into unit strips and that the restraints exercised by the supporting beams at right angles to one another cause the outer strips to rotate and twist. We may therefore

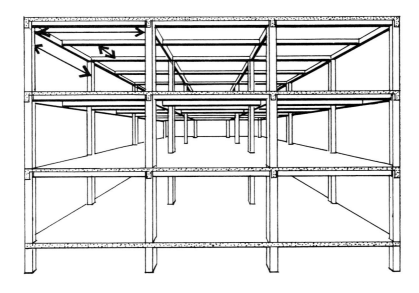

4.4

Reinforced concrete frame with one-way slabs. The columns support the girders (primary beams), the girders support the joists (secondary beams) and the joists support the floor slab. The arrows indicate the direction of the span.

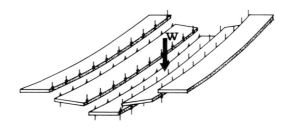

4.5

Deformation of two-way slab under load. Because of restraints of the supporting beams at right angles to one another the outer strips rotate and twist.

consider that the slab is made up of a number of interwoven strips (Fig. 4.6), each unit square forming a part of two strips at right angles. We can solve this problem by considering the central unit square of a slab spanning L_1 in one direction, and L_2 in the other at right angles.

Modern deflection theory dates from 1741 when Daniel Bernoulli determined the deflection of simple beams (Ref. 1.1, Section 8.4). Using modern notation, the deflection y due to a uniformly distributed load w (per unit area) over a simply supported beam of span L is

$$\frac{5}{384} \frac{wL^4}{EI}$$

where E is Young's modulus and I is the second moment of area.

The central unit square of the slab forms part of two strips, one carrying a load w_1 and spanning L_1, the other carrying a load w_2 and spanning L_2. The deflection of both strips at the center must be the same so that

$$\frac{5}{384} \frac{w_1 L_1^4}{EI} = \frac{5}{384} \frac{w_2 L_2^4}{EI}$$

which gives

$$\frac{w_1}{w_2} = \left(\frac{L_2}{L_1} \right)^4 \tag{4.1}$$

This equation enables us to determine how the total load w (= $w_1 + w_2$) carried by the slab is distributed between the two strips at right angles, and thus we can calculate the amount of reinforcement required in each direction. Two authoritative books (Ref. 4.10, p. 58, and 4.56, p. 698) credit this method to F. Grashof, Professor of Applied Mechanics at Karlsruhe Technical University from 1863 to 1893, and W. J. M. Rankine, Professor of Civil Engineering at the University of Glasgow from 1855 to 1872; however, I have been unable to find this method in their better known

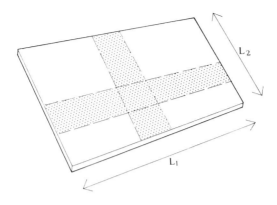

4.6

A two-way slab may be considered as made of interwoven strips at right angles to one another. Each unit square is therefore part of two strips at right angles.

writings. The Grashof-Rankine method was included in the first American national concrete code,* drafted in 1908 by a Joint Committee of representatives from various interested groups (Ref. 2.16, p. 445) and has been embodied in the German, British, and Australian concrete codes.

The two-way action was cleary demonstrated by tests on concrete slabs carried out by Bach and Graf in the early twentieth century (Section 2.8). We noted that in a slab bending in two directions the outer strips rotated (Fig. 4.5). As might be expected, the tests by Bach and Graf showed that the corners curled up if unrestrained (Fig. 4.7).

In actual fact the concrete slab is cast in one piece with the supporting beams so that curling of the corners is prevented; this induces torsion in the corners of the slab, a fact ignored in the Grashof-Rankine method; therefore a correction is needed in (4.1). A modified method, proposed by H. Marcus (Ref. 4.12), gave a series of constants that depended not merely on the ratio of the two spans but also on the restraint of the supporting beams. This method was specifically referred to in the German concrete code of 1925 (Ref. 4.13) and subsequently included in the first British code of 1934 (Ref. 4.10).

* In addition to being represented on the first and second Joint Committees on Reinforced Concrete, the American Concrete Institute (ACI) also issued its own concrete building code. In 1917 the ACI and the Joint Committee Codes differed in several important particulars, including the flat-slab design constants. The Chicago and New York Building Codes have also at times differed from those of the ACI. Because of the importance of these two cities, their codes had far-reaching effects on design practice. In addition, there are several regional codes (Ref. 4.11). Thus the *equivalent frame method* was included in a Californian building code (see below) before it was admitted by the ACI Code. In recent years, however, the ACI Code has acquired the status of a national American code, and a new edition of the ACI Code usually influences changes in the national codes (Section 4.10) of other countries.

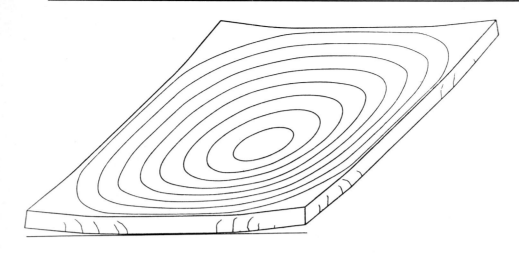

4.7

The tests by Bach and Graf at the Stuttgart Technical University showed the curling up of the corners which occurred when the slab was not held down at the supports (Ref. 4.56).

The Marcus method, although more accurate than Rankine's, neglects the transverse (shear) forces between adjacent strips, shown in Fig. 4.5. A precise elastic solution is obtained by considering the elastic deformation of the plate, taken as a whole, instead of dividing it into interweaving strips. This problem had already been examined in 1826 by Navier in *Résumé des leçons* (Section 2.3). The general solution was obtained by Gustav Kirchhoff in 1877 (Ref. 4.24) as a fourth-order partial differential equation which can be solved in the form of a Fourier series, that is, an infinite series containing sines and cosines. The specific solution for elastic slabs, cast in one piece with the supporting beams and spanning in both directions, was derived by H. M. Westergaard (Ref. 4.14) in 1921. The coefficients calculated by this analysis, with some empirical adjustments, were first used in the American code of 1925 (Ref. 2.16, p. 459).

4.4 FLAT SLABS AND FLAT PLATES

In timber and steel structures assembled from individual members beams supporting the floor are essential. In a concrete structure cast in one piece the beams can be omitted, although this makes the floor structure more flexible for the same slab thickness. Without supporting beams, however, the column is liable to punch or shear through the slab and cause a diagonal tension failure (Fig. 4.8). The earliest beamless slabs therefore had enlarged column heads and are called flat slabs (Fig. 4.9*b*). A patent for flat-slab construction was registered by Orlando Norcross in the United States in 1902, and the first recorded flat-slab structure was the five-story

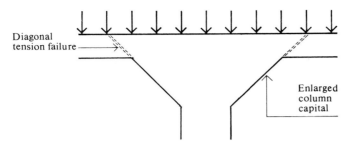

4.8

Flat slabs are liable to fail in diagonal tension due to shear around the periphery of enlarged column capital.

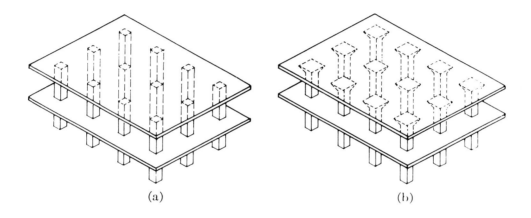

(a)   (b)

4.9

Flat plate and flat slab.

The flat slab (b), which was first used in 1905, has enlarged column heads to increase the circumference at the joint between the column and the slab and thus spread the punching shear over a larger concrete area. The enlarged column heads were first omitted in 1940, and this is now called a flat plate (a).

Bovey-Johnson Building in Minneapolis designed by C. A. P. Turner (Ref. 4.15, p. 2). By 1914 Turner claimed to have built 200 million dollars worth of flat-slab structures which proved to be a particularly popular form of construction for industrial buildings and warehouses with live loads of 100 psf (4.8 kPa) or more.

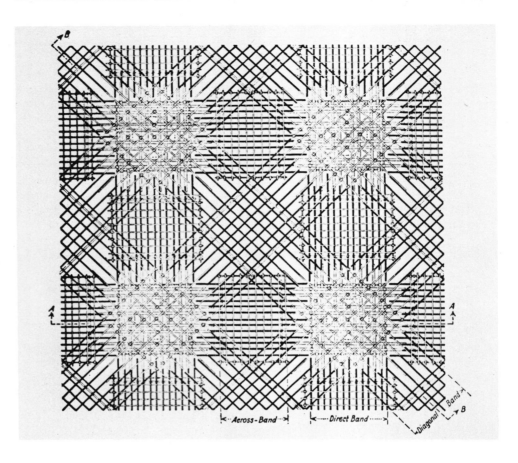

4.10

Reinforcement for a flat slab arranged in four bands; this was the system employed by C. A. P. Turner. The bars shown solid are at the bottom face of the slab, and the bars drawn in outline are at the top face.

The invention of flat-slab construction has frequently been claimed for Robert Maillart (Ref. 3.13); however, Maillart did not start to experiment with flat slabs until 1910 and built the first building to employ them in 1912.

The enlarged column heads are a prominent feature of flat slabs and for this reason they were also known as mushroom floors. Maillart's mushrooms gradually curved into the slab, which made them more elegant but also more expensive than the straight column heads normally used in America. Occasionally American column heads were decorated with Corinthian acanthus or Egyptian lotus leaves, but fortunately this practice did not establish itself (see Section 3.1).

The reinforcement employed by Turner was normally arranged in four bands, two running parallel to the two spans and two running diagonally (Fig. 4.10). Maillart

Tall Buildings

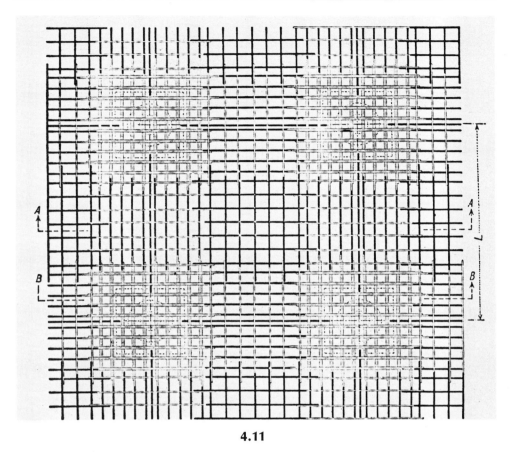

4.11

Reinforcement for a flat slab arranged in two bands; this was the system employed by Robert Maillart, and it is the system normally used at present. The bars shown solid are at the bottom face of the slab, and the bars drawn in outline are at the top face.

used only two bands of reinforcement running parallel to each span as in a two-way slab supported on beams. This method is simpler and ultimately became the standard procedure (Fig. 4.11). An interesting variation was the circular system patented by Edward Smulski (Fig. 4.12). It anticipated the reinforcement patterns employed later by Torroja and Nervi (Fig. 6.15), but in spite of some saving in reinforcement the extra labor cost was not warranted.

The flat slab is arbitrarily divided into two strips of equal width: a column strip and a middle strip (Fig. 4.13). Over the columns the bending moment is negative (i.e., it produces tension on top of the slab), but it changes to positive at midspan. The sum of the positive and negative moments is statically determinate (Fig. 4.14). Let us consider half a flat slab—that is, from the column center to midspan. If an entire panel of the slab carries a load W, then the half-slab carries a load $\frac{1}{2}W$ and the

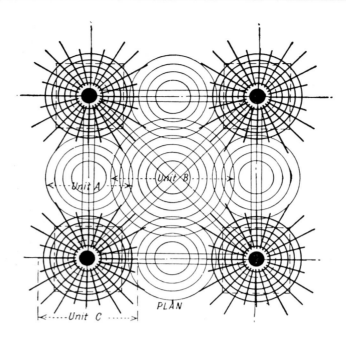

4.12

Reinforcement for a flat slab arranged radially and concentrically, a system patented by Edward Smulski. The thin lines indicate bottom reinforcement, and the heavy lines top reinforcement.

reaction of the column is also ½W. The half-slab is held in equilibrium by these two forces by the total positive bending moment M_+ and the total negative bending moment M_-. By taking moments about the column center line the total bending moment

$$M_o = M_+ + M_- = \frac{1}{8} WL \left(1 - \frac{2}{3}\frac{c}{L}\right)^2 \qquad (4.2)$$

This equation was derived by J. R. Nichols in 1914 (Ref. 4.17). The total moment M_o must, for design purposes, be divided into the maximum positive and negative bending moments in the column and middle strips. This part of the problem is statically indeterminate.

In 1878 the problem was solved by Grashof for the firebox of a steam locomotive (Ref. 4.18). There was a similarity between the interaction of the steam pipes and the boiler plate and between the interaction of the concrete slab and mushroom columns. Grashof's solution was adapted to flat-slab design by a number of people, although their interpretations differed.

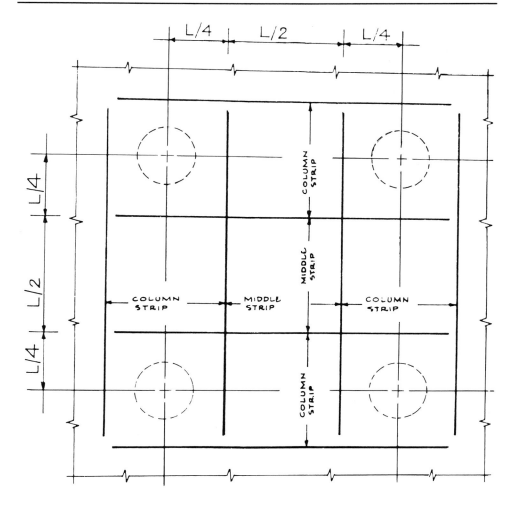

4.13

Division of flat slab or flat plate into column strips and middle strips in accordance with the 1917 ACI Code.

An alternative approach considered the concrete slab as a cantilever up to the point of contraflexure (see Fig. 3.13) of the slab and regarded the remainder of the slab as an inset girder, as in a Gerber beam (Fig. 3.4).

The third and most economical approach was employed by Turner. Although he quoted Grashof as a source, it was essentially empirical, based on his extensive experience and on data obtained from tests to destruction, the first of which was carried out in 1910 by Arthur R. Lord on a flat-slab floor in Minneapolis. Several other tests were carried out between 1910 and 1917 (Ref. 4.15, p. 7, Ref. 4.9, pp. 381–440).

Turner's method gave a maximum bending moment (which determined the thickness of the slab) of $WL/50$, where W is the total load and L is the span. In 1878 Grashof obtained $WL/25.6$ for the moment that determined the thickness of the boiler plate.

By 1917 Nichols' formula (4.2) had been adopted in the American Joint Committee and the ACI codes (Ref. 2.16, p. 454) for the design of flat-slabs but with the coefficient reduced below ⅛ because lower bending moments had already been accepted in the Chicago building code. Rules were also given for apportioning this total bending moment between the positive and negative moments and between the column and middle strips.

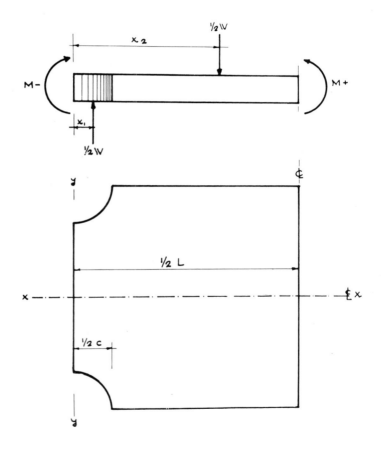

4.14

Nichol's method of determining the sum of the positive (M_+) and negative (M_-) bending moments for a flat slab: W is the total load carried by one panel of the slab; L is the span; c is the diameter of the column head or the equivalent diameter of a square of rectangular column head. The distances x_1 and x_2 depend on the ratio c/L.

A more accurate assessment, based on thorough theoretical and experimental investigations by Westergaard and Slater (Ref. 4.14), was introduced in 1925. These rules have been adopted by building codes throughout the world.

The methods discussed so far were based on the concept of an elastic plate supported on mushroom columns. As we noted in Section 3.2, the columns were regarded as supporting members rather than a part of the frame. The alternative approach, which is now preferred by most designers, treats the flat slabs and their supporting columns as a series of elastic frames interacting at right angles. This method was permitted by the German concrete code of 1925, and in 1926 a design procedure based on the Maxwell-Mohr method (Section 3.6) was derived by Viktor Lewe, a lecturer at the Berlin Technical University (Ref. 4.20, pp. 128–141).

In 1933 the *equivalent frame method* of flat slab analysis was included in the Californian edition of the Uniform Building Code (which was used mainly on the West Coast). The work of the subcommittee that investigated the feasibility of the method throws an interesting light on the structural thinking of the time (Ref. 4.21). In 1941 the method was embodied in the ACI code and has since been adopted by most building codes. It is now generally preferred to the Westergaard-Slater method because of its greater flexibility, but without the earlier development of Cross's moment distribution method (Section 4.2) the labor of computation would have been excessive.

The enlarged column heads were first omitted by Joseph Di Stasio (Ref. 4.22) in 1940. This reduced the area available to resist the punching action of the column through the slab (Fig. 4.8), and it was therefore necessary to introduce some shear reinforcement or to use a column somewhat larger or a slab somewhat thicker than otherwise required. The last two alternatives are usually cheaper than the use of shear reinforcement. In order to distinguish the slab without column capitals from a flat slab with column capitals, the former is called a *flat plate* (Fig. 4.9a).

The flat plate represents the ultimate in structural simplicity, for it consists merely of concrete plates of uniform thickness and columns of uniform section. The formwork is simple and so is the placement of the reinforcement and concrete. The flat-plate structure is economical provided that no false ceiling is required for services such as air conditioning. When there is need for a false ceiling, the economy of the flat-plate structure, which requires more concrete than a beam-and-slab floor because of its smaller overall depth, is open to question. The flat plate has acquired a particular popularity in Australia (Ref. 4.23) where many buildings are without heating or air conditioning. The flat-plate structure does not fit easily into the Westergaard-Slater rules and is usually designed by the equivalent-frame method.

Because flat plates have no projecting column heads, they can easily be combined with walls and partitions between the columns and thus are suitable for office and residential buildings. The flat-slab structure, because of the projecting column heads, was limited almost entirely to buildings such as warehouses and garages, with open floors.

Flat plates, however, have a deflection problem that is more pronounced than that of flat slabs. Because of their small overall depth and the absence of stiffening column heads, the elastic deflection of flat plates is relatively high. This elastic deflection could be allowed for in the design, but there is an additional long-term deflection (Section 4.4) due to creep that is caused by water in the concrete being squeezed from the pores by the sustained load of the concrete slab itself. Thus additional creep deflection is at least double the elastic deflection. This problem was not fully appreciated until the late 1950s when a number of flat-plate buildings showed signs of distress such as cracked partitions or jammed windows. There are two remedies: one is to have shorter spans to reduce the deflection that is proportional to (span)^4; the other is to prestress the reinforcement and use load balancing (Section 3.4).

4.5   THE FACTOR OF SAFETY AGAINST FAILURE

Failure is a term that requires definition in relation to a specific problem. If the plumbing of an old building fails, we may decide that it is worthwhile to install new pipes or to demolish the building. If it is of historical interest, we may turn it into an uninhabited museum piece, in which case we accept its failure as a habitable building. Many famous old buildings and monuments have, in fact, failed in that sense.

The Pyramids of Gizeh are often held up as an example of durability, but in a strict sense they failed long ago. They no longer serve the purpose for which they were designed and their surfaces have greatly deteriorated. The first cannot be remedied, but it would be possible, if it were considered a reasonable expenditure of money, to restore the surfaces.

There are two types of failure for which this alternative of renovation or restoration is rarely practical: one is destruction by fire and the other, structural failure. Both have been the subject of safety regulations from an early time. Quite specific building regulations relating to fire existed in ancient Rome (Ref. 1.1, Section 3.8), and in the City of London they go back as far as the twelfth century.

By contrast, early regulations on structural safety were generally vague. Thus Hammurabi (Ref. 1.1, Section 3.1) ruled that a builder who built a house badly so that its owner was killed should himself be slain. A rule of this type naturally encouraged conservative structural design without giving any guide to structural sizes.

It was not possible to legislate specifically on structural design until a structural theory had been accepted. The elastic theory of structural design was established in the midnineteenth century, largely due to the influence of Navier (Ref. 1.1, Section 8.4). It was necessary only to specify a maximum permissible or working stress to turn this theory into a legally binding method of structural design. This was first done about 1840 when the (British) Board of Trade fixed the working stress for wrought iron in railway bridges as 5 tons/in^2 (11.2 ksi or 77.2 MPa) (Ref. 4.25, p. 114). It was obtained by dividing the average ultimate strength recorded in various tests on wrought iron (a nominal 20 tons/in.^{2}) by 4 to provide a margin of safety.

Rankine (Section 2.4) formally defined the factor of safety as the ratio of the ultimate strength of the material to the maximum stress permissible under the action of the actual or working loads acting on the structure. Toward the end of the nineteenth and the early twentieth centuries this *working stress* was laid down in building regulations for various materials and structural applications. Rankine also distinguished between the dead loads, the loads always acting on structures that could be determined with some precision, and live loads whose magnitude was not known with the same accuracy (Ref. 4.26).

The factor of safety of 4 was still considered appropriate for the first British regulations on the design of steel-framed buildings, promulgated in 1909 by the London County Council (established in 1888 as the administrative authority for the metropolitan boroughs); the working stress for steel was given as 7½ tons/in² (16.8 ksi or 115.8 MPa).

A higher factor of safety was specified for columns to allow for the inevitable imperfections that introduced some bending in addition to the compression (Section 2.4). The magnitude of this factor depended to a large extent on the method used for determining the ultimate strength of the material. If it was based on laboratory tests for which specimens had been carefully prepared, a much higher factor was needed because this degree of accuracy could not be reproduced on the building site. E. H. Salmon in his classical treatise on columns (Ref. 4.27) recommended that the factors of safety should be 10 for dead loads and 20 for live loads for cast-iron columns if the constants in the Rankine formula [which included the ultimate strength (Section 2.4)] were determined from carefully prepared laboratory specimens. These figures were more than double those recommended by Rankine in 1866 when laboratory specimens could not be made and tested with the same precision. The problem of determining the factor of safety was therefore not so simple as it had appeared in Rankine's time (Ref. 4.25, p. 18).

Another problem was the determination of the working loads. The dead load could be calculated from the dimensions of the structure and the known weights of the material. The live load, however, was initially based on guess work. It was possible to determine the weight of a tightly packed crowd of people, which is about 150 psf (7.2 kPa). This, however, would happen only under conditions of panic, and the factor of safety could in part be expected to allow for this situation. Evidently the working load could not be less than (load of a tightly packed crowd/factor of safety). Thus live loads of 40 to 80 psf were considered reasonable for buildings other than warehouses, in which loads might be higher.

During the nineteenth century columns were invariably considered as carrying the full live load of all the stories they supported. After 1900 the building bylaws of both Chicago and New York permitted reduction of the live loads because it was improbable that all floors would carry the full live load at the same time. This was done by specifying the full live load only for the top floor, 95% for the floor below, 90% for the floor below that, and so on until the live load had been reduced to 50%, which was then used for all lower floors. The Times Building in New York, completed in 1909, was the first tall building to be designed by this procedure. The same rule was adopted by the London Buildings Acts in 1909 (Ref. 4.25, p. 62) and is still used, with some variations in the percentage reduction, in most building codes.

The problem of a combination of loads is also important in relation to wind. We noted in Section 3.5 that the earlier determinations of the wind load were associated with the design of railway bridges following the Tay Bridge disaster. It was argued that a bridge was unlikely to carry a full vertical live load when a strong wind was applying a high lateral load. The railway operators would stop trains in a severe storm or at least reduce the vertical loads. Thus the custom arose of allowing the stresses due to a combination of wind and vertical loads to exceed the normal working stresses by 20%. This type of rule spread to the design of tall buildings and was embodied first in American and then in British building codes (Ref. 4.25, p. 61); its validity, however, is open to question.

The elastic theory, in conjunction with working stresses and working loads which were increasingly specified by building codes, became the basis of structural design from about 1870 on and was unquestioned until about 1920.

This theory had one incidental implication for masonry structures of the conventional type. The strength of a Gothic or Renaissance structure depended only on the line of action of the thrust and not on the strength of the material (Ref. 1.1, Sections 6.5 and 7.6). If the tensile strength of the mortar in the masonry joints is neglected, as it should be, the elastic theory requires that the thrust fall within the middle third. In fact, the structure remains stable as long as the resultant lies within the section. In the first case the thrust may be displaced by $\frac{1}{2} \times \frac{1}{3}$ of the thickness of the masonry section; in the second case by $\frac{1}{2}$ the thickness. Thus the elastic theory provides a factor of safety of $\frac{1}{2} / \frac{1}{2} \times \frac{1}{3} = 3$ against collapse for most masonry structures. The working stresses of the material do not enter into the calculations.

4.6   CONCEPTS OF STRUCTURAL SAFETY

In the 1920s the new aircraft industry stimulated the search for a more accurate assessment of the factor of safety. A reduction in the factor which does not endanger the safety of the structure evidently saves material and money. In building this is important, but in airplanes it is essential, for a plane that is too heavy may not be able to fly at all. Thus factors of safety for aircraft structures are much lower than for architectural structures; these lower factors, however, are associated with careful control of the quality of the materials, frequent inspection of structures, replacement of damaged parts, and accurate methods of structural design based on a highly mathematical theory. The design methods originated by the aircraft industry in due course affected the design of buildings.

In addition, the search for a theory of the ultimate strength of buildings, abandoned in the nineteenth century (Ref. 1.1, Section 8.3), was resumed in the 1920s (Section 4.10). The elastic theory was both improved and challenged by these developments.

Little thought had been given to the useful life-span of a structure before that time. The pyramids had been built to last forever and so had the medieval cathedrals and perhaps the palaces of the Renaissance. An aircraft had a limited life—about 30,000 flying hours or about ten years (Ref. 4.25, p. 87). By 1920 it had become evident that most buildings had a finite life. Sanitary facilities considered excellent in 1870 had become barely adequate by 1920, and it was sometimes simpler to demolish and replace a building than to improve its services. Industrial processes and office procedures were slowly changing. Thus factories, office buildings, and blocks of apartments had a limited lifespan, and no purpose was served in building a structure that would outlast its services or the purpose for which it had been designed. The useful life of a building, depending on the degree of innovation to be expected, might be fifty to one hundred years. Only a few monumental buildings were expected to last for a longer time.

An acceptable probability of failure could also be defined. The value placed on human life still varies greatly throughout the world, but on the whole a death caused by the collapse of a building is far less acceptable than a death caused by a traffic accident. I calculated from official statistics that in the State of New South Wales in 1960 it was 130 times more likely that the designer of a building would be killed in an automobile accident during a fifty-year period than that one of his buildings would suffer structural damage in the same span of time. It might be argued that we spend too much money on the safety of buildings and too little on preventing traffic accidents. On the other hand, building bylaws are enacted by elected bodies, and a city councilor or member of parliament is unlikely to win on a platform of making buildings less safe, however justified his arguments might be.

After 1946, the probability calculations developed during World War II were applied to the structural design of buildings (Refs. 4.28 to 4.30). We must compromise because the safer the structure, the greater its cost. A reasonable probability for a local failure, that is, structural damage that will not cause collapse of the structure, is 1 in 10,000 or 1×10^{-4}. This figure strikes an acceptable balance between the social acceptability of a failure and the cost of structure.

It is now necessary to determine the probability that the loads for which the building is designed may be exceeded and the probability that the structural material used in the building will fall below the design strength. The first is set by load surveys (Sections 4.7 and 4.8); the second is ensured by an appropriate testing program.

When materials are tested, the shape of the curve recording the distribution of a large number of results ranges from a few very low strengths through the average to a few very high strengths. The shape of this curve was established about 1800 and is still called after Carl Friedrich Gauss (whose name has also been given to a unit of magnetic measurement: Ref. 4.57). The smaller the spread of the curve, the better the degree of quality control. Once this curve is known it is possible to assign an average strength that will ensure that only a given proportion, say one in ten or one in twenty, will fall below the acceptable minimum strength (f'_c in Fig. 4.15) specified for the material.

Although quality control had been used by American industry since the early years of this century, statistics became a major element in production control only during World War II. In 1940 the American Standards Association initiated a project that led to the publication of the American Defense Emergency Standards, which specified methods of quality control based on statistical theory. The British Standards Institution adopted them in 1942 and the Standards Association of Australia, in 1943.

A statistical method for evaluating compression test results for concrete, adopted by the American Concrete Institute in 1957, resulted in a considerable economy of the most widely used structural material (Fig. 4.15). This method has since been adopted by the European Committee on Concrete and incorporated in the British and Australian concrete codes.

The program for testing concrete presently prescribed in the ACI code is intended to ensure that no more than 5% of defective material is used in the structure. From this figure, and a 5% probability that the loads may exceed the design loads, it is possible to determine from theory the factors of safety against a probability of 1×10^{-4} for local failure and 1×10^{-7} for total collapse (Ref. 4.25). Since 1950 factors of safety in building codes have to an increasing extent been based on considerations of probability.

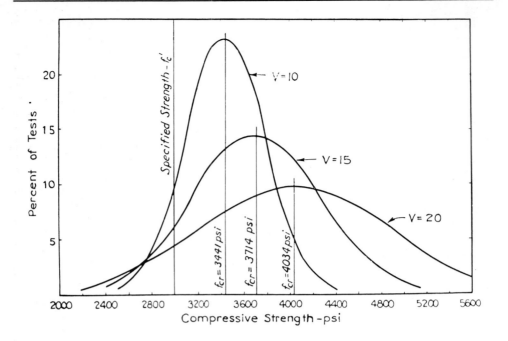

4.15

Normal frequency curve for coefficients of variation of 10, 15, and 20%. The required average compressive concrete cylinder strength f_{cr} is based on a probability of 1 in 10 that a test will fall below a specified strength, f'_c, of 3000 psi (20.7 MPa). The coefficient of variation is a statistical measure of the degree of dispersion of the experimental data (from *Recommended Practice for Evaluation of Compression Test Results of Field Concrete*, Standard 214, American Concrete Institute, Detroit, 1957).

4.7 NORMAL LOADS ON BUILDINGS

The earliest detailed surveys of live loads were undertaken by the U.S. Department of Commerce Building Code Committee (Ref. 4.58) in 1924 and by the British Steel Structures Research Committee (Ref. 4.58) in 1931. More recently office loads have been surveyed in the United States by Dunham, Brekke, and Thompson (Ref. 4.58) and Bryson and Gross (Ref. 4.31) and in England by Mitchell and Woodgate (Refs. 4.32a and b and 4.59).

Weighing all the movable items in an office and recording the data is tedious and time consuming, and recent surveys have been made more thorough by the introduction of automatic weighing equipment and data-recording systems. Thus Mitchell's survey (Ref. 4.32a) consisted of a sample of thirty-two buildings (chosen from a total of 473), more than 100 occupying organizations, and a total area of 1,750,000 ft.² (160,000 m²).

Other investigations dealt with retail stores (Refs. 4.32*b* and 4.58), storage occupancies (Ref. 4.58), industrial occupancies (Ref. 4.58), churches and theaters (Ref. 4.58), and residential buildings (Ref. 4.33).

Weighing the furniture, movable equipment, filing cabinets, books, and other contents is an objective activity and, provided that the sample is sufficiently large, the result should be reliable. Problems may develop in warehouses and retail stores in which the quantity of goods stored can vary considerably, but even they can be overcome by sampling with reasonable accuracy.

Investigators differed in their assessment of the weight of people. Dunham (Ref. 4.58) gave one entry for "unit live load as surveyed" that included the weight of the people actually present at the time the observations were made. Under the heading "unit live load, with aisles crowded" he gave the weight that would result if the aisles were filled with people at the rate of 60 psf (2.9 kPa), which is the kind of fairly crowded condition that might occur at Christmas, at a sale, or in a fire. Thus the unit live load surveyed in the men's shoe section of a New York retail store in 1950 was 10.6 psf (0.5 kPa); with "aisles crowded" it increased to 43.6 psf (2.1 kPa). Yet in the storage section of the same department, which was not open to the public, the load remained unaltered at 57.0 psf (2.7 kPa).

Mitchell (Ref. 4.59), whose investigation was made twenty years later, was able to arrive at a more accurate estimate of the effect of crowding by relying on observations of the numbers of people at lunchtime, at the weekend, at the time of sales, or just before Christmas, depending on the nature and location of the store. This led him to an allowance of 50 psf (2.4 kPa) for crowding at stairheads and at shop exits. Thus more precise observations have tended to reduce previous estimates.

Snow loads are a particular problem only on high ground and in countries near the Arctic, notably Canada, Scandinavia, and the USSR. Assessing the weight of snow is a comparatively simple matter, but recent research has concentrated rather on modifying the design to prevent accumulations of snow (which can impose very heavy loads) resulting from monitors, interior roof valleys, and changes in roof level (Ref. 4.60). Solar radiation can reduce snow loads by melting even when the air temperature does not rise above 0°C if suitable drainage is provided, but melting snow followed by freezing can produce an ice barrier at the edge of a sloping roof that will increase the depth of the snow retained.

As we noted in Section 3.5, estimates of wind loads were made as far back as the eighteenth century (Ref. 4.61), but accurate data on the wind loads acting on buildings did not become available until the subject received intensive study in the 1950s and 60s (Ref. 4.62).

The static design of buildings is now based on the average wind speed determined from anemometer measurements over a period of years, usually at airfields or on the roofs of meterological offices. In recent years a number of measurements have been made on actual buildings in North America and various European countries (Ref. 4.62) which allow comparison with the wind velocities recorded by the meterological bureaus.

Very high wind velocities occur over very short periods (less than a second) and these gusts can be more than double the average wind speed. They are important in the design of roof coverings and curtain walls, particularly if suction is likely to result. Once a fastening has been loosened by the suction of a gust of wind the roof or wall covering may roll up, thus presenting a front to the wind that will lead to rapid failure.

Architectural structures have too much inertia to be able to fail under the action of a gust, at least if we exclude light domestic buildings, and average wind speeds are therefore used for the static design of structures. It is, of course, essential that multistory buildings be prevented from collapsing because of wind loading, but this has never occurred. Few cases of serious structural wind damage to multistory buildings have been reported, and in only one instance (the Meyer-Kiser Building, damaged by a hurricane in Miami in 1926) has yielding of the steel frame been claimed. On the other hand, collapse of domestic buildings and minor damage to multistory buildings by cyclones (hurricanes, typhoons) are frequently reported. More could be done to minimize it (Ref. 4.34), but the complete elimination of cyclone damage is probably not economically feasible.

Although average wind speeds are used for static design, short-term gusts seem to determine the dynamic behavior of the building (Ref. 4.35). The very tall (more than 1300 ft or 400 m) steel-framed buildings with lightweight fireproofing erected in the United States since the 1960s have appreciable deformation, with frequencies of the order of 0.1 hertz (0.1 cps), and viscoelastic dampers may be needed in the structural frames. The aerodynamic behavior of such buildings has only recently received attention (Refs. 4.36 and 4.62).

The aerodynamic criterion for building frames is the perception of motion by its occupants. It is not sufficient to prevent collapse by aerodynamic instability like that of the Tacoma Narrows Bridge near Seattle (Ref. 4.37). If the building moves to a disturbing extent more frequently than once in ten years, it may be difficult to let the space at an economical rent. How much motion is, in fact, acceptable to people remains to be determined at the time of writing (Ref. 4.63, Vol. 1a, pp. 165–175).

Because of the complexity of aerodynamic analysis, many problems are solved by model investigations in a boundary-layer wind tunnel in which the roughness or smoothness of the ground, and particularly the effect of surrounding buildings, is imitated with considerable accuracy (Ref. 4.63, Vol. Ib, pp. 335–365). In addition, the buildings under investigation must be modeled aerodynamically with regard to their geometry, stiffness, and distribution of mass to produce movements and forces in the wind tunnel from which the behavior of the building can be predicted by dimensional analysis (Ref. 4.64).

4.8 EARTHQUAKES

Small earth tremors occur in most parts of the world at some time, but in regions in which earthquakes are common they occur many times each year. They probably went unnoticed before the development of seismographs (instruments for recording earth tremors) and do little harm to buildings.

The main zones in which earthquakes occur are the Circum-Pacific Belt, which passes through Chile, Central America, California, Alaska, Japan, Indonesia and New Zealand and the Alpide Belt, which passes through Portugal, Italy, Yugoslavia, Greece, Turkey, Iran, and Afghanistan. Major earthquakes in the Alpide Belt figure prominently in the histories and legends of the Mediterranean and Middle Eastern countries, but the explanations offered are mostly in terms of special divine intervention in human affairs. Several scientists from Aristotle to Robert Hooke (who wrote a *Discourse on Earthquakes* for the Royal Society in 1668) have offered rational explanations without obtaining a correct or even a useful solution.

On November 1, 1755, Lisbon, the capital of Portugal, was destroyed by an earthquake that lasted about six minutes. About 12,000 buildings were demolished and more than 60,000 people died. This was the first great earthquake in modern times and it attracted the attention of many physicists and geologists. The Rev. John Mitchell, Professor of Mineralogy at Cambridge University, published an essay in the *Philosophical Transactions of the Royal Society* that stated that the earth had a liquid interior covered by a comparatively thin crust. The waves generated in the subterranean liquid shook this flexible cover and, if the shocks were strong enough, produced an earthquake. In 1807 Thomas Young in his *Lectures on Natural Philosophy* interpreted earthquakes as elastic waves in solid material; this was the beginning of a useful seismic theory. The first seismographs were built in the 1840s, and by mid-century earthquakes were being recorded and classified according to their intensity (Ref. 4.65, Vol. VII, Article *Earthquake*).

Further great earthquakes helped to formulate theories and design methods. In 1811–1812 one of the greatest earthquakes, consisting of three separate shocks, occurred in southern Missouri. It was felt from Canada to Mexico and did much damage to nature but caused small loss of life because the region was sparsely inhabited.

On April 18, 1906, the San Andreas Fault in California slipped over a segment of 270 miles (435 km), which included San Francisco (Ref. 3.7). Approximately 700 people were killed and the subsequent fire caused damage of about 400 million dollars.

On September 1, 1923, Tokyo was struck by an earthquake and fire that destroyed about 700,000 houses. Estimates of casualties ranged from 74,000 to 143,000 people.

The building failures during the San Francisco earthquake of 1906 clearly demonstrated the significance of the horizontal component of the quake's motion (Section 3.5). The earth moved rapidly and because of their inertia the buildings were left behind. The forces created by the vertical component of the motion were less damaging, for the buildings had been designed to carry vertical loads, and unless those due to the earthquake exceeded those for which the buildings had been designed, multiplied by the factor of safety, the buildings were not seriously damaged by them. Because most of them had not been designed to resist horizontal forces, or only relatively small horizontal forces due to wind, the horizontal component of the earth's motion produced great damage and even caused the buildings to collapse. After 1906 tall buildings in earthquake zones were designed to resist a static horizontal force equal to C times their own weight, where C was an empirical coefficient ranging from 0.02 to 0.14, depending on height and location. The value of C was to an increasing extent specified in building codes in earthquake zones.

In San Francisco in 1906 masonry buildings performed badly, but the small number of steel-frame buildings sustained little damage (Ref. 3.7); this was before the time of reinforced concrete. Tokyo in 1923 witnessed the collapse of 54% of all brick buildings but only 10% of those with reinforced concrete frames (Section 3.5). It was held that the stiffer the building, the greater its resistance to horizontal forces, and the better its performance in earthquakes. Thus buildings with stiff reinforced concrete frames and massive shear walls were regarded as particularly suitable.

In the 1930s a different approach emerged. A rigid structure has good resistance to a static horizontal force but cannot absorb much energy because of its small elastic deflection. A flexible structure, on the other hand, is capable of appreciable deformation and thus of storing strain energy (Section 3.6). The horizontal force, which is actually not static but dynamic, is dissipated by the backward and forward movement of the building instead of being largely absorbed by a rigid frame. In the Los Angeles earthquake of 1971 flexible steel frames performed better than rigid reinforced concrete frames; both were greatly superior to unreinforced brick buildings. In recent years, therefore, steel frames have been preferred to reinforced concrete frames in earthquake zones (Ref. 4.63, Volume Ib, pp. 151–334).

4.9 EXPLOSIONS AND IMPACT

War damage had not until recently been considered in the design of buildings, yet even in Roman times they could be destroyed by missiles hurled in a siege. Nonetheless, the damage inflicted was a minor problem compared with the wholesale plunder that frequently followed when a city fell to a conquering army.

Even the bombardment of Paris by German guns and the first air raids of World War I did not do enough harm to cause structural engineers to consider it in the design of their buildings. During the Spanish Civil War (1936–1939) however, aerial bombardment did great damage and the collapsing buildings caused much loss of life. In the short period before the outbreak of World War II in 1939 and during the war years much thought was given to the design of buildings that would have better resistance to air attack; the advantages of rigid frames and reinforced concrete slabs were clearly demonstrated both by experiments and actual performance. In 1945 the effect of the two atomic bombs dropped on Japan was so devastating that clearly no structure could be designed to resist this type of warfare. It may be possible to build shelters to protect people from radioactive fall-out (Ref. 4.38) but that is not a structural problem.

The recent troubles in Northern Ireland (Ref. 4.39) have produced information of considerable interest on the stability of structures damaged by explosions. During World War II the damage was generally too great and professional time too scarce to make such studies possible. It may seem heartless to say that we can learn something of interest from senseless terrorism, but the same objection could be raised to any study of damage caused by warlike acts. Peter Rhodes, the Chief Structural Engineer of the Department of Finance in Northern Ireland, pointed out that in traditional construction stability frequently depended on weight to stop a diagonal tension failure (Fig. 4.16). The same principle applies in Gothic construction (Ref. 1.1, Section 6.5 and Fig. 6.10). Modern framed structures were capable of resisting tension and thus withstood explosions much better than the nineteenth-century buildings designed to be mainly in compression.

Many precast concrete structures depend for their lateral stability on the frictional forces within the joints caused by the comparatively heavy weight of the components

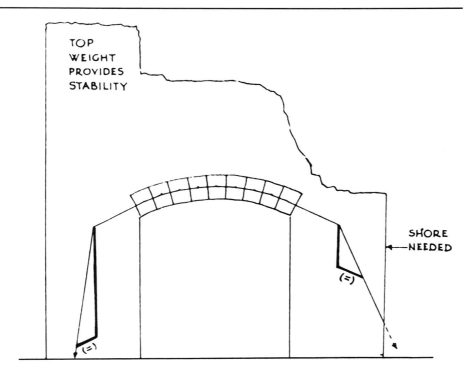

TOP
WEIGHT
PROVIDES
STABILITY

SHORE
NEEDED

(=)

(=)

4.16

Damage by explosion to a nineteenth-century masonry wall with an arched opening. The weight on the masonry at the left is sufficient to cause the resultant to fall inside the wall, but on the right-hand side the resultant falls outside, so that shoring is required (Ref. 4.39, p. 331).

and the high friction between two concrete surfaces. Although these forces are adequate under normal conditions, an explosion may cause a collapse.

Explosions are not necessarily associated with warlike acts. At about 5:45 A.M. on May 16, 1968, a gas explosion occurred in a one-bedroom apartment on the eighteenth floor of a twenty-four-story building at Ronan Point in Canning Town, a suburb of London. The explosion blew out the nonloadbearing face walls of the kitchen and the external flankwalls of the living room and bedroom, thus removing the southeast corner supports for the floors and walls above; this caused their collapse. The weight and impact of the debris from these walls and floors falling on the floors below resulted in the progressive collapse of the floor and wall panels in this corner of the tower down to the cast-in-place podium level. Four people were killed and seventeen injured. Because of the early hour most people were still in their bedrooms. Had the explosion occurred a little later, when more people were in their kitchens and living rooms, the loss of life would have been much heavier.

Explosions and Impact

The explosive was ordinary town gas and the explosion was not of exceptional violence. At the subsequent inquiry (Ref. 4.40) the presssure was estimated between 2 and 12 psi (13.8 to 82.7 kPa). Based on 1966 British statistics, the frequency of explosions involving town gas was approximately 8 per million dwellings per year, of which 3.5 per million were violent enough to cause structural damage. Ronan Point had 110 flats and an estimated life of sixty years. From these data the chance of a gas explosion heavy enough to cause structural damage occurring during the lifetime of the block was 1 in 50, which is unacceptably high.

The Ronan Point disaster caused much discussion (Ref. 4.41) and an amendment to the British Building Regulations (Ref. 4.42). The reliance on friction and weight, which modern precast concrete had inherited from Gothic construction, came under criticism, and most systems are now designed to have sufficient reinforcement to provide continuity at the joints.

Progressive collapse had, before this disaster, been considered a possible consequence of war damage or earthquake. The amendment (Ref. 4.42) required buildings to be designed even in nonearthquake zones by one of two alternatives:

1. A combined dead load, live load, and pressure of 5 psi (34.5 kPa) in any direction to allow for a gas explosion,
2. A consideration of progressive collapse; this implied that if any structural member was removed, the damage would be confined to the two adjacent floors and would not cause total collapse.

The amendment met with considerable criticism, both in Britain and elsewhere (Ref. 4.43). The concept of designing a structure against progressive collapse is still under discussion. At the time of writing the issue has not been resolved.

As buildings grew taller the possibility of a collision between an aircraft and a tall building had to be considered (Ref. 4.44). One must accept the loss of the aircraft and local damage to the building; however, the building must be designed to prevent collapse as a result of the impact. Such collisions have occurred, the most notable being in July 1945 when a B25 Mitchell bomber flew into the twenty-seventh floor of the Empire State Building in New York. Part of the fuselage passed right through the building, one engine fell down an elevator shaft, and local damage was considerable. There was no consequential collapse (Ref. 4.45), however.

4.10   THE PLASTIC THEORY AND THE FAIL-SAFE CONCEPT

We noted in Section 4.6 that the search for an ultimate strength theory, abandoned in the nineteenth century, was resumed in the 1920s, and most building codes now include some elements of ultimate strength design.

The stress-strain diagram of wrought iron had been established, in general terms, in the eighteenth century. The material first deformed elastically in accordance with Hooke's law. It then yielded. This phenomenon was clearly visible when wrought iron was tested by a machine of the type used by Musschenbroek (Ref. 1.1, Section 8.2) because the greatly increased rate of deformation at yield produced a drop in the lever arm. The physical nature of the *plastic deformation* of iron was not at that time understood.

The yielding of the iron did not reduce its strength. On the contrary, there was stiffening above the yield point and the ultimate strength at which the iron broke was higher than its yield strength; but the yield point marked the end of the elastic deformation and the beginning of unacceptably high plastic deformations. Navier therefore (Section 2.3) based his elastic theory on the elastic range of deformation. The upper limit of this range was the yield stress and all working stresses were kept well below it.

In mechanical engineering this approach was never fully accepted. The plastic deformation was used to form metals at normal temperatures in a press, and it was well known that stress concentrations at sharp corners caused local yielding without necessarily endangering the safety of the machine part.

In 1868 H. Tresca presented two notes on plastic deformation to the French Academy. They were submitted to Saint-Venant, who in 1864 had produced the third edition of Navier's book (Section 2.3). He became interested in the concept of plasticity, and in 1871 published the first of a number of papers on the strength of iron beams above the yield point (Ref. 2.8, p. 242). He called the new subject plasticodynamics; we now describe it as the theory of plasticity.

In the 1920s E. C. Bingham coined the term *rheology* for the science of viscous flow and plastic deformation. Rheologists use physical models (Ref. 4.46) to illustrate their concept, and the model for plastic deformation is called a Saint-Venant body (Fig. 4.17). It consists of a heavy block lying on a rough surface, pulled through an elastic spring. As long as the pull is less than the frictional force the deformation of the system is purely in the spring. The spring extends elastically as the load is increased and contracts elastically as the load is reduced. When the pull on the spring is increased, the frictional force is overcome. The system deforms under this constant pull and goes on doing so indefinitely. This model therefore describes a perfectly elastic deformation, followed by a perfectly plastic deformation, the assumption made by Saint-Venant in his papers during the 1870s.

In actual fact the deformation of wrought iron and low-carbon steel is a little more complicated, but this simple model proved a satisfactory basis for the theory of plasticity, as Hooke's model of an elastic spring had done for the theory of elasticity. The marked yield point is a phenomenon confined to wrought iron and low-carbon steels. In particular, it is not found in high-strength steels and aluminum alloys. It is possible, however, to define an equivalent yield point, called a proof stress, and use the plastic theory, provided the material shows sufficient ductility to deform without fracture until the structure actually collapses.

In the 1940s the new science of metal physics (Ref. 4.47) explained the plastic deformation of metals in terms of their atomic structure (Fig. 4.18). A metal crystal was described as an assembly of atoms held together by electrostatic attractions. Musschenbroek's concept (Ref. 1.1, Section 8.2) of a *vis interna attrahens* (internal attracting force) is thus supported by modern research. When the crystal is loaded, these forces permit small deformations and restore the original shape as soon as the load is removed. The crystal then deforms elastically. When the load is increased, the atoms are displaced sufficiently to cause some to jump by one position. The crystal is thus deformed permanently. This plastic deformation is not recovered, but the crystal

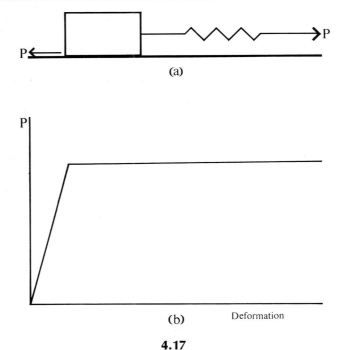

(a)

(b)   Deformation

4.17

The rheological model for a plastic material and the corresponding stress-strain diagram.

(a) Rheological model, consisting of a heavy block lying on a rough surface, pulled through an elastic spring.
(b) The idealized stress-strain diagram, which follows from the model in (a). The plastic deformation is assumed to continue indefinitely. Because structural steel is never deformed sufficiently to break (except in a fire, earthquake or explosion), this is a realistic assumption.

is capable of further plastic deformation at the same load. The stress-strain diagram of this idealized crystal structure thus corresponds to the idealized elastoplastic stress-strain diagram in Fig. 4.17b which follows from the rheological model in Fig. 4.17a

A plastic body can be deformed indefinitely in compression; for example, we can squeeze a mild steel bar into a flat disk without causing any rupture in the metal. In tension the plastic deformation eventually leads to a thinning out of the material so that it breaks; however, this requires so much plastic deformation that it is unlikely to occur in any architectural structure except by fire, earthquake, or explosion.

All pure metals have the type of ductility that produces high plastic deformations. As impurities are introduced, for example, carbon atoms into iron, the plastic deformation is partly blocked by foreign atoms and the elastic strength is consequently increased. On the other hand, the foreign atoms provide a starting point for a crack so that rupture occurs more easily and ductility is reduced. Thus low-carbon steel is stronger but less ductile than pure iron and high-carbon steel is stronger but less ductile than low-carbon steel. Eventually the amount of plastic deformation prior to rupture is so small that the plastic theory can no longer be applied. Steels with high carbon content (useful for toolmaking) are thus not suitable for structural steelwork.

For the same reason cast iron (which has an even higher carbon content) is a brittle material; that is, its fracture is not preceded by plastic deformation.

Natural stone, concrete, and brick are all brittle materials. The plastic theory can, however, be used to a limited extent for reinforced concrete because the steel reinforcement is plastic (Section 2.8).

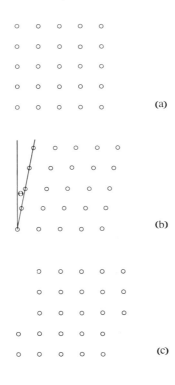

(a)

(b)

(c)

4.18

Deformation of a metal crystal.

(a) The unstrained crystal.
(b) The crystal deformed elastically. The angle θ is a measure of the stiffness, usually given in the form of the elastic modulus E (which is the hypothetical stress that would cause unit deformation). When the load is removed, the atoms spring back to their original positions.
(c) The crystal deforms plastically. When the elastic limit is reached, the atoms jump one position, producing permanent or plastic deformation.

These physical explanations became available only with the development of metal physics in the 1940s and materials science (Ref. 4.48) in the 1950s, but the results had been known in general terms since the latter part of the nineteenth century from laboratory tests on iron, steel, stone, and concrete (Ref. 4.66, pp. 45–65).

The implication of plastic deformation on the postelastic behavior of built-in beams was first discussed by Gabor Kazinczy (Ref. 4.49), who showed that the maximum bending moment for which a built-in beam needed to be designed (Fig. 4.19) was $WL/16$, not $WL/12$, as obtained from the elastic theory (Fig. 3.1). The plastic theory thus allowed a saving in material of about 25% in this instance.

In 1928 the British steel industry became concerned about the growing competition from reinforced concrete and set up the Steel Structures Research Committee as a cooperative effort with the government. John (later Sir John) F. Baker was appointed its technical officer. The Committee published three reports (Ref. 4.50), the last of which contained *Recommendations for Design* based on the conventional elastic theory (see also Section 2.4). These were an improvement on the previous building regulations, but the laboratory investigations and test on actual buildings convinced Baker that no really economical method for the design of steel structures that ignored postelastic behavior could be derived. He therefore commenced a research program in 1936 to investigate the behavior of ductile structures when loaded beyond the elastic range until plastic hinges were formed and collapse occurred. In 1933 Baker had been appointed to a chair at Bristol University, and 1943 he moved to the mechanical sciences chair at Cambridge where the work is still in progress under his successor.

In 1941 Baker developed the Morrison indoor table air-raid shelter, named after the Minister of Home Security of the day. The design of air-raid shelters for individual families had been based mainly on the experience of the Spanish Civil War (Section 4.9) when holes in the ground covered by plenty of soil proved satisfactory. Shelters of this type were structurally sound, but the climate of England was cool and damp and the pattern of air raids was less predictable than it had been in Spain. The underground outdoor shelters thus caused much illness and discomfort. The Morrison shelter consisted of a thick corrugated sheet of steel, bent to fit under the living room table. It was strong enough to resist the partial collapse of the building and did so by absorbing the energy of the collapse through plastic deformation; that is, the steel distorted plastically. Thus the whole of the area under the stress-strain curve in Fig. 4.17b was available to absorb the potential energy of the falling materials; in an elastic design only the elastic strain energy could be utilized (i.e., the small triangle at the left-hand side).

In the early 1950s Baker predicted that in a few years' time all steel frames would be designed by plastic theory, but further research uncovered new problems, some of which still remain to be solved.

The problem of the collapse of a rigid frame by the formation of plastic hinges lacks a unique solution. Even a simple portal frame can collapse by the formation of plastic hinges in at least six different ways (Fig. 4.20). In multistory frames the number of possibilities is greatly increased. The critical mechanism is the one produced by the lowest load and may require much calculation.

The frame may buckle (Section 2.4) before all the plastic hinges can form. The buckling of frames partly deformed in the plastic range is a complicated problem that

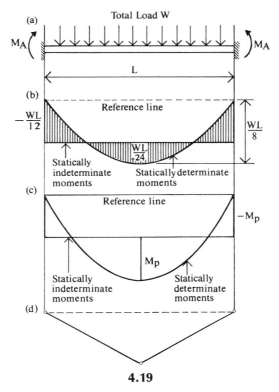

(a) Total Load W

M_A ⟩ M_A

L

(b) Reference line

$-\dfrac{WL}{12}$

$\dfrac{WL}{8}$

$\dfrac{WL}{24}$

Statically indeterminate moments

Statically determinate moments

(c) Reference line

$-M_p$

M_p

Statically indeterminate moments

Statically determinate moments

(d)

4.19

Built-in steel beam of uniform section carrying a uniformly distributed load.

(a) Load diagram
(b) Variation of bending moment according to the elastic theory
(c) Variation of bending moment according to the plastic theory
(d) Formation of plastic hinges prior to collapse

As the load is increased beyond the elastic range, plastic deformation begins at the points of highest bending moment (i.e., at the supports). This process continues until the entire steel section at the supports has turned plastic and forms a "hinge" which allows continued rotation at the constant moment M_p (the resistance moment of the fully plastic section). The two hinges counteract the two statically indeterminate restraining moments M_A. A third hinge, needed (Section 3.3) to turn the beam into a mechanism, forms at the section with the next highest bending moment which occurs at midspan. When the bending moment at midspan reaches M_p, the third plastic hinge forms and the beam becomes a mechanism and collapses. This is shown in (c) and (d). As in Fig. 3.1, the statically determinate bending moment is $\frac{1}{8}WL$, and, from 4.19(c),

$$2M_p = \text{⅛}WL,$$

which gives $M_p = \frac{1}{16}WL$. This is only 75% of the maximum bending in the elastic state which, from (b) is $\frac{1}{12} WL$. There is a corresponding saving of material if the plastic theory is used for design.

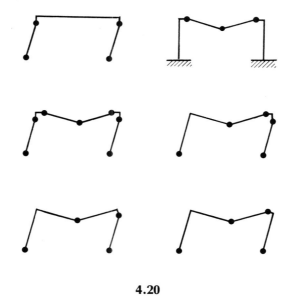

4.20

Possible collapse mechanisms for a single-bay, single-story portal frame, built-in at the supports. The circles indicate the possible location of plastic hinges.

has not yet been completely solved. Unless the designer can be certain that buckling will not precede the formation of the plastic collapse mechanism, the plastic theory cannot be used for the frame as a whole.

One-story rigid steel frames for industrial buildings are now frequently designed by the plastic theory; but it can be used only for multistory buildings that are relatively stocky (Ref. 4.54).

Since the late 1950s Lehigh University in Pennsylvania has been the main center for research on plastic steel design.

Although the plastic theory has not replaced the elastic theory, it has served an important purpose by drawing attention to the visible signs of distress that in structural steel frames precede the total collapse of the building. The plastic theory thus provides a "fail-safe mechanism"; that is, it offers a warning of possible failure while the structure is still safe. This technique, not necessarily associated with plastic deformation, originated in Germany and was quickly accepted by the aircraft industry. In conjunction with plasticity it became an important factor in the safety of structures.

One of the classical problems posed by M. Gruening in 1926 (Ref. 4.51) is shown in Fig. 4.21. A structure consisting only of two inclined bars, AC and BC, is statically determinate (Section 2.2). If one bar fails, the structure fails. If we add a bar CD, the structure becomes statically indeterminate, and one bar can fail without necessarily causing collapse of the structure. We can design this structure by one of a number of methods (e.g., Section 3.6).

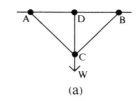

(a)

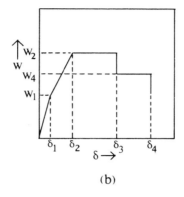

(b)

4.21

A simple fail-safe structure.

(a) The pin-jointed structure consists of three bars AC, BC, and CD.
(b) Relation between load W and deformation δ of point C. At a load W_1 and deflection δ_1 the bar CD reaches its elastic limit. At a load W_2 and deflection δ_2 the bars AC and BC reach their elastic limit. At a deflection δ_3 the bar CD snaps. The load must then be reduced to W_4, and the structure fails completely with a deflection δ_4 when the other two bars break.

Let us make the bar CD small enough to yield plastically at a load W_1 while the bars AC and BC are still well below the yield point. Because the stress in the yielded bar CD cannot be further increased (Fig. 4.18c), any increase in load must come from the other two bars and the deflection increases more sharply. This could be instrumented to provide a warning signal to indicate failure of part of the structure; but the structure itself is still safe.

At the load W_2 the other two bars reach the yield point, which is the maximum load that the structure can carry. Failure does not, however, occur instantly because the plastic deformation takes time. At the deflection δ_3 the bar CD snaps, but the structure does not actually collapse until the deflection reaches δ_4.

There are several warnings. At δ_1 the structure is still safe, but the yielding of the bar CD gives warning of a load greater than the normal working load. At δ_2 the

---

The Plastic Theory and the Fail-Safe Concept

maximum load is reached, but there is still time before failure. At δ_3 the life of the structure can be increased a little if the load can be reduced. Fail-safe structures are frequently used in aircraft design because they give warnings that may enable a pilot to land a damaged plane safely.

The fail-safe concept has also been introduced into the design of architectural structures. The formation of plastic hinges (Fig. 4.20) offers the same sort of warning as the yielding of the bar CD in Fig. 4.21. It does not matter whether the structure has been designed by the elastic or the plastic theory: its failure occurs by the formation of plastic hinges. Thus even an approximate determination of their formation enables us to estimate the margin between the first warning and the ultimate collapse.

In Section 2.8 we discussed the development of the elastic theory of reinforced concrete design. In 1922 H. Kempton Dyson proposed a theory that dispensed with elastic moduli (Ref. 4.52). In 1937 Charles S. Whitney (Ref. 4.53) published an ultimate strength method that was, with some modifications, made an optional design method in the 1956 Building Code of the American Concrete Institute. In 1971 it became the standard method of design in the United States. Since then similar methods have been introduced in Britain, Australia, and several European countries.

The ultimate strength method is based on the assumption that failure of a rein-forced concrete *section* is initiated by plastic yielding of the steel. The plastic extension of the steel causes a rise in the neutral axis (Fig. 2.15b) and its consequent plastic deformation produces a plastic hinge, the rotation of which before the disintegration of the concrete is far less than for steel. It is difficult therefore to prevent the concrete in some of the hinges from disintegrating before all the hinges required by the plastic theory for rigid frames have formed. Thus the plastic theory cannot be used for reinforced concrete frames—only reinforced concrete members.

Nevertheless the plasticity of the hinges provides a fail-safe margin that would be missing if the hinges did not form at all. This could happen if there were so much reinforcement in the concrete beams that the concrete crushed before the steel yielded (Fig. 2.15a). The ACI Code of 1971 therefore limits the amount of reinforce-ment to 75% of that required to produce a concrete failure while the steel is still elastic. Similar restrictions have been adopted in other codes.

4.11 LIMIT STATES DESIGN

As ultimate strength methods were being introduced in the 1950s, it became evident that they provided no information on some problems that the elastic theory had solved satisfactorily. One was the deflection under the action of the working loads, now renamed *service loads*. This can be calculated only by elasticity from the actual service loads, and a structure that is quite satisfactory on an ultimate strength basis may have excessive deflection. This is a particular problem in flat plates (Section 4.4), but excessive deflection is likely to occur in concrete slabs generally. Because of the time-dependent creep deformation of concrete (Section 3.4), it cannot be fore-stalled by an upward camber. Thus the gradual increase in the deflection of the concrete slab might damage brittle finishes, cause cracks in brick or block partitions, or jam doors or windows in the walls below.

Related to deflection is the possibility that the structure might vibrate excessively in a high wind (Section 4.7).

Concrete structures may crack excessively under the service loads, which would cause deterioration of the concrete or rusting of the reinforcement.

In the 1960s the term *limit states design* was coined to denote a multiple check of the ultimate limit states, discussed in the preceding section, the abovementioned serviceability limit states, and the durability of the structure (Ref. 4.63, Vol. III, pp. 843–953). The current American (1971), British (1972), and Australian (1973) concrete codes specify serviceability limit states. In structures that incorporate reinforced concrete slabs the limit on deflection is often more critical than the ultimate strength; the elastic theory is then restored to its former dominant position.

4.12 TWO-DIMENSIONAL AND THREE-DIMENSIONAL STRUCTURES FOR TALL BUILDINGS

The great advances made in the theory of structural frames during the 1940s and 1950s were to a large extent stimulated by the effort put into the design of structures for ships, aircraft, and civil defense during World War II. When the war finished many highly trained people were released for other work, and the building industry acquired engineers who knew more structural theory than had ever been considered necessary for architectural structures.

The actual construction of tall buildings had been inhibited first by the Depression in the 1930s, then by the war in the early 1940s, and thereafter for another decade by essential reconstruction. The Empire State Building, conceived before the Depression, was completed in 1931 (Section 4.1) and retained its height record for forty-two years, which is a long time for a record in the twentieth century. The Empire State Building was designed as a series of plane frames. When the construction of tall buildings was resumed in the 1960s this simple two-dimensional concept was replaced with the more economical concept of a three-dimensional tube (Ref. 4.63, Vol. Ia, pp. 403–656).

Tall buildings require a service core to house the elevators. In recent years this core has also accommodated the vertical ducts of the air-conditioning system, the plumbing, and the electrical services. It needs appreciable depth, particularly for elevators. Furthermore, it must be surrounded by concrete to protect the elevators against fire. The taller the building, the more substantial the core and its contribution to the structure of the building as a whole.

A building requires external walls whose supporting columns and spandrel beams constitute another tube. The core and the structure of its facade thus form a tube within a tube which resists the vertical loads as a composite column and the horizontal loads as a tube-shaped cantilever.

The tube concept was used by Fazlur Khan in the design of the John Hancock Center (Fig. 4.22). Six sets of diagonal bracing give the outer tube shear resistance in the same way as diagonals do in simple trusses (Section 2.1). The giant diagonals, however, form a striking visual feature that does not please everybody.

4.22

The John Hancock Center, Chicago, completed in 1968. Its height is 344 m (1127 ft). *Photograph by Hedrich-Blessing.*

There is a considerable saving in material (Fig. 4.23) even when the diagonals are omitted. The outer structure is conceived as a perforated tube rather than an assembly of beams and columns forming a rigid frame (Fig. 4.24). This concept was used by Leslie Robertson in the World Trade Center, completed in New York in 1973. Its height (411 m or 1350 ft) was the first to top the Empire State Building. In the Sears Tower Fazlur Khan used bundled tubes that were discontinued at different levels (Fig. 4.25).

The tube concept for the design of tall buildings produced a considerable increase in the span of the floor structure. The MLC Centre in Sydney (Fig. 5.4) measures 140 × 140 ft (43 × 43 m) in plan, and the sixty-four levels are vertically supported only by the service core and by eight outer columns. These are large [12 ft (3.6 m) deep at the base], but because they are placed outside the building they do not take up interior floor space. The relatively heavy floor structure spanning between the outer columns and the service core, without inner columns, provides the horizontal

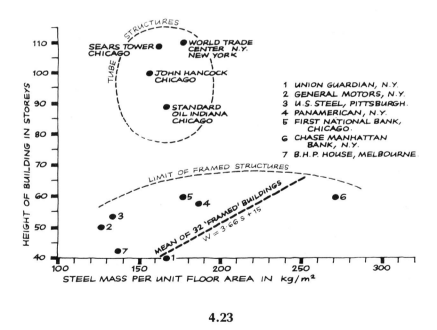

4.23

Relation between the height of the building and the amount of steel in its structure for a number of American and Australian tall buildings. (By courtesy of the Australian Institute of Steel Construction).

Two-Dimensional and Three-Dimensional Structures for Tall Buildings

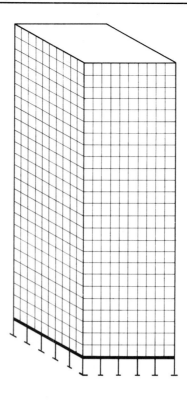

4.24

The tall building as a perforated tube, formed by the vertical columns and the horizontal spandrel beams at each floor level, with spaces between for window and other openings.

members of the outer tube which resists the horizontal forces. This concept, developed in the United States in the 1960s has been employed in most of the recent tall buildings.

Every multistory building requires some services on the roof, such as water tanks and hoisting machinery for the elevators. In many of the early buildings these service units were left exposed because they were the highest at the time. As taller structures were erected, the untidy arrangements on the roofs of the lower "skyscrapers" detracted from the prestige of the once highly acclaimed buildings. It therefore became common to enclose all the services in an upper story. In the 1960s this service floor was utilized to provide the structure with a "top hat" which gave it extra stiffness and wind resistance. Because windows could be made small or omitted on the service floor, the service core was extended laterally to give additional restraint to the columns in the outer tube (Fig. 4.26). The same principle was applied to intermediate service floors that were commonly used in very tall buildings.

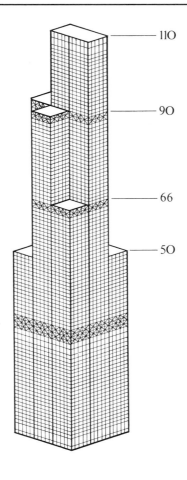

110

90

66

50

4.25

The "bundled-tube" concept of the Sears Tower in Chicago, completed in 1974. At the time of writing, this building is the tallest (442 m or 1450 ft). Two of the nine perforated tubes terminate at the 50th floor, two at the 66th floor, three at the 90th floor, and two go to the full height of 110 storys. The diagonals are extra stiffeners surrounding the service floors.

In the 1930s it was regarded as axiomatic that the steel frame was the only suitable form of construction for tall buildings, and this was accepted throughout the world. By 1960 the dominance of steel was being challenged by reinforced concrete, and by 1970 reinforced concrete had become the cheaper material for tall buildings in most countries. At the time of writing the tallest buildings are steel-framed only in the United States, Canada, and Japan. In all other countries they have a reinforced concrete frame (Fig. 5.4). The tallest concrete building, however, is still much lower

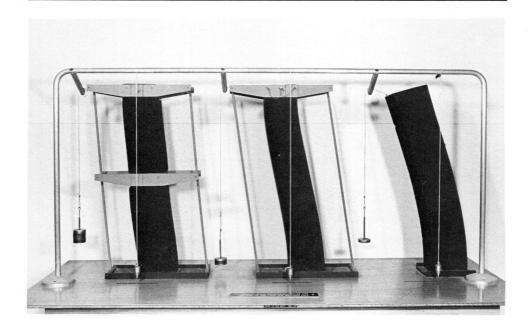

4.26

A "top hat," consisting of an extension of the service core into the topmost story, greatly increases the stiffness of a tall building.

than the tallest steel-framed building; at the time of writing it is Chicago's Water Tower Place, which has a height of 262 m (860 ft).

The extent to which the service core has assumed a major structural function is illustrated by the fact that the outer structural tube can be omitted. There are advantages in certain circumstances in hanging the floors and the outer walls from the roof (Fig. 4.27). This means that the loads have to be transferred by structural material to the roof and then conveyed by more structural material through the core to the ground (Fig. 4.28). Because the cost of the structure of a tall building is usually only one-fourth to one-sixth of the total cost, this method is not necessarily ruled out on economic grounds if there is an appreciable saving in construction cost. By

4.27

Standard Bank Centre, Johannesburg, South Africa. The building is supported by the reinforced concrete service core, cast with sliding formwork. This carries prestressed concrete cantilever brackets at three levels; nine precast concrete floors are hung from each set of brackets by means of prestressed concrete hangers.

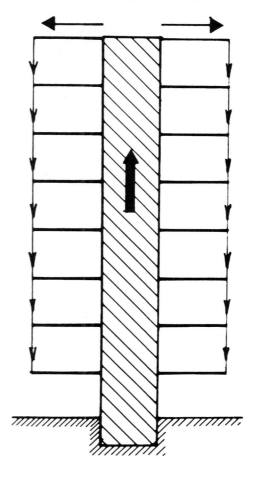

4.28

Structural principle of a building with floors supported by hangers from roof cantilevers.

building from the top down the ground can be used for storage and subassembly and the components can then be hoisted directly into position.

4.13   LIMITATIONS ON THE HEIGHT OF BUILDINGS

We noted in Section 2.6 that the new iron-framed buildings of Chicago in the 1880s were known as "elevator buildings" because their height was made possible only by the installation of elevators. Before the advent of the elevator height had been limited by the amount of stair climbing people were willing to do; although in ancient Rome buildings went above 20 m (65 ft) (Ref. 1.1, Section 3.8). Only in primitive societies was height restricted by the structure.

In the early days of the Chicago School, while people were still uncertain about the design of iron-framed buildings, the structure may have set the limit but soon the speed of the elevators became the controlling factor.

Nineteenth-century elevators were slow, and thus a comparatively large number were needed to service a tall building. The ratio of the space taken up by the service core to the rentable space, which depended to some extent on the size and shape of the available site, determined the maximum economic height. As elevators became faster, buildings became taller, but without the income from the radio and television transmitters in its tower the Empire State Building might have been above the economic limit.

When the invention of the sky lobby (Section 8.7) again raised the economic height, transportation to and from the tall building became a cause for concern as its population increased.

If tall buildings were connected, as Le Corbusier envisaged them, (Section 4.1), to a specially designed rapid-transit system, vertical transportation might still remain the limiting factor. In fact, most tall buildings, except for a few prestige structures, are located on existing streets, usually in the old part of a city whose communication system was designed for a much smaller town. In the City of London some of the main roads still follow their alignment in the days of the Roman occupation, and their width is possibly no greater than it was in Roman times. High land values would make the realignments or widening of these roads excessively costly. In Sydney, where the roads in the central business district date only from the nineteenth century, the problem is precisely the same because the streets were laid our for a small town with little traffic. Even in cities like Chicago or Melbourne, where the streets are wider, there is insufficient parking and an inadequate public transport system. In Los Angeles, which has good freeways and fairly adequate parking, there are fewer very tall buildings. Contrary to Le Corbusier's *City of Tomorrow* (Ref. 4.1), tall buildings are associated in practice with traffic congestion made even worse by the large number of people they attract.

Another ideal concept, the city within a city, has also remained unrealized. The Hancock Center contains apartments, shops, offices, and recreational facilities. It was thought that the traffic to and from this building would be reduced but few people who live in the building work there and vice versa.

We could build structures much higher than the Sears Tower and we could provide adequate vertical transportation within them. However, getting people in and out of these buildings would require a great improvement in horizontal transportation. At present this seems to be neither politically nor economically feasible. The problem is beyond the scope of this book. Until it is solved there is little point in considering how high we could build with our present technology.

CHAPTER FIVE

The Mechanization

of Structural Design

*The challenge of the computer:
how to manage its sophistication
without going broke.*

FAZLUR KHAN

In the first section we review the development of arithmetic and the early calculating machines. In the next section we survey the evaluation of structures with the help of models: first those that model the structural theory and then those that model the structure and are used for the direct measurement of strains and deflections.

In the third section we look at some of the physical analogs used for the solution of structural problems. Finally we trace the development of structural analysis with the digital computer which has so profoundly changed structural design practice.

5.1 PROGRESS IN ARITHMETIC

Symbols for numbers are almost as old as symbols for letters, syllables, and words. In Egypt the oldest identified numbers go back to 3000 B.C., that is, about the same time as the oldest pyramids. In Mesopotamia the oldest identified numbers date from 2400 B.C., that is, the period of the Sumerian-Akkadian Empire (Ref. 5.23). We are better informed about arithmetic than about most other aspects of ancient science because numerous papyri and tablets in clay and stone of accounts of goods received, payments made for them, and astronomical calculations have been found. There are also some exercises, of which those in the Rhind Papyrus are particularly interesting. In 1858 Henry Rhind, a Scottish antiquary, bought this roll of papyrus, 1 × 18 ft, believed to date from 1650 B.C.; most of it is now in the British Museum. The Rhind Papyrus contains the rules of proportion, later known in England from medieval to Victorian times as the "Rule of Three." ("If 258 loaves of bread cost 3 pounds 4 shillings sixpence, what is the cost of 13 loaves?") It gives a surprisingly accurate value for π, obtained by squaring the circle: "the area of a circle with a diameter of nine units equals a square with a length of eight units." Thus $\pi = 4 \times 8^2/9^2 = 3.1604$, an error of less than 1% (Ref. 4.57).

We are not so well informed about Mesopotamian arithmetic but apparently it also reached a high standard. The Greeks were therefore building on a solid foundation. Their particular development was geometry. They also made some advances in calculations. Ptolemy about A.D. 150 gave the value of $\pi = 377/120$, which equals 3.1416 and is, to five significant figures, the value still used today.

Virtually all ancient arithmetic was based on the decimal system derived from counting on the ten fingers of the two hands; the word digit is derived from the Latin *digitus*, which means a finger. However, this applied only to whole numbers; decimal fractions were not known before the Renaissance, and this greatly complicated even the simplest calculations. The addition of vulgar fractions is today easily accomplished by converting to decimals:

$$\frac{29}{16} + \frac{43}{38} = 1.81 + 1.13 = 2.94$$

For the Greeks, even at the height of their arithmetic achievement in Alexandria, this was a lengthy calculation because it was necessary to find a common denominator for the two fractions.

Calculating devices were in use in China and the Mediterranean countries by the third century B.C. The word abacus is derived from the Phoenician word *abak*, a board strewn with sand on which marks could be drawn; but the Egyptians were already using pebbles for counting and later mounted them to slide in grooves or on bars. A Roman abacus now in the British Museum looks quite similar to a modern Chinese abacus. Chinese and Japanese trained in the use of the abacus from childhood can perform simple addition, subtraction, and multiplication on an abacus almost as fast as on an electronic pocket calculator, and it is likely that in the ancient world there were people who could use an abacus with comparable speed.

Greek and Roman arithmetic was limited by its numerals and the need to work with vulgar instead of decimal fractions. The Greeks used the letters of the alphabet as numerals; Roman numerals, still in use today, were a definite improvement, but they could not be used for decimal fractions.

Arabic numerals were invented in India and adopted by the Arabs in the eighth century A.D. when the Hindu mathematical works were translated into Arabic. They were first used in Europe about 1200 by Leonardo of Pisa, generally known as Fibonacci. Leonardo had traveled with his father, a Pisan merchant, in North Africa and had studied there under a Muslim teacher.

The first recorded use of zero is in an Indian inscription dated A.D. 876; it was also introduced independently by the Maya of Mexico (Ref. 5.23, p. 335).

The decimal point was adopted to denote division of a number by ten in *Compendio do lo abaco,* by Francesco Pellos, published in Turin in 1492 (Ref. 5.23, p. 307), but decimal fractions were not used in calculations to any significant extent before the seventeenth century. Simon Stevin, who first published the parallelogram of forces, was a prominent advocate of decimal arithmetic.

The introduction of decimal fractions greatly increased the effectiveness of the abacus as a calculating device for addition, subtraction, and multiplication, but it had limitations even for simple division and for squaring numbers and could not be used at all for drawing square roots.

Logarithms were invented by John Napier, Laird of Merchiston in Scotland, in 1614. They greatly simplified division, squaring and cubing of numbers, and drawing square and cube roots. In 1620 Edmund Gunter plotted logarithms on a 2-ft straight line. Using dividers to transfer the numbers, multiplication and division could thus be performed graphically. In 1633 William Oughtred took two of Gunter's lines and performed the calculation by sliding them past one another. He thus invented the first analog computer (Section 5.3). In 1654 Robert Bissaker made a slide inside a fixed stock. This type of instrument was still in use in James Watt's Soho works.

The first modern slide rule was produced in 1859 by Amédée Mannheim, a French artillery officer. It had scales of logarithms from 1 to 10 and 1 to 100 on the slide and on the fixed stock, an additional scale from 0 to 1000, and a reciprocal scale at the center. This arrangement is still used on most slide rules today. He also invented the cursor which greatly increased the accuracy and speed of the slide rule. Mannheim's instrument performed with ease calculations involving division, squares, and square roots and could be used for cubes and cube roots.

The slide rule, like most analog calculators, is continuously graduated and its accuracy depends on the fineness of the graduations. The longer the slide rule, the more precise the answer; yet there are obvious limits to its practical length. The electronic pocket calculator achieves greater accuracy in a smaller instrument by using digits.

The first digital calculator was produced by Blaise Pascal, a French mathematician who is best known for his pioneer work on the theory of probability. In 1641, when he was only 18 years old, he built the first adding machine to use rotating gear wheels and later built and sold another fifty machines. They did not, however, do anything that an abacus could not do as well.

In the 1670s Gottfried Wilhelm Leibniz, better known as a philosopher and the coinventor of the infinitesimal calculus, devised a digital calculating machine that performed multiplication by repeated additions. It was exhibited to the Académie des Sciences in Paris and to the Royal Society in London.

Only minor improvements were made in digital calculators during the eighteenth century. In 1813, at the age of 21, Charles Babbage, later Lucasian Professor of Mathematics at Cambridge, proposed the production of mathematical tables by machine and built a small machine that could tabulate a function whose second differences were constant (Fig. 5.1). It had several vertical columns that contained "figure wheels" engraved with numbers 0 to 9, and in 1876 Scott Lang, Professor of Mathematics at the University of St. Andrews, described it:

The mechanism was so contrived that whatever might be the numbers placed respectively on the figure wheels of each of the different columns, the following succession of operations took place as long as the handle was moved. Whatever number was found upon the column of first differences, would be added to the number found on the table column. The same first difference remaining on its own column, the number found upon the column of second differences would be added to that of first differences (Ref. 4.65, Vol. IV, Calculating Machines).

In 1822 Babbage sent a description of this machine to Sir Humphry Davy, as President of the Royal Society, and asked for assistance to build one for twenty-decimal numbers, working up to seventh differences. He obtained a grant in 1823, but at his death in 1871 the machine was still incomplete. The first of a number of difference machines was built by George Scheutz in 1853, on one of which the *English Life Tables* of 1864, used by insurance companies, were calculated.

In 1833 Babbage conceived a more complex and versatile machine he called an analytical engine. He described it as follows:

The analytical engine is therefore a machine of the most general nature. Whatever formula is required to develop, the law of its development must be communicated to it by two sets of cards. When these have been placed the engine is special for that particular formula. The numerical constants must then be put on wheels, and on setting the engine in motion it will calculate and print the numerical results of that formula (Ref. 4.65, Vol. IV, Calculating Machines).

Babbage's grant for the construction of the difference engine was terminated by the government of Sir Robert Peel in 1842, although the Royal Society had recommended further support, and he never built either the difference engine or the analytical machine; he left some fragments and many drawings.

One reason why Babbage's machine encountered so many engineering problems was his use of decimal arithmetic. In 1937 Howard H. Aiken proposed the construction of a computer at Harvard University that, in retrospect, has much in common with Babbage's analytical machine. His design included a conversion of decimal to binary numbers, which employ only two numerals, 0 and 1:

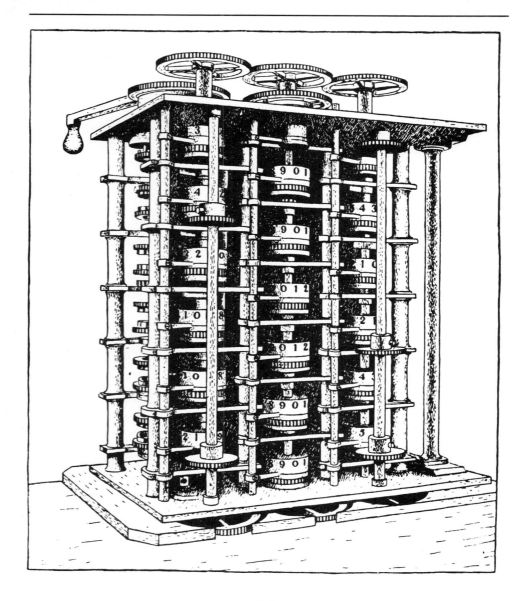

5.1

A model of a difference engine built by Charles Babbage in 1820. Some of the gear wheels of which it consisted had the figures 0 to 9 on their rims. One column carried the (constant) second differences, the central column carried the first differences, and the last column carried the numbers to be tabulated. The turning of the handle performed the additions.

The Mechanization of Structural Design

In decimal numbers	0	1	2	3	4	5	6	7	8	9	10	11
In binary numbers		0	1	10	11	100	101	110	111	1000	1001	1010	1011

and so on. Although binary arithmetic contains a lot of digits, it required only a two-way mechanism in the calculator. Aiken used an electric relay similar to those employed in automatic telephone exchanges. When the relay was open and no current passed, the digit was 0; when the relay was closed and an electric current passed, the digit was 1. The Harvard Mark I Automatic Sequence Controlled Calculator was completed in 1944 in association with International Business Machines (IBM), and was used to carry out computations for the US Navy (Ref. 5.22, p. 122).

During the 1940s J. Prosper Eckert and John W. Mauchly started work at the University of Pennsylvania on a computer that used electronic circuits; the Electronic Numerical Integrator and Calculator (ENIAC) was first shown in 1946 (Ref. 5.1, p. 9).

It is doubtful whether electronic calculating machines would have been built without the boost that World War II gave to scientific research. The magnitude of the calculations needed for ballistics and atomic research required high speeds, which could be supplied only by electronic circuits. After the war the new electronic digital machines were applied to a variety of scientific problems, but they were soon found to be even more useful in routine business applications. Few people in the 1950s expected the growth in the use of digital computers that occurred in the 1960s and 70s.

Although the developments in arithmetic were important for accounting and other commercial calculations, they had hardly affected structural design. All structural design in the ancient world, the Middle Ages, and the Renaissance was based on geometry (Ref. 1.1, Sections 3.9, 6.3, and 7.6). The new structural theory devised in the early nineteenth century required only simple calculations. Graphical methods (Section 2.1) were available for most statically determinate problems and were widely used in the late nineteenth and early twentieth centuries. The perfection of the slide rule in 1859 (discussed earlier in this section) provided a calculator adequate for the structural calculations of the time.

Problems developed in the 1920s, however, when statically indeterminate structures of some complexity were introduced (Section 4.2). This started the search for mechanized methods of structural analysis.

5.2 STRUCTURAL MODELS

In 1922 G. E. Beggs, Professor of Civil Engineering at Princeton University, published a description of his deformeter (Ref. 5.2). It was based on a theorem derived in the 1880s from Castigliano's work (Section 3.6) by H. Müller-Breslau, a professor at the Berlin Technical University.

Let us consider a beam with one statically indeterminate reaction at A (Fig. 5.2). Let us remove this reaction and displace A by δ_A, thus producing a displacement δ_B at B. Müller-Breslau's theorem states (Ref. 5.21, p. 19) that the reaction R_A at A produced by a force W_B at B is given by

$$R_A \, \delta_A = W_B \delta_B \qquad (5.1)$$

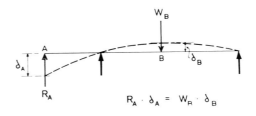

$$R_A \cdot \delta_A = W_R \cdot \delta_B$$

5.2

Müller-Breslau's theorem states that the reaction R_A produced by the force W_B is given by

$$R_A \delta_A = W_b \delta_B$$

where δ_A and δ_B are the deflections without the redundant reaction R_A.

This theorem had been used, particularly in Germany, for solving statically indeterminate structures by calculation. Beggs devised equipment for the accurate production of very small displacements. He then used a micrometer microscope to measure the resulting deformations along the members of a cardboard model of the structure to be analyzed. Thus the redundant reaction or reactions could be determined for any location of the load. Once this was known the structure could be designed by simple statics.

The theory of elasticity applies only to small elastic deformations, and Beggs therefore used small displacements and a microscope to measure them. This caused considerable eye strain, and it was not practicable for one operator to use the equipment for any length of time. The use of cardboard for the models, on the other hand, led to errors that were later avoided by metal or plastics.

In fact, large displacements cause only a relatively small error. Otto Gottschalk's Continostat, the first device based on large displacements, appeared in 1926; it included prefabricated sections from which the models were assembled. Like Begg's apparatus it was produced commercially, but neither became important as a design tool. The main defect of the Continostat was the restriction that the number of available sections, with a limited range of sectional properties, imposed on the design.

William Eney's Deformeter (Ref. 5.4) employed celluloid models that could be, made to fit any sectional properties and large displacements, but it was not developed until the late 1930s.

The 1920s were a decade of rapid development in the design of tall buildings (Section 4.1); they culminated in the design of the Chrysler Building and the Empire State Building in New York. The choice was between simple but rather approximate design methods and more accurate but complicated procedures based on strain energy (Section 3.6) or slope deflection (Section 4.2). Because of the Depression and

World War II, there was little building during the 30s and the early 40s, and in 1930 Hardy Cross published a comparatively simple method for *calculating* rigid frames (Section 4.2). Eney's Deformeter, which therefore came too late, remained primarily a useful laboratory tool for teaching structures.

Improvements of structural-model-testing equipment deriving from Begg's Deformeter continued to be made throughout the 1930s, 40s, and 50s (Ref. 4.64, Chapter 2); as the laboratory equipment improved, however, so did the techniques of calculation.

The early twentieth century was also a period of revival of interest in curved masonry and concrete (Chapter 6). Mathematical solutions were produced for a limited range of shells (Section 6.3), but they did not satisfy some designers, and in the 1930s engineered reinforced concrete structures were built for which there were no precise theoretical solutions, notably by Pier Luigi Nervi in Italy and Eduardo Torroja in Spain.

This led to the construction of accurately scaled models as an aid to design. The loads on these models were determined by the theory of dimensional similarity, and measurements of the resulting strains and deflections were made from which the strains and deflections in the actual structure could be calculated (Ref. 4.64, Chapter 3).

Galileo in the seventeenth century had already outlined the principles of dimensional similarity for structures (Ref. 1.1, Section 7.4). Models designed in the 1920s in accordance with dimensional theory were used for the structural analysis of dams (Ref. 5.5), which have an irregular shape because of the topography of the valleys in which they are raised; mathematical solutions therefore are difficult (calculations have recently been greatly simplified by the finite element method—Section 5.4). These models, however, were large and heavy by comparison with those used later for architectural structures; consequently, strains could be measured with fairly heavy instruments over gauge lengths varying from 5 to 6.5 in. (127 to 165 mm).

Instruments capable of measuring small elastic strains (i.e., deformations of about 10^{-5} mm/millimeter of length, or 10^{-5} in./inch), developed rapidly after 1870, but they were too heavy or required a long gauge length. Lightweight tensometers with a 1-in. (25-mm) gauge length were developed in the 1920s by A. U. Huggenberger, a Swiss instrument maker; they had an accuracy of about 1×10^{-4}.

The first architectural structure designed with the help of strain measurements on a scale model was an aircraft hangar at Orvieto, Italy (Fig. 6.24). It was designed and constructed by Nervi. A structural model made of celluloid was tested in 1935 by Guido Oberti, professor at the Milan Technical University, who used Huggenberger tensometers for strain measurement and dial gauges for deflection measurement. Several other models of architectural structures were tested by Oberti in Italy and Torroja in Spain before 1945.

During the 1939–1945 war a vast amount of structural testing done on aircraft structures, which developed rapidly during that time, led to great advances in strain-measuring techniques. The electric resistance strain gauge invented in the late 1930s consists of a length of wire glued to the structure through a paper or resin backing. As the short length of structure to which the gauge is glued (the gauge length) extends, the wire also extends and its diameter is reduced. This increases its electrical resistance. Conversely, if the gauge length is compressed, the electrical resistance is decreased. If certain alloys, glues, and electrical circuits are used, the strain can be measured with an *electric resistance strain gauge* to an accuracy of about 10^{-5}. When these gauges became available for general use at the end of the war, they made the model analysis of architectural structures a more attractive proposition. It was then possible to attach numerous gauges to the model and read them all at a central switchboard. The gauges weighed only a fraction of a gram, and the gauge length could be reduced to a few millimeters. Models were particularly useful for complex structures whose theory was either laborious or, in the current state of knowledge, impossible to evaluate numerically. Since the 1960s, however, progress in computer application (Section 5.4) has narrowed the useful field of model analysis.

It is still employed for preliminary design; for example, an architect with an unusual concept for a structure wishes to know whether it can be built at a reasonable cost; structures that cannot be built at all are now exceptional. A small model (Fig. 5.3) can normally establish the structural feasibility of the project at a small cost; getting results sufficiently accurate for design purposes is more expensive. If the test shows that the concept would involve too great an expense, it naturally receives little or no publicity. Hence the use of model analysis for this purpose may be more common than the literature suggests.

At the other extreme models are particularly useful for architectural structures so expensive that the cost of model analysis is small by comparison (Fig. 5.4). Structures of unusual complexity also offer scope for model analysis (Fig. 5.5).

In most theoretical analyses it is necessary to make some simplifying assumptions; end restraints on structural members and the complex curvature of a structure can often be reproduced more accurately in a model. On the other hand, model measurements are not so precise as the results obtained with a digital computer. The model thus serves as a check on the calculation and vice versa.

In addition, models help to clarify the structural concept. Even structural engineers find it easier to understand the behavior of a model than to read a computer printout. Structural behavior can be explained with the help of models to the architect and client and it is thus of considerable assistance in decision making on projects that involve large expenditures of money.

We have so far considered only those models tested within the elastic range to determine the stresses and deflections. As we noted in Sections 2.4 and 4.10, a structure may also fail by buckling before it has reached its limiting stresses, and failure is easily checked with a model (Fig. 5.6) if elastic buckling predominates.

The models shown in Figs. 5.3 to 5.5 were made from suitable elastic material for testing in the elastic range only. There is an ultimate-strength theory for some frames and for some slabs, but there is still no satisfactory solution for the ultimate-strength theory of shell structures. Model tests for these structures are particularly interesting beyond the elastic range. It is a relatively simple matter to model a steel structure for ultimate strength. The iron and carbon crystals are of microscopic size, and it is sufficient to use the same steel in the structure and model. In the case of concrete, however, the particles of aggregate need to be scaled down to produce *microconcrete* (Ref. 5.6).

Evidently the limits on the reduction of scale possible with ultimate strength models are more severe. For the frames of a tall building a reduction of scale between

5.3

A small perspex model of a hyperbolic paraboloid for a church roof. The model showed that this structure would cost more than the client intended.

5.4

A structural model for the frame of the MLC Centre in Sydney, a 64-level reinforced concrete structure tested in 1972 at the University of Sydney. This perspex model has a scale ratio of 1 : 95. Strains were measured with electrical resistance gauges and deflections with dial gauges.

5.5a

A perspex model of a club in Sydney, tested at the University of Sydney in 1960. The strains were measured with electric resistance gauges and the deflections with dial gauges. The scale ratio is 1 : 32.

1:100 and 1:50 is suitable for an elastic model. Thus the model of a 400-m (1310-ft) tall building can be reduced to between 4 and 8 m. If the microconcrete is to model the real concrete, it is difficult to use a scale reduction greater than 1:15 (Ref. 5.7), so that only structures of moderate size, or parts of larger structures, can be successfully modeled for ultimate strength.

Although computer technology has restricted the scope of model analysis, it has also greatly aided the evaluation of the results. In the 1940s the use of a hundred strain gauges would have required a larger staff to record and reduce the data than most laboratories possessed. It is now possible by means of a digital converter to obtain an automatic typewritten printout and to record it directly on computer tapes or cards which can be fed into a computer to perform the calculations necessary to obtain principal stresses or reduced data.

5.5*b*

An architectural model of the same club in Sydney. It has a reinforced concrete folded plate dome and was completed in 1961.

5.6

Perspex model of a rectangular frame which has just reached its buckling load *(Architectural Science Laboratory, University of Sydney).*

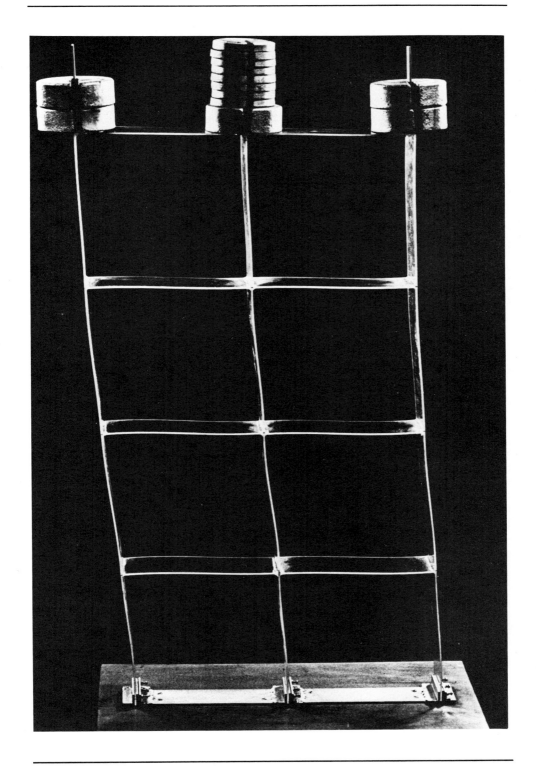

5.3   ANALOGUES

There is a great variety of analogues, although few have been particularly useful in the design of buildings. The most successful analogue computer of all time is the slide rule (Section 5.1). If we add the logarithm of two numbers A and B, we get the logarithm of the product AB; thus the multiplication of numbers is analogous to the addition of their logarithms. It is easier to *add* the logarithms than to *multiply* the numbers.

   All analogues are based on this principle. If we have a problem too complicated to solve directly, we may look for another with a similar mathematical equation that is easier to evaluate.

   We shall examine the torsion problem as an example of the analog technique. Barré de Saint-Venant in 1853 derived the general theory for the elastic torsion of a shaft or structural member which has a simple solution for a circular section, rather more complex solutions for certain geometric shapes (e.g., a rectangle), and no exact mathematical solutions fog irregular shapes (e.g., structural steel I-sections). Before the advent of the electronic computer, analogues were used for a number of torsion problems (Ref. 5.8, pp. 107–111 and 120–122). Saint-Venant's torsion theory produces a partial differential equation:

$$\frac{\partial^2 \phi}{\partial x^2} + \frac{\partial^2 \phi}{\partial y^2} = -2G\theta \tag{5.2}$$

where G = the shear modulus of elasticity of the material,
   θ = the angle of twist per unit length of the member,
   x, y = the coordinates of the cross section of the member,
   ϕ = the torsion function from which the elastic shear stresses in two perpendicular directions are derived by differentiation:

$$v_{xz} = \frac{\partial \phi}{\partial y} \quad \text{and} \quad v_{yz} = \frac{\partial \phi}{\partial x}$$

In 1903 Ludwig Prandtl, then Professor of Engineering Mechanics at Hannover Technical University, pointed out that (5.2) was mathematically similar to the equation for a membrane under pressure:

$$\frac{\partial^2 z}{\partial x^2} + \frac{\partial^2 z}{\partial y^2} = -\frac{p}{T} \tag{5.3}$$

where p = the pressure acting on the membrane (in excess of atmospheric pressure),
   T = the surface tension in the membrane,
   x, y = the coordinates of the section at the base of the membrane,
   z = the transverse deflection of the membrane.

We can solve the torsion problem by a simple experiment instead of a complicated calculation. First we make a box and at one end cut holes equal to the sections we wish to solve (Fig. 5.7). We glue a rubber membrane to close the hole. We then increase the pressure inside the box, for example, through a bicycle valve with a bicycle pump, and contour the shape of the deflected membrane. This can be done with a micrometer needle or by photography. The elastic shear stresses at any point in the section are given by the differential coefficient of ϕ (or of z); that is, by the slope of the membrane, which is obtainable from the contours. The elastic torsional resistance moment, according to Saint-Venant, is

$$2 \int \phi \, dA$$

(where A is the cross-sectional area of the member), so that we can obtain it by analogy from the volume displaced under the membrane by the pressure; the volume is calculated from the contours.

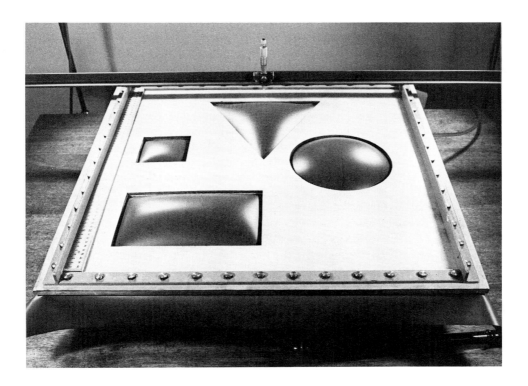

5.7

Prandtl's membrane analogue for the torsion of a square, a rectangular, a circular, and a triangular section. The contours of the membranes are determined with the micrometer needle shown at the top end of the photograph. The same method can be used for more complex shapes, such as structural-steel I-sections.

In 1923 A. Nadai derived the theory of plastic torsion and also produced an analogue for its numerical solution (Ref. 5.9). When a section is twisted and becomes fully plastic (Section 4.10), the shear stress v is by definition uniform across the entire section and equal to the shear stress when the material is yielding:

$$\left(\frac{\partial\phi}{\partial x}\right)^2 + \left(\frac{\partial\phi}{\partial y}\right)^2 = v^2 \tag{5.4}$$

If we heap dry sand on a horizontal plane surface of any shape, it settles at a constant slope m, and therefore the depth of the sand z is given by

$$\left(\frac{\partial z}{\partial x}\right)^2 + \left(\frac{\partial z}{\partial y}\right)^2 = m^2 \tag{5.5}$$

If we equate the slope of the heap of sand m to the constant plastic shear stress of the structural material v, the plastic torsional resistance moment is proportional to the volume of the sand, which is obtained by weighing. Thus we can determine the plastic torsional moment by cutting a flat shape out of plywood equal to the section we wish to solve; for example, a structural I-section. We heap dry sand on it and weigh the sand that is retained.

Nadai pointed out (Ref. 5.10, pp. 494–526) that we can also investigate the partially plastic section in torsion by a combination of the membrane and the sandheap analogues, although this requires rather more work. When a steel section is twisted, it deforms elastically until at some point in the section the yield stress is reached. This corresponds to the membrane analogy. With increase in torsion the shear stress at that point remains constant, for the material has turned plastic. The plastic zone gradually spreads until the entire section is plastic; this occurs at the plastic torsional resistance moment and corresponds to the sandheap analogy.

Having determined the shape of the sandheap, we make a "roof" from a transparent material such as perspex or plexiglass of the same shape and place it over the elastic membrane. As we increase the pressure on the membrane, it eventually touches two or more points of the roof. This contact can be shown up more easily by moistening the membrane so that it forms a line where it touches the roof. This part of the section in contact with the roof corresponds to the section that has turned plastic. The elastoplastic torsional resistance moment is proportional to the volume displaced by the membrane restrained by the "roof" of the sandheap analogy.

There are numerous analogues for structural problems based on various physical phenomena, notably fluid flow and electricity (Refs. 5.8 and 5.11). At a time when the numerical evaluation of differential equations without an exact solution was laborious they offered a quicker method of evaluation that could generally be made as accurate as the problem required by increasing the size of the apparatus and the precision of the measuring equipment. The torsion analogues were, in fact, used for practical design problems in the 1930s, 40s, and 50s. By 1960, however, it was simpler and quicker to make a numerical evaluation with the help of an electronic computer (Section 5.4).

The most important structural problem since the 1930s has been the solution of the rectangular rigid frame (Section 4.2), and several analogues have been employed

(Ref. 5.8, pp. 132–141), mostly based on electrical circuits. The analogue proposed in 1934 by Vannevar Bush, the inventor of numerous successful analogues (Ref. 5.12) consisted of adjustable resistors and tapped transformers that could be connected to simulate trusses and rigid frames of different configurations. The properties of the individual members were simulated by suitable setting of the resistors and transformer taps. Currents were introduced to simulate applied loads, and the internal forces and moments were represented by other currents that could be measured easily. The instrument was essentially an electrical analogue for Maney's slope-deflection method (Section 4.2).

In 1953 Frederick Ryder proposed an analogue based on strain energy (Section 4.6). He described it as follows:

The labor involved in the solution of statically indeterminate structures may be greatly reduced by the use of electrical analog, especially in those cases where various choices of dimension are being investigated. There is a similarity between the structure and its analog which often permits the construction of the analog on a pictorial basis without regard to the governing equations.

In the electrical circuit, force and moment are simulated by current, and deflection and slope by voltage drop. The analogy is based on the fact that Castigliano's theorem and the principle of least work have their counterparts in certain electrical networks (Ref. 5.13, p. 1).

The elements of the analogue are shown in Fig. 5.8. Electric consoles based on the Ryder analogue were used for the analysis of rigid frames in the 1950s. In the late 1960s, however, digital computers were performing the calculations quicker, more accurately, and cheaper.

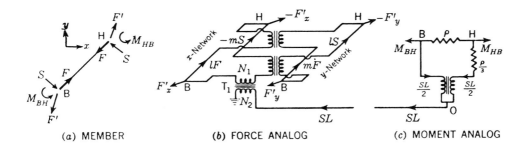

| (a) MEMBER | (b) FORCE ANALOG | (c) MOMENT ANALOG |

5.8

The Ryder analog consisted of adjustable resistances and tapped transformers which simulated the forces and the moments.

5.4 DIGITAL COMPUTERS

We noted in Section 5.1 that the first electronic digital calculating machine became available in the United States in 1946. The first practical electronic computer in Britain (EDSAC) was completed at Cambridge University in 1949.

The late 1940s were not, however, a period in which the demand for improved methods of calculating building frames was pressing. Since the design of the very tall buildings in the late 1920s (Section 5.2) the moment distribution method had been developed (Section 4.2), and electrical calculating machines, which were the conventional adding machines of the 1890s (Section 5.1) with the addition of electric motors, had become available at reasonable cost. They also performed multiplication and division, not quite so fast as a slide rule, but with greater accuracy; some machines produced a printout that made it easy to check the input.

The first paper on the use of electronic digital computers for structural analysis was published in 1953 by R. K. Livesley, then a lecturer in engineering at the University of Manchester (Ref. 5.24). Electronic computers are particularly well adapted to the solution of matrices, and Livesley formulated the conventional equations of the elastic deformation of structural frames (Section 4.2) in terms of matrix algebra.

In a later review (Ref. 5.14) Livesley credited John Bennett, now Professor of Electronic Computing at the University of Sydney, with having performed in 1950 the first structural calculations done on a computer described in a Ph.D. thesis submitted to Cambridge University in 1952. Bennett, in turn, based his method on a paper by G. Kron (Ref. 5.15) which described an electrical analogue for simulating a structure. In 1968 Livesley commented as follows:

Although not the first paper on structural analysis to use matrix algebra, it is one of the first to provide a systematic description of the topology of a general structure—an idea borrowed from the use of connection matrices in electrical network analysis. Kron realized that there are two quite separate systems of equations which together define the properties of a structure. The first system describes the behaviour of the individual elements, while the second defines the manner in which the elements act together. This second system provides the coefficients for both the "equilibrium" and the "compatibility" equations of classical structural analysis. This implicit distinction between element properties and topological form was an important structural advance (Ref. 5.14, p. 177).

The electronic digital computer would have been of some help if it had been used as a particularly fast electrical calculator to speed up the existing methods of structural analysis (Sections 3.6 and 4.2). To get the real benefit from it, however, it was necessary to take the "machine's-eye view" of the problem. In the late 1950s the analysis of rigid frames was therefore rewritten in terms of matrix algebra, and the first textbook of structural design written specifically for computer applications appeared in 1959.

The first commercial application, reported by Livesley (Ref. 5.14, p. 178), was in 1953 to the rectangular five-bay frame of a British power station, but it is possible that there were earlier commercial applications in the United States that have not been published. At that time the structural analysis could have been done faster by conventional methods. By 1960 computerized structural analysis was quicker and cheaper for tall frames and competitive for frames of medium height, but many consulting engineers lacked satisfactory access to computers. By 1970 computerized structural analysis was in general use in America, western Europe, and Australia for all but the smallest frames. Whereas model analysis and analogue computers have influenced structural analysis only marginally, digital computers have transformed it.

The earliest computers required formulation of their programs in mathematical terms, so that designers could make no assessment of the results until printouts were produced. In 1956 the FORTRAN (FORmula TRANslation) programming language was introduced in the United States and, at the same time, Autocodes were developed in Europe which made possible the use of ordinary words within a restricted vocabulary. These methods overcame the language barrier that had restrained engineers and architects from using computers.

Designers were also enabled to communicate with a computer on an on-line input-output device in the form of an electric typewriter. Later this communication was further improved by graphical input-output devices, such as a cathode-ray tube equipped with a light pen. Because of the speed of the computer, the main delay was the reaction time of the designer.

This paved the way in the late 1950s for time sharing which enabled a designer from an input-output unit in his office to communicate by telephone with a computer, possibly hundreds of miles away. Thus large computers were brought within the reach of small design offices.

Although computer technology had advanced most rapidly in the United States, the early work on computer-based structural analysis was carried out in Britain. In the late 1950s, however, a number of American universities, backed by computer firms, set to work to simplify structural programs which resulted in problem-oriented languages that permitted the engineer to write a special-purpose program quickly for any of a great variety of structural problems in words familiar to him (Ref. 5.16). The first of these programs was STRESS (STRuctural Engineering Systems Solver), which became fully operational in 1964; it was developed by the Massachusetts Institute of Technology in cooperation with IBM (Ref. 5.17). STRESS employs the stiffness method of analysis; that is, the elastic deformations of the joints in the frame are regarded as the unknown quantities. The stress resultants in the members are expressed in terms of these unknown displacements, and the joint equilibrium equations are set up in matrix form and solved for the displacements. The STRESS program is written so that its user need not be familiar with the technique of analysis employed.

With a program of this type it became possible, even with a modern minicomputer, to print out in rapid succession the solutions to every one of the hundreds of rigid frames for which Kleinlogel had so painstakingly derived the design formulas (Ref. 4.8 and 4.9). Only fifteen years earlier this had been regarded as a monumental task.

One of the great advantages of computer solution of structural problems is that it provides a practicable method of obtaining optimum designs for statically indeterminate structures. For a statically determinate structure the forces and moments acting on the structural members can be obtained without assuming the size of the members (Section 2.1 and 2.3). Hence the sizes of the members that are optimal for strength can be obtained without difficulty. For a statically indeterminate structure it is necessary to assume the sizes of the members in order to determine their elastic stiffness. Without a computer optimization is therefore a laborious process. With a computer the same program can easily be rerun with different structural sizes to obtain members that are stressed to the permissible stresses, deform to the permissible limits, or satisfy other design requirements. Computer programs have recently been devised to produce automatically structural sizes that are optimal for some simple criterion, such as the least weight of the structure.

The weight of the structure is, however, only one item in the total cost of the building. Fabricating cost for steelwork and formwork for concrete enters into the structural cost. Even if these costs are included, optimization may not be achieved because the cheapest structure does not necessarily produce the cheapest building. At the time of writing the problem of optimizing the cost of a building remains to be solved.

The analysis of plane stress problems and surface structures (Chapter 6) was also transformed by computer analysis. In 1935 Sir Richard Southwell, Professor of Engineering Science at Oxford University, had developed the relaxation method for pin-jointed space trusses used in aircraft. Later this was generalized and adapted to a variety of other problems (Ref. 5.18), including plane stress analysis. The plane stress system was replaced by a network or lattice of trusses, which was solved by "relaxation" or successive redistribution of the finite differences (as in the moment distribution method; Section 4.2). To obtain reasonably accurate results it was necessary to use a fine-mesh lattice of *bars,* which increased the arithmetic labor. The finite difference method was thus a laborious process, although it was made much easier when electronic computers became available.

In 1954 W. J. Argyris in England and R. W. Clough in the United States developed a method for the solution of complex aircraft frames (Ref. 5.19) which replaced the continuous plane stress system with an assemblage of plane stress *plate* elements interconnected at the corners. Clough later coined the name finite element method and adapted it to plane stress analysis (Ref. 5.20). The finite element method is thus essentially the stiffness method for frames applied to continua, which are idealized into a large series of interconnected elements of various geometric shapes. It is more accurate and versatile than the method of finite differences and has been used in the design of architectural structures (Chapter 6) and other surface structures for which an exact mathematical solution is not available.

Computers have also been applied to the design of building services and to general architectural design problems (Sections 9.10 and 10.9).

The effect of the computer may be compared to the invention of power-operated handling and earthmoving equipment, the lack of which did not stop the ancient Romans from building hundreds of kilometers of aqueducts; today we can lay water mains without the hard labor and at a lower cost. The absence of calculating machines did not stop the compilers of mathematical tables a hundred years ago from performing prodigious feats of arithmetic with the help of hundreds of people doing routine calculations. It is now possible to perform comparable calculations much cheaper and to dispense with at least one unsatisfying type of work. What was an exceptional task has now become a matter of routine.

Structural theory had by the 1950s become so complex that the choice was between an approximate design, on the one hand, and an amount of calculation that few engineers were willing or could afford to undertake. The computer has not entirely solved this problem because the structural analysis of a frame for a modern tall building would still be too expensive without some simplifications. It is now possible, however, to obtain a much more accurate solution and to run a program several times to determine the most appropriate design. The structural systems briefly described in Section 4.12 would not have been practicable before suitable computer programs became available.

CHAPTER SIX

Long-Span Structures

and

Curved Structures

The method employed I would gladly explain
While I have it so clear in my head,
If I had but the time and you had but the brain—
But much yet remains to be said.

LEWIS CARROLL

The Hunting of the Snark

The last two chapters have been devoted mainly to the problem of height. We now consider designs suitable for long spans. There was probably no span exceeding 50 m (164 ft) before the eighteenth century; thereafter spans increased rapidly, and at present the longest bridge spans 1300 m (4260 ft) and the longest span for a building is 219 m (718 ft). There is no limit to the useful span of a bridge, but it is doubtful whether we would benefit from a further increase in the span of buildings.

In this chapter we examine first the recreation of the traditional dome and barrel vault as thin reinforced concrete shells and the invention of new concrete shell forms such as the hyperbolic paraboloid. We then examine the economic advantages of folded plates for moderate spans.

In Sections 6.6 and 6.7 we consider long-span steel structures, such as braced domes, space frames, and suspension structures, and finally we take a brief look at pneumatic membranes.

6.1 LONG-SPAN BRIDGES

The longest spanning structures are, and probably have always been, bridges. It is a relatively simple matter in a building, if the spans get too big, to insert extra columns. Thus the larger temples of ancient Egypt, India, and Greece contained veritable forests of columns.

Palladio, describing a bridge he designed in the sixteenth century (Ref. 1.1, Section 7.7), explained that a single span was used because the river "is very rapid . . . a resolution was taken to make a bridge without fixing any posts in the water, as the beams that were fixed there were shaken and carried away by the violence of the current." Undoubtedly this argument also influenced the design of a number of Roman and medieval bridges.

The longest spanning Roman bridge to survive is the Puente Trajan at Alcantara in Portugal. It was built by Caius Julius Lacer in the reign of the Emperor Trajan (second century A.D.) and its longest masonry arch spans 98 ft (30 m) at a height of 170 ft (52 m) above the River Tagus. It has been claimed that the longest span of the timber bridge built at about the same time by Apollodorus over the Danube was 170 ft (52 m) (Ref. 4.65, *Bridges*); assuming that the illustration on Trajan's column is diagrammatic only, this is not impossible, but it cannot be proved and most estimates are much lower.

A semicircular masonry arch bridge at Ceret, over the River Tech in France, near the Spanish border on the old Roman road from Nimes to Spain, was built in the fourteenth century on a Roman foundation. It has a span of 45 m (149 ft), but because there is no evidence of an intermediate support it is likely that it replaced a Roman bridge of the same span.

This span was not exceeded until the eighteenth century. From 1756 to 1758 Johann Ulrich Grubenmann, a local carpenter without theoretical training, built a timber bridge over the Rhine at Schaffhausen in Switzerland. He had originally proposed a single span of 110 m (360 ft) and made a model to illustrate the soundness of his design which easily carried his own weight. The city council, however, insisted on an intermediate support and the bridge was built with two spans of 59 m (193 ft) and 52 m (171 ft). Fritz Stüssi (Ref. 6.1, p. 11) relates that Grubenmann designed the bridge to cross the river in one or two spans. Before the official opening he knocked a block from under the intermediate support to make the bridge span the whole length, saying to his journeyman, *"Da habt ihr euren Pfeiler, aber ich habe meine Brücke."* (You have your prop, but I have my bridge.)

The bridge consisted of top and bottom chords with inclined diagonals converging on the end supports and additional diagonals converging on the central support. Since it was destroyed after a life of only forty-one years, this story cannot be checked, although even as a two-span bridge it set a record. Thomas Tredgold (Section 2.3) referred to it:

The French army, in 1799, destroyed the celebrated bridge across the Rhine at Schaffhausen; but the fame of Grubenmann the carpenter will long continue; and the form of that excellent specimen of the art will only cease to be remembered, when carpentry itself no longer exists (*Elementary Principles of Carpentry, 1820*).

In 1812 Louis Wernwag built the Colossus Bridge over the Schuykill River at Fairmont, near Philadelphia, a timber bridge with a span of 340 ft (104 m).

Spans increased rapidly after iron came into use as a structural material, and since 1826 the world record for span has always been held by an iron or steel bridge. Table 6.1 is a chronology of the longest spans to date.

Although suspension bridges have an ancient history, Thomas Telford's wrought-iron bridge over the Menai Strait (Fig. 6.1) was the first with a really long span. Like all preceding and some subsequent suspension bridges, it was too flexible and has since required stiffening to prolong its life. Twenty years later Stephenson built parallel to it the tubular Britannia railway bridge, discussed in Section 2.5, which no longer carries traffic. John Roebling was the first to design a stiffened suspension bridge, the Grand Trunk Bridge just below the Niagara Falls, completed in 1855 with a span of 820 ft (236 m). Roebling was also the first to use spun cables of high-strength steel wire with an ultimate strength of 143 ksi (986 MPa) (Brooklyn Bridge, 1883).

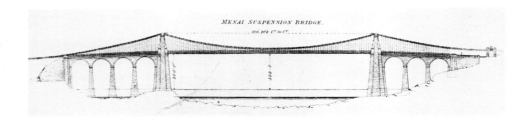

6.1

Thomas Telford's wrought-iron suspension bridge over the Menai Strait in North Wales, with a span of 580 ft (177 m), was completed in 1826. It was the first iron suspension bridge. (From Ref. 4.65, Article *Bridges*)

Table 6.1

Span		Type of	Name and Location	Year of
m	ft	Bridge	of Bridge	Completion
177	580	Suspension	Menai Bridge, Menai Strait, North Wales	1826
265	870	Suspension	Fribourg Bridge, Fribourg, Switzerland, D	1834
308	1010	Suspension	Ohio River Bridge, Wheeling, West Virginia	1849
318	1043	Suspension	Niagara River Bridge, Lewiston, New York, D	1851
322	1057	Suspension	Ohio River Bridge, Cincinnati, Ohio	1867
387	1268	Suspension	Niagara-Clifton Bridge, Niagara Falls, New York, D	1869
486	1595	Suspension	Brooklyn Bridge, New York	1883
521	1710	Cantilever	Forth Bridge, Edinburgh, Scotland	1889
549	1800	Cantilever	Quebec Bridge, Quebec City, Canada	1918
564	1850	Suspension	Ambassador Bridge, Detroit, Michigan	1929
1067	3500	Suspension	George Washington Bridge, New York	1931
1280	4200	Suspension	Golden Gate Bridge, San Francisco, California	1937
1298	4260	Suspension	Verrazano Narrows Bridge, New York	1964

D = Demolished

Othmar H. Ammann almost doubled the longest span with the George Washington Bridge, completed in 1931, contemporary with the Empire State Building (Section 4.1). The span of bridges, like the height of buildings, has not greatly increased since that time. Joseph Baermann Strauss's Golden Gate Bridge has acquired special fame because of its magnificent physical setting.

All the record-making bridges since 1826 have been suspension bridges, except for the Forth Bridge (Section 3.3) and the Quebec Bridge, which are both cantilever types. Yet suspension structures (Section 6.7) have not played an important part in architectural design.

Since the eighteenth century an ever-growing gap has developed between the longest spans of bridges and buildings. In ancient Rome, the Middle Ages, and the Renaissance the longest bridge was only a little longer than the greatest span in a building. At the present time the longest span in a building is 718 ft (219 m), which is about one-sixth of the span of the longest bridge (Section 6.3). This gap is likely to increase (Section 6.9).

6.2 THE REVIVAL OF CURVED ROOF STRUCTURES

In studying the structures of the past one can observe the progress from post-and-lintel construction through the corbeled and barrel vaults to the ribbed (Ref. 1.1, Sections 3.1 and 6.1). Each step reduced the unit weight of the structure and each increased the complexity of design. A similar development can be traced for steel and reinforced concrete structures.

The first-ever complete iron structure, the bridge at Coalbrook Dale (Ref. 1.1, Section 8.6), built in the late eighteenth century, was an arch, but at that time there was neither an empirical nor scientific basis for its design. Within a few decades, however, the theory of bending for straight elastic beams was perfected and it was soon used for the design of fireproof iron structures (Ref. 1.1, Sections 8.3 and 8.6). The development of the theory of the steel frame (Section 2.6) established the linear iron structure in the early twentieth century as firmly as the linear stone structure had been established in Ancient Egypt and Greece. Thus the cities of Europe and the United States acquired an unprecedentedly large number of multistory buildings, most of which are shaped like big rectangular boxes (Section 6.9).

Iron was used from the midnineteenth century on for a few spectacular, curved, long-span structures (Section 2.6 and 3.3), the most noteworthy being St. Pancras Station (Fig. 3.7) built in 1886 with a span of 244 ft (74 m). The span of iron and steel structures was made much more reliable when Schwedler invented the third pin (Fig. 3.8). This rendered the structure statically determinate so that accurate calculations could be made and spans increased. *The Galerie des Machines* (Section 3.3) erected in 1889 with three-pin steel portals had a span of 113 m (370 ft); this was more than twice the biggest span of antiquity and the Renaissance and more than four times the span of the widest Gothic cathedral.

Long-span steel structures had, however, one important limitation. Unlike traditional masonry structures, they provided merely the skeleton. This did not matter for a structure like the Eiffel Tower, which was built as a vertical feature for the Paris Exhibition of 1889 and only much later found a practical use as a radio transmitter; but a building needed a roof and walls. In temporary exhibition halls, such as the Crystal Palace (Section 10.1), and industrial buildings this could take the form of glass and sheet iron, but these materials were unsuitable for monumental buildings. Since the turn of the century, therefore, curved structures have been designed in reinforced concrete to provide a roof surface as well as the skeleton.

Traditional structures were limited by the lack of tensile strength of unreinforced masonry (Ref. 1.1, Sections 3.9, 5.2, 6.5, 7.2, 7.3 and 7.6). Reinforced concrete presented no problem.

In 1897 M. A. de Baudot, architect to the Paris diocese, designed the Church of Jean de Montmartre in the style of the Art Nouveau; it had a dome with reinforced concrete ribs and brick shells. There is no record of a complete reinforced concrete dome before 1900, but in the next decade at least six were built (Refs. 2.11, 2.15, and 6.2); all had ribs.

The use of ribs is probably attributable to the precedent set by steel structures and the influence of the Gothic Revival. Gothic structures, still the lightest form of masonry construction in existence, provided a natural prototype for curved reinforced concrete structures. The Gothic cathedrals had in the late nineteenth century been analyzed in accordance with the new principles of structural mechanics and interpreted in the light of the experience gained from iron structures. This produced a theory, espoused notably by Viollet-le-Duc (Ref. 1.1, Section 6.7), which considered ribs to act like a structural framework supporting the stone shells and suggested that reinforced concrete might also be designed in this manner. Concern for the buckling of thin sheets of reinforced concrete, following the problems encountered with iron (Section 2.5), may have been a contributory factor.

Several large ribbed reinforced concrete domes were built before 1914, notable among which are Wesleyan Hall in Westminster, England, completed in 1910, the main reading room of the Public Library in Melbourne, Australia, 1911, and Centenary Hall in Breslau, Germany (now Wroclaw, Poland), 1913.

The dome of the Melbourne Public Library (Fig. 6.2), although much smaller than Breslau's, is perhaps the more remarkable because it was the first reinforced concrete

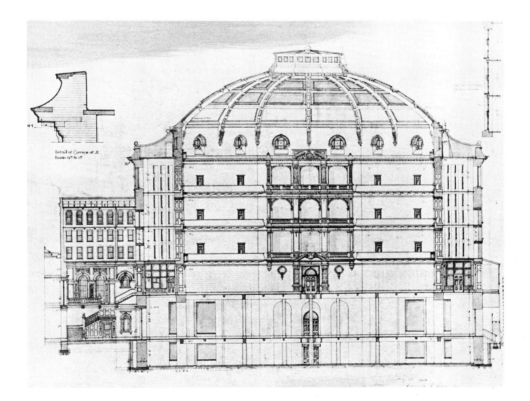

6.2

Working drawing of section of the Melbourne Public Library Reading Room by Bates, Peebles, and Smart, 1909 (Ref. 6.3).

6.3

Airship hangars designed by Freyssinet under construction at Orly near Paris. The completed structure had a clear span of 80 m (262 ft), a height of 56 m (184 ft), and a length of 300 m (984 ft). The corrugations were 5.4 m (18 ft) deep, but the maximum thickness of the concrete was only 90 mm (3½ in.).

building of any size constructed in Australia at a time when a reply to a letter to Europe or America took three months. The architects were Bates, Peebles and Smart (Ref. 6.3), and the first sketch plans were prepared in 1906 by N.G. Peebles from rough sketches drawn by the chief librarian who presumably had in mind Smirke's 140-ft (42.6-m) iron-framed reading room in the British Museum (Section 2.6). John Monash (later General Sir John Monash, commander of the Australian army during World War I), then director of a construction company, was apparently responsible for the choice of reinforced concrete for the 115-ft (35.1-m) dome and advised that it would be cheaper than steel. His company did not win the contract, however; the lowest bid was submitted by the Trussed Steel Company of London which had just completed the 114-ft (34.7-m) dome for Wesleyan Hall. Thus the Melbourne Public Library gained the record as the longest spanning reinforced concrete dome by 1 ft and it held it for a year.

Centenary Hall in Breslau had the first concrete dome to exceed the span of the Pantheon, built 1800 years earlier. It was designed by the city architect, Max Berg, and in accordance with German practice the contractors, Dyckerhoff und Widmann, were responsible for both structural design and construction; the structural calculations were made by Willy Gehler (later professor at the Technical University of Dresden) and Eugen Schulz. The dome had a clear span of 65 m (213 ft).

In 1916 construction started in Paris on two reinforced concrete airship hangars, each with a clear span of 80 m (262 ft) (Fig. 6.3). They were designed by Eugène Freyssinet (Section 3.4), constructed by Enterprise Limousin, and completed in 1924, but were destroyed in 1944 in an air raid. Because of the shape of the structures, the stresses due to dead weight were, as in the Ctesiphon arch (Ref. 1.1, Section 5.3), almost wholly compressive, but because of the height and exposed location of the structures and their small weight there were appreciable bending moments due to wind loads which were resisted by corrugations in order that the concrete itself could be made quite thin. The airship hangars at Orly were much admired not so much for the lightness of their construction but for their elegant shape, although Freyssinet denied any attempt to produce an artistic design and claimed that the shape resulted logically from the conditions of equilibrium.

6.3 CONCRETE DOMES AND CYLINDRICAL SHELLS

Although the airship hangars at Orly and Centenary Hall in Breslau were subject to bending stresses, they required only a small fraction of the material employed in classical domes and vaults; the weight could be further reduced by using membrane shells.

It has been clearly understood for at least two centuries that a flexible material such as leather or thin sheet metal can, as a membrane, resist an appreciable fluid pressure. The problem was solved by G. Lamé and E. Clapeyron (Section 3.2), two French engineers who were at the time professors at the Institute of Engineers of Ways of Communication in St. Petersburg. Their solution of the membrane under pressure is contained in a one-hundred-page *mémoir* on the general theory of elasticity, submitted to the French Academy on an unrecorded date; it was reviewed by Navier in 1828 and published in the *Mémoires présentés par divers Savants* in 1833. It has been abstracted in English by Todhunter and Pearson (Ref. 6.35, Vol. 1).

A membrane resisting forces only within its surface (Fig. 6.4) can form a structural member. Because a membrane does not require thickness to resist bending or twisting moments or transverse shear forces, it can be made as thin as constructional limitations and waterproofness of the structure permit. A membrane structure is therefore very light.

A really thin shell cannot be produced if there are appreciable flexural stresses (Fig. 6.5) because an internal resistance moment requires a moment arm a (Fig. 3.12) and the shell must be thick enough to accommodate it. The alternative solution, employed by Freyssinet in the Orly hangars, is to use corrugations in which the tensile and compressive forces act in alternate hollows, whose depth forms the lever arm. This can be done only if the bending is limited to one direction and if the corrugations are functionally and aesthetically acceptable.

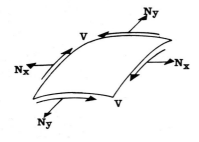

6.4

A membrane can resist only those forces that are within its surface. The only possible membrane forces are two tensile or compressive forces N_x and N_y at right angles and shear forces V, as shown.

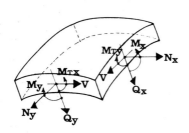

6.5

A "thick" shell can also resist bending moments at right angles to one another, M_x and M_y, twisting moments at right angles to one another, M_{Tx} and M_{Ty}, and transverse shear forces, Q_x and Q_y.

A thin shell can exist without bending. This is easily demonstrated by blowing a soap bubble which is quite stable under the membrane forces within its surfaces but bursts as soon as it is subjected to transverse shear or bending. A closed sphere is not, however, a useful architectural structure, for as soon as we cut the sphere we introduce edge disturbances that give rise to bending stresses. Nevertheless, there are a number of shell forms in which the stresses under normal loading conditions are close to the theoretical membrane stresses over most of the surface.

The membrane theory for reinforced concrete shells was devised by Franz Dischinger, chief engineer of the construction company Dyckerhoff und Widmann, and by Walther Bauersfeld of the Carl Zeiss optical company, who had developed an instrument for showing the movement of the stars. Bauersfeld required a large "celestial" hemisphere on which his instrument could project the moving image of the sky. The membrane theory was thus used to design the world's first planetarium completed in Jena in 1924. The design and construction method was patented by Dyckerhoff und Widmann as the *System Dywidag,* and until 1945 thin-shell construction remained a patent in several European countries and North America.

The hemispherical dome of the Jena Planetarium (Ref. 6.36, p. 66) had a span of 24.9 m (82 ft) and a thickness of only 60 mm (2⅜ in.) without ribs. Larger shells designed by the Dywidag System soon followed, and most of the larger shell structures had ribs, notable among which are the three octagonal 90-mm (3½ in.) thick) domes of the Market Hall at Leipzig (East Germany) with spans of 65.8 m (216 ft), completed in 1927, the cylindrical shell of the sports stadium for the Hershey Chocolate Factory near Philadelphia (Fig. 6.6), and the indoor stadium for the 1936 Olympic Games (Fig. 6.7).

The membrane theory for shells is statically determinate. The solution can be obtained by resolving the forces acting in the x-, y-, and z-directions and taking moments. Although this produces some quite complicated equations (Ref. 6.37), they

6.6

Sports arena for the Hershey Chocolate Factory near Philadelphia designed by Roberts and Schaefer Co. in 1936. The span of the "short cylindrical shells was 225 ft (69 m), and the shell thickness varied from 3½ in (90 mm) to 6 in (152 mm) at the edges, stiffened with 22 in by 60 in (0.56 m by 1.52 m) arches.

are due to the differential geometry needed to describe the shell surfaces in mathematical terms suitable for equations of equilibrium (Ref. 6.4). Dischinger and Bauersfeld obtained solutions for spherical, parabolic, and elliptical domes, cones and cylindrical shells. Later F. Aimond derived the membrane stresses for the hyperbolic paraboloid (Section 6.4) and M. Soare, those for the conoid.

In a complete sphere floating in space the membrane conditions are completely satisfied. The membrane theory is also adequate for small hemispherical shells like the Jena Planetarium and other planetaria built subsequently as thin hemispherical concrete shells. In a hemisphere the shell is restrained by hoop tension (Ref. 1.1, Fig. 3.30), and there is no need for the external horizontal reactions required in semicircular arches. The reactions of the hemispherical shell are therefore purely vertical and easily resisted by a wall or thin ring supported on columns. Because the hoop stresses at the base of the shell are tensile, the shell expands slightly, but unless the shell is tied to a rigid supporting structure only slight bending stresses are produced.

6.7

Sports arena for the Berlin Olympic Games was designed and constructed by Dyckerhoff und Widmann in 1935. The elliptical dome had major and minor axes of 45 m (148 ft) and 35 m (115 ft), one of the few domes built on an elliptical plan since the Baroque domes of the eighteenth century.

Problems arise, for example, in a shallow spherical dome if it subtends an angle of 60° at the center of curvature (Fig. 6.8). This was the type used in Byzantine and Muslim (Ref. 1.1, Section 5.2) and also in modern architecture. The reason for the preference of shallow domes in twentieth-century buildings is not primarily aesthetic but functional. The dome in Classic and Renaissance architecture had a monumental function; its height was intended to be impressive and the building was not heated. The planetarium was a special case, for the celestial hemisphere was needed to project the movement of the stars. For other modern buildings, however, a ratio of 1:2 for height to span (which is inherent in the hemisphere) was unnecessarily high; it increased the volume of air that required heating, it increased the cost of the formwork and materials used in the shell, and concreting the steep portion of the hemisphere was especially difficult.

The shallow dome has, however, an inclined reaction (60° to the vertical in Fig. 6.8). In Byzantine architecture this was absorbed by buttresses or additional semi-

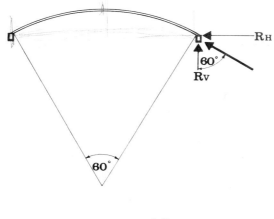

6.8

Reactions required at the supports of a shallow spherical dome.

domes (Ref. 1.1, Section 5.2), which is also possible in reinforced concrete domes and has been done a few times, the best-known structure being the smaller indoor stadium built in 1957 by Pier Luigi Nervi for the Rome Olympic Games (Fig. 6.9). This was an expensive way of absorbing the inclined reaction, for it is much simpler and cheaper to cast a reinforced concrete ring monolithically with the shell (Fig. 6.10). Because the ring absorbs the horizontal reaction R_H shown in Fig. 6.8, it expands as the shell is loaded; this happens when the formwork is lowered and the dome carries its own weight. The hoop stresses in a shallow dome are entirely compressive and the lower edge of the shell contracts. Consequently the ring induces bending stresses in the shell.

The same sort of thing happens in all cylindrical shells (Fig. 6.11), because they require a tie or edge frame to prevent the cylindrical sheet from straightening out into

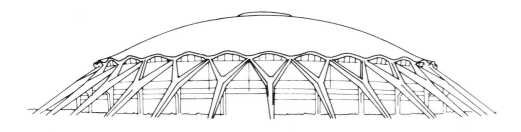

6.9

Buttresses transmit the horizontal reactions to the ground in the Palazetto dello Sport, built in Rome in 1957 by P.L. Nervi. The dome is reinforced inside with diagonally intersecting ribs that converge on the buttresses; its diameter is 58.5 m (192 ft).

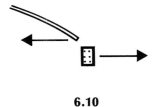

6.10

When the shallow dome shown in Fig. 6.8 is loaded, the hoop stresses are compressive throughout and the base of the shell contracts. The ring that absorbs the horizontal reaction R_H is in tension and consequently expands. These two deformations are incompatible in a membrane shell. Bending stresses are unavoidable in the lower part of the dome.

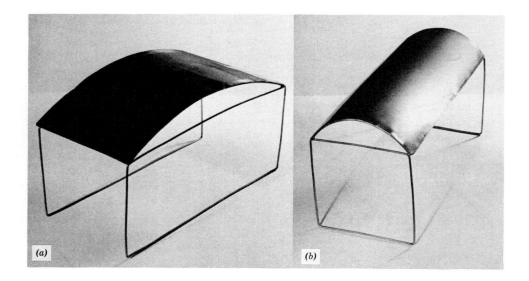

(a)

(b)

6.11

Unless buttresses are used, as in Gothic architecture, ties or rigid arches are required at the ends of cylindrical shells to prevent the curved structure from flattening under load. We call a shell "short" (a) if the straight lines run in the shorter direction; we call it "long" (b) if they run in the longer direction.

6.12

Shallow spherical domes on a near-square plan covering a rubber factory in Bryn Mawr, Wales. The shell is 3½ in. (90 mm) thick, and the columns are spaced 90 ft (27 m) and 70 ft (21 m). Architects: Architects' Cooperative Partnership. Structural Design: Ove Arup and Partners.

a flat sheet. The interaction of the tie and shell produces edge disturbances, which set up bending, torsional and shear stresses; the bending stresses are often the largest.

In small shells it is sufficient to thicken the shell near the supports and place steel bars or welded mesh on both faces to provide flexural reinforcement. In shells with longer spans, the bending stresses must be determined.

The general formulas for the bending theory of thick shells were published in the form of second-order partial differential equations by A. E. H. Love, Professor of Natural Philosophy at Oxford, in 1888 (*Philosophical Transactions of the Royal Society*, Vol. A.179, p. 491). The solution for the spherical shell was derived by H. Reissner in 1912 (*Müller-Breslau Festschrift*—presentation volume for Professor Müller-Breslau, p. 192); but it proved to be too complicated for practical design, and Bauersfeld invited J.W. Geckeler to produce a simpler approximate theory for domes. This was printed in Berlin in 1926 as *Forschungsarbeiten des Ingenieurwesens*, No. 276, a series of research papers published by the German

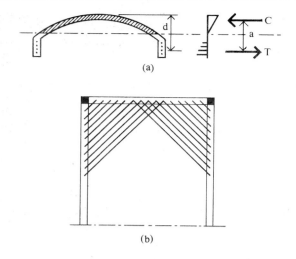

(a)

(b)

6.13

A "long" cylindrical shell resists vertical loads like a beam with a curved cross section. Although ties are essential only across the curved ends, they are frequently used also at the straight edges because this type of shell requires a great deal of tension reinforcement. The resistance to the bending moment (a) is formed by the force C provided by the large volume of concrete in the upper curved portion of the shell, the force T provided by tension reinforcement which is placed as low as possible, and the moment arm a (Fig. 3.12). Shear reinforcement is provided near the supports (b) by bending this reinforcement up diagonally to resist the diagonal tension.

Society of Engineers. It included a theory for the buckling of shells which was subsequently checked by tests on models made of sheet metal (Section 5.2).

Since the 1930's Russian engineers have taken a particular interest in the theory of shells, and *A General Theory of Shells* by V. Z. Vlasov, published in Moscow in Russian in 1949 (Ref. 6.5), became the basis of most subsequent elastic theory.

Square or rectangular spaces are best suited to the functional requirements of most commercial and industrial buildings. Domes have been erected on square plans (Fig. 6.12), and rectangular spaces are easily roofed with cylindrical shells. The reason for the great popularity of shells in the late 1940's and early 50's in both eastern and western Europe and North and South America was partly due to the steel shortage following World War II which limited the use of traditional steel trusses. The damage done to uncased steel structures by fire started by aerial bombardment may have been another factor that caused designers to choose concrete structures for garages

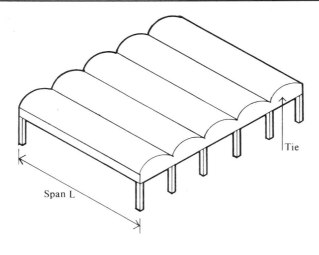

Span L

Tie

6.14

A multiple "long" shell forms a flat roof consisting of individually curved shell elements.

and factories. In the 1960's rising labor costs made curved concrete structures less attractive in western Europe and North America, and the perfection of welding and high-strength bolting made rigid steel frames more economical.

There are basically two types of cylindrical shell. One is called a *short shell* because the straight lines run in the shorter direction (Figs. 6.6 and 6.11a). In this type of shell the arch action, restrained by the horizontal ties (Figs. 6.11a) or stiffening arches (Fig. 6.6), dominates the design, and the membrane theory gives a good approximation for shells with small spans. Such shells also provide roof structures that have a pronounced curvature in the same way that Roman and Romanesque vaults had a pronounced cylindrical curvature.

In the second type, called a *long shell,* because the straight lines run in the long direction (Fig. 6.11b), the arch action is subordinate to the bending action and the membrane theory cannot be used for its design. A good approximation to the precise bending theory of shells is obtained by considering that the shell acts like a straight beam of curved cross section (Fig. 6.13).

The distinction between "long" and "short" shells lies in the direction in which the straight lines run and does not imply classification of size. Thus a roof formed by "long" cylindrical shells tends to give the impression of a *flat roof* formed by elements that are themselves curved (Fig. 6.14), whereas a "short" shell, because the curvature is in the long direction (Fig. 6.6), looks more like a vaulted roof.

The design of cylindrical shells was greatly simplified in the 1950's by the publication of two design manuals that included detailed design tables (Ref. 6.6 and 6.7). In the late 1960's the problems of shell analysis of all types were largely resolved by the development of suitable computer programs. Using, first, the method of finite differences (Ref. 6.8) and then the method of finite elements (Section 5.4), the design of

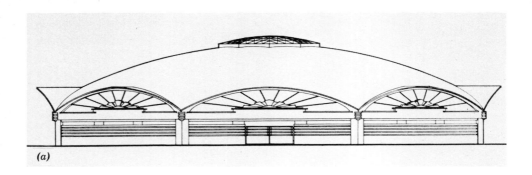

(a)

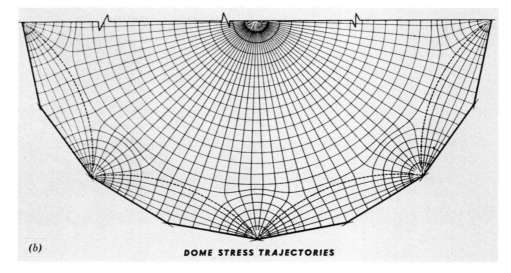

DOME STRESS TRAJECTORIES

(b)

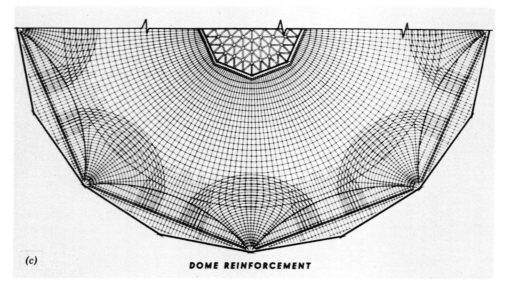

DOME REINFORCEMENT

(c)

6.15

Market hall at Algeciras, Spain, designed and built by Eduardo Torroja in 1933 and destroyed during the Spanish Civil War. The dome had a span of 48 m (157 ft) and a thickness of 90 mm (3½ in.) increasing to 450 mm (18 in.) at the supports. The horizontal reactions were resisted by an octagonal hoop joining the column heads, prestressed with sixteen 300 mm (12 in.) diameter bars. (Ref. 6.52, p. 24). (a) Elevation; (b) stress trajectories due to vertical load; (c) plan of reinforcement.

continuously curved shells of almost any geometric form has become a practical proposition (Ref. 6.9).

The membrane stresses in domes are quite small, particularly near the crown (Ref. 3.6, pp. 271–276), and large openings must be cut to admit natural light. Reinforcement is only nominal over the greater part of the shell; however, if the dome is on point supports, high stress concentrations require heavy reinforcement. Both aspects are illustrated in Eduardo Torroja's design of the market hall in Algeciras, Spain, built in 1933, one of the first shallow thin concrete domes without ribs and with a large skylight (Fig. 6.15).

Another interesting point-supported structure is the airport terminal building in St. Louis (Fig. 6.16), which is virtually a Gothic cross vault in thin-shell form.

The CNIT Exhibition Hall in Paris, the longest spanning building so far erected, is also a cross vault, although on a triangular plan (Fig. 6.17). This shell structure has an enormous span of 219 m (718 ft) which created buckling problems and its convergence on only three supports created bending moments. The structure was therefore designed as a double shell (Ref. 6.38), with intermediate stiffening diaphragms at 9-m (30-ft) centers, a return, in essence, to Brunelleschi's concept of the dome in Florence Cathedral (Ref. 1.1, Section 7.2). In this structure, however, the shells are 60 mm (2.36 in.) thick, and therefore the total thickness of the double shell is only 120 mm. The space between the shells is 3.75 m (12 ft 3 in.), which is an ample allowance for maintenance and provides a substantial moment arm to resist bending. The horizontal reactions are absorbed by prestressing cables in the floor structure.

When this shell was built at the Rond-Point de la Defense on the main arterial road from Paris to St. Germain in 1957 and 1958, it was surrounded only by buildings it dwarfed, and its crown could not easily be seen except from the air and the Eiffel Tower. It was also questionable whether so large a clear span served any useful purpose in an exhibition building (Section 6.9). The structure seemed altogether too big for its purpose and surroundings. Since then the construction nearby of a group of the tallest buildings in Europe has provided a visual balance.

6.16

Airport terminal at St. Louis, Missouri, completed in 1954. Each of the three point-supported barrel vaults has a span of 122 ft (37 m); shell thickness varies from 4½ in. (115 mm) to 8 in. (200 mm). Deep ribs are required at the junctions of the shells and are reinforced with 10 No. 11 (35 mm) bars on top and 10 more at the bottom. Architects: Hellmuth, Yamasaki, and Leinweber. Structural design: W. C. Becker and A. Tedesco.

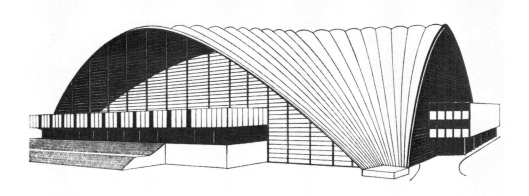

6.17

Palais du Centre National des Industries et des Techniques (C.N.I.T. Exhibition Hall) in Paris, a double shell cross-vaulted over a triangular space. Completed in 1958, it is still the longest spanning architectural structure (219 m or 718 ft). The combined thickness of the two concrete shells is only 120 mm (4¾ in.). Architects: Camelot, de Mailly, and Zehrfuss. Structural design: Nicholas Esquillan.

6.4 HYPERBOLIC PARABOLOIDS (HYPARS), HYPERBOLOIDS, AND CONOIDS

The revival of the dome and vault, which are traditional masonry forms, was a natural first step in the design of reinforced concrete shells. The economics of concrete structures is largely determined by the cost of the formwork. A masonry dome was composed of blocks of stone cut to the required shape, and the cost of cutting blocks with nonparallel faces was, before the development of mechanical saws, not so different from that of cutting rectangular blocks.

It is known that Brunelleschi built the dome of Florence Cathedral freehand, and masonry vaults have been built in the same way (Ref. 1.1, Sections 6.4 and 7.2). Concrete must be cast on a form that gives it its shape, and formwork is made either from straight pieces of timber or steel.

The geometry of the dome was an advantage to Brunelleschi because the rings of masonry blocks were stable until the angle of inclination of the joints became too great for friction to hold the blocks in position. On the other hand, the geometry of the dome is a hindrance in concrete construction because it is impossible to generate a dome with straight lines and therefore there is no simple way to build the formwork from straight pieces of timber (see also Section 6.8). The cost of formwork has reduced the attractiveness of domes for small and medium spans; for large spans the distances are so great that the linearity of the timber has less effect on the economics of the geometric form.

Only two surfaces can be formed entirely by two families of straight lines. One is the hyperboloid, which is generated by rotating a line inclined at an angle to the vertical about a vertical axis to produce the familiar "cooling tower" shape.

Christopher Wren published his discovery of the hyperboloid (Ref. 4.57) in the *Philosophical Transactions of the Royal Society* in 1669 during his term as Savilian Professor of Astronomy at Oxford before he acquired fame as an architect.

The other surface, called a hyperbolic paraboloid, or hypar for short, is generated by one straight line moving over two other straight lines at an angle to one another (Fig. 6.18). It was discovered in the late nineteenth century by a systematic investigation to determine the shape of a number of equations, including the very simple

$$z = kxy \tag{6.1}$$

where x, y, and z are the three rectangular coordinates in space and k is a constant.

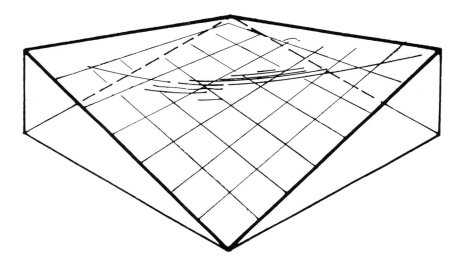

6.18

Geometrically, a hypar shell is formed by one straight line moving over two other straight lines inclined to one another. Structurally, it acts like a network of interlaced arches and cables at 45° to the straight lines. Because of the interaction of tensile and compressive stresses, this surface has good buckling resistance.

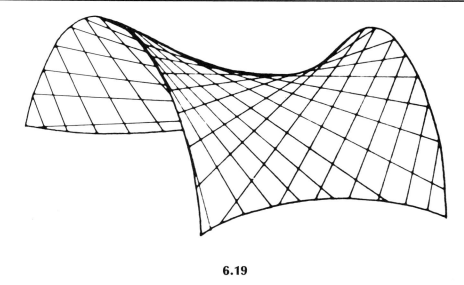

6.19

If the surface in Fig. 6.18 is cut at 45°, a shell surface with curved boundaries results.

Because of the simplicity of (6.1), the membrane theory for the hypar shell has a simple solution. It was derived in the early 1930s by F. Aimond (Ref. 6.10), an engineer with the French State Railways, who built several engine sheds in the new form.

The architectural versatility of the hypar shell (Figs. 6.19 and 6.20) was discovered by Felix Candela, a Spanish architect who moved to Mexico after the Civil War (Ref. 6.39). He demonstrated it in structures ranging from industrial buildings, noteworthy for their low cost and elegance, to spectacular churches reminiscent of Gothic architecture.

Candela condemned the use of complex mathematical analysis and designed his shells mainly by considering the membrane stresses in them and the conditions of equilibrium at their boundaries. Most of his shells have moderate spans.

Since the late 1950s hypar shells have also been used by other designers (Ref. 6.40), notably in the United States (Fig. 6.21). These had initially small spans because of the lack of an adequate bending theory, but since the development of the method of finite differences (Section 5.4) hypars with larger spans have been built; for example, the roof of a hall in Puerto Rico, which seats 10,000, has cantilever spans of 138 ft (42 m) (Ref. 6.11).

The conoid, a related type of shell (Ref. 6.41, Vol. 2, pp. 103–133), is generated by a straight line moving over a straight and a curved line (Fig. 6.22).

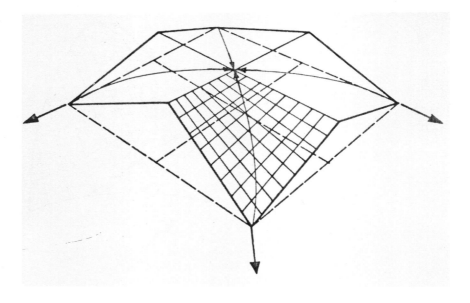

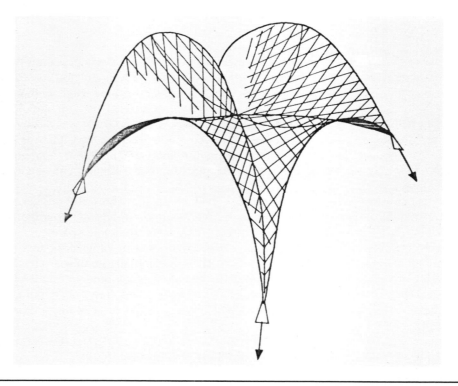

Complex roof surfaces can be formed by combining the hypar shells in Fig. 6.18 or 6.19. There are numerous possible combinations that produce a variety of different roof forms (Ref. 6.40).

6.21

Priory of St. Mary and St. Louis, St. Louis, Missouri. Although the church has a diameter of 140 ft (43 m), the individual hypar elements have a maximum span of only 21 ft (6.4 m) and a height of 21 ft (6.4 m). Architects: Hellmuth, Obata, and Kassabaum.

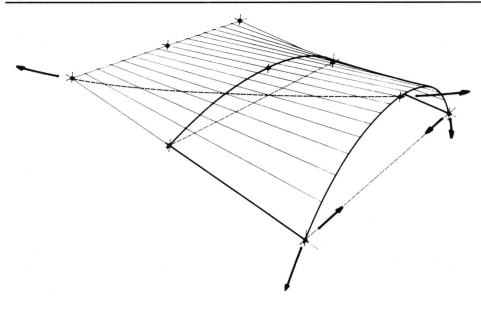

6.22

The parabolic conoid is generated by a straight line moving at one end over another straight line and at the other over a parabola. The flat portion of the conoid cannot be utilized because it is liable to buckle, and the practical conoid is bounded by two parabolas, which usually face north and south to admit mainly indirect daylight. The concrete conoid shell originated in Eastern Europe in the early 1950s, and its membrane theory was published by Mircea Soare in 1958 (Ref. 6.12).

6.5   LINEAR SHELLS

Reducing the weight of curved structures by constructing them as reinforced concrete membranes had occupied the attention of structural engineers for about forty years, from 1920 to 1960. Toward the end of that period the emphasis changed to reducing the labor cost by linearizing the shells.

For spans of about 20 m (65 ft) the thickness of a shell is determined less by the stresses in it than by constructional limitations. The minimum thickness of a curved or sloping concrete sheet that a contractor is able to cast at a reasonable cost with the requisite dimensional accuracy and the requirements regarding waterproofing vary in different parts of the world. In countries with a high standard of living labor costs are high and clients insist on perfect waterproofing. In developing countries thinner shells are acceptable, the cost of concreting them is less, and shells have remained popular. Developed countries are now using shells mainly for long spans.

Our concepts of what constitutes a long span have also changed. The great medieval masonry spans are small compared with those in use today, for few Gothic cathedrals have naves wider than 15 m (50 ft); we can span this distance today with a folded-plate roof. In the United States, Australia, and northern Europe the minimum economic thickness for a concrete shell is about 80 mm (3¼ in.), and a structure of this depth is capable of resisting some bending.

A folded-plate structure is composed of flat plates connected at various angles. Because it is not continuously curved, it is necessarily subject to some bending (Section 6.3). The cost of the formwork, however, is less than that of any shell,

6.23

Addition to Casey Junior High School, Boulder, Colorado. The maximum span of the north-light folded plates is 56 ft (17 m) and their thickness is 3½ in. (90 mm). The folded plates perform essentially the same function as north-light cylindrical shells, and for this span the thickness would be no different. The cost of the formwork is less. Architect: H. D. Wagener. Structural design: Ketchum, Konkel, and Hastings.

6.24

Aircraft hangar at Orvieto spanned 100 × 40 m (328 × 13 ft), designed and built by Pier Luigi Nervi in 1935 and destroyed by the German army in 1944. This cylindrical vault was formed by a combination of linear elements. This is the structure whose model analysis is discussed in Section 5.2.

including one generated by straight lines. Folded plates fulfil functions similar to domes and vaults and have comparable, if different, aesthetic qualities. The first reinforced concrete folded plates (Faltwerke) were built in Germany in the 1920s as coal bunkers; these had previously been designed with supporting beams (Ref. 6.2, p. 357).

The first theoretical analysis was presented by Schwyzer in a doctoral dissertation to the Zurich Technical University in 1920 (Ref. 6.13, p. 218), and several papers based on the equality of slopes and deflections (Section 4.2) in two adjacent plates along the lines at which they are joined by the folds (Ref. 6.13, p. 247) appeared in German-language periodicals during the 20s and 30s. The theory was greatly simplified when in 1947 George Winter, Professor of Structural Engineering at Cornell University, adapted moment distribution to the design of folded plates (Ref. 6.14). Each flat plate forming a part of the folded plate structure was considered rigidly clamped. Each of the joints at the folds was then released in turn and the unbalanced moments distributed. This provided a reasonably economical method of analysis.

The theory of shell and folded-plate analysis before 1939 was thus radically different, although in fact both are surface structures that require some depth and both resist bending moments caused by transverse loads (Fig. 2.6). In the 1950s unifying elements were found (Ref. 6.15), and the distinction became less significant in computer analysis (Ref. 6.16).

In appearance, also, we now see less distinction between shells and folded plates, and it is possible to form roofs from folded plates that resemble cylindrical shells (Fig. 6.23), domes, or hypars. Like the shell, the folded-plate structure requires ties or rigid frames as boundary restraints.

Folded plates became notably more economical with the development of water-proof plywood and hardboard (Section 9.8), which made available a readily cut sheet material suitable for concrete formwork at a reasonable cost.

The alternative approach to the linearization of the shell form was mainly due to Pier Luigi Nervi who has for more than forty years combined both design and construction. An early example is an aircraft hangar built in 1935 in which the roof covers a space frame composed of linear members (Fig. 6.24). In his recent structures the frame has become integrated with the roof covering and broken up into elements precast under factory conditions and assembled with concrete joints cast on the site.

6.6 SPACE FRAMES

We noted in Section 2.1 that triangulated trusses were designed in the nineteenth century as an assembly of tension and compression members that resisted bending. In the twentieth century parallel-chord trusses, like the one shown in Fig. 2.1, were sometimes interpreted as solid girders with holes cut in them where the material is not required. Thus the top chord resists the compressive component of the bending moment (Fig. 3.12), the bottom chord resists the tensile component, and the connecting members resist the shear forces produced by the loads.

In the same way a space frame, like that in Fig. 6.24, can be considered as a three-dimensional assembly of linear members or as a shell with holes cut in it. This argument can be extended to metal domes. The forces in the members can be derived by equating their horizontal and meridional components to the hoop and meridional forces required for the stability of a membrane dome. Since the development of space-frame computer programs this is no longer the simplest method of design, but it provides the link between the structural behavior of concrete domes and metal domes.

The simplest plane or space frame formed by straight linear members consists of an assembly of triangles (Fig. 6.25), and this structure can be worked out by statics alone if the members are pin jointed; that is, if the connection is not rigid. The concept of building triangulated domes originated with J. W. A. Schwedler (Ref. 6.17) who in 1863 also built the first triangulated iron dome with a span of 30 m (98 ft) (Ref. 6.18, p. 7) over a gas tank in Berlin. Schwedler's domes were formed by vertical ribs corresponding to the meridians of longitude on the earth's surface, horizontal members following the circles of latitude (which are "small circles" in spherical trigonometry), and diagonals that performed the triangulation (Fig. 6.25).

For symmetrical loading a simple analysis could be obtained by the method of

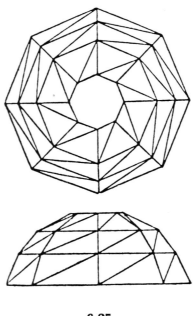

6.25

Schwedler domes are formed by ribs following the circles of latitude and longitude, with additional diagonal members.

resolution at the joints (Section 2.1), used by Schwedler and subsequent designers. It was, however, a tedious method unless the triangles forming the dome were few, in which case the dome had an angular form and the roof covering presented problems.

A graphical method by Karl Culmann (Section 2.1) was also laborious because it involved the determination of a three-dimensional stress diagram by projection on horizontal and vertical planes (Ref. 6.18, p. 13).

The first successful method was developed by Sir Richard Southwell (Section 5.4) for the design of airship frames in 1920 (Ref. 6.19). Southwell's method of tension coefficients was a modification of Jourawski's method (Section 2.1). The latter was expressed in terms of sines and cosines and could therefore not be used conveniently in three dimensions. Southwell expressed the trigonometric functions as ratios of the coordinates ($x, y,$ and z) to the length of the member and thus obtained the equations in cartesian coordinates. These were easier to solve in three dimensions. Moreover, the equations could be expressed in matrix form, which has more recently made them suitable for computer evaluation.

Schwedler himself built a dome of 63-m (207-ft) diameter in Vienna in 1874 (Ref. 6.42, p. 118). In 1954 a Schwedler-type dome was built for the Coliseum or Civic Center at Charlotte, North Carolina, with a 300-ft (91-m) diameter. This is at present the largest of its kind.

The lamella dome in which the triangles are further subdivided into lamellae, or smaller triangles, is a modification of Schwedler's. When this pattern is built up of deep, trussed members, large spans can be achieved. The two largest domes of this type are the Harris County Stadium, commonly known as the Astrodome, in Houston, which was designed by Roof Structures Inc. of St. Louis, and built in the 1960s with a span of 642 ft (196 m), and the Louisiana Superdome in New Orleans, designed by Sverderup and Parcel of St. Louis, and completed in 1973 with a span of 678 ft (207 m); this, however, is smaller than the largest concrete dome (Fig. 6.17). These large domes have stiff joints and are thus statically indeterminate (Ref. 6.20); a large computer is required for their analysis.

The third principal type is the dome formed by the intersection of regularly spaced great circles (or geodesic lines). It differs from the Schwedler and lamella domes which are formed by a combination of great and small circles. The most notable of the early great-circle domes was the Dome of Discovery, an aluminum structure with a steel tension ring erected for the 1951 London International Exhibition and later demolished (Fig. 6.26).

In 1954 Richard Buckminster Fuller patented the great-circle dome under the name *geodesic dome* (Ref. 6.43, p. 181). The majority of geodesic domes in existence at present are based on his patents. They range from children's small play-frames to large industrial structures, such as the Union Tank Car Company's car-rebuilding plant at Baton Rouge, Louisiana, built in 1958 with a span of 384 ft (117 m).

An arrangement of tubes in a single layer is adequate for the smaller domes. For large spans two interconnected layers are required to provide the depth necessary to resist the bending moments due to unsymmetrical loading which results from wind forces and to reduce the tendency of individual tubes to buckle (Fig. 6.27).

The concept of linearization can logically be extended to the overall form of the frame. Straight pieces of steel can be assembled into a space frame which is dome-, vault-, or hypar-shaped: they can also be assembled to form a flat roof arranged as a double layer grid (Fig. 6.28). The term space frame on its own usually means a frame for a flat roof.

Although the space frame has a machine aesthetic which makes it appropriate for technical exhibitions and industrial buildings, its main justification is its lightness. A plane frame is more economical than a beam because it uses greater depth and therefore less material; similarly, a space frame uses less material than a plane frame because it can be made deeper without creating buckling problems, but this is frequently outweighed by the cost of the joints. Site-welded joints are too expensive for most space frames.

If mechanization on the scale customary in the automobile or aircraft industry could be introduced into building construction (Sections 10.4 and 10.5), the space frame would immediately become a more attractive proposition. Many of the proponents of the modern school of architecture thought in the 1920s that the time had come for buildings to be produced under factory conditions, particularly in Germany, where the first space frame systems were devised.

The MERO (MEngeringhausen's ROhr-Bauwerke) joint was developed in Germany in 1937 (Ref. 6.44, p. 133) and is still in use today. It consists of a steel casting with faces at 45° to one another which can accommodate eighteen circular steel members. The joint has eighteen flat surfaces, predrilled and tapped to receive a threaded end-connector device which is prewelded to the ends of the circular member. The

6.26

Dome of Discovery constructed in 1950–1951 for the 1951 London Exhibition and subsequently demolished. It was formed by the intersection of great (geodesic) aluminum ribs, restrained by a steel tension ring. The span was 365 ft (111 m). (a) Dome under construction; (b) after completion, showing lattice struts supporting tension ring. Architect: Ralph Tubbs. Structural design: Freeman, Fox, and Partners.

6.27

The Climatron, a greenhouse in St. Louis, Missouri, designed by Buckminster Fuller and built in 1960, with a span of 175 ft (53.3 m). It is an aluminum tubular geodesic dome covered with plexiglass. Groups of six triangles combined into hexagons give the dome its distinctive appearance.

joint is compact, attractive in appearance, and expensive. With it the structure can be quickly assembled from steel tubes of any length and with any geometry that produces 45° angles.

In the 1950's the University of Michigan at Ann Arbor, located near the heart of the American automobile industry, undertook a research project sponsored by Charles W. Attwood, which produced the Unistrut system (Ref. 6.21). The Unistrut joint consisted of a plate pressed to allow bolted assembly of nine members with angles at 45 and 90°. Although a cheaply produced joint, it was not sufficiently flexible in application to find wide acceptance.

At the same time A.E. Fentiman, in Canada, developed the triodetic joint, sliced from an aluminum extrusion with serrated cuts into which flattened aluminum tubes were pressed (Ref. 6.45, pp. 1073–1082). The joint could accommodate nine members, and by cutting and flattening the tubes at appropriate angles it was possible to make connections at any angle. Because of the high cost of aluminum, steel tubes were later substituted, but the connectors continued to be made of

6.28

Plan of a double-layer triangulated space frame, with diagonals in three directions.

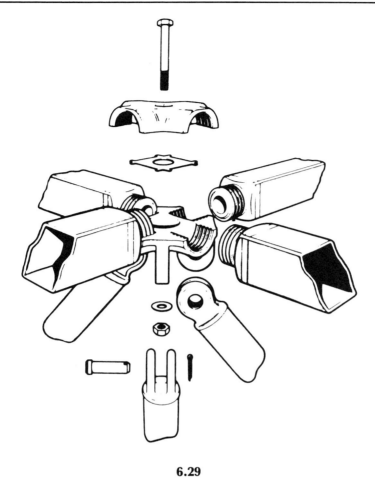

6.29

The parts of the Nodus joint for space frames developed by the British Steel Corporation.

aluminum because the complex shape was economically produced by extrusion. The triodetic joint has been particularly successful for small geodesic domes and other relatively small structures in which high deflection resulting from the usual fabrication tolerances is acceptable.

Since the middle 1960s there has been a revival of interest in space frames and several new patent joints have been produced. The Nodus joint (Fig. 6.29) is typical of the new generation of elaborately machined joints which have high strength and are expensive to make but easy to assemble.

The recent popularity of three-dimensional construction is partly due to research by Z.S. Makowski, now Professor of Civil Engineering at the University of Surrey, and partly to the development of computer programs that make it possible to take full advantage of the economy to be achieved by accurate design. There is now also a greater need for long spans in conjunction with flat roofs (Fig. 6.30).

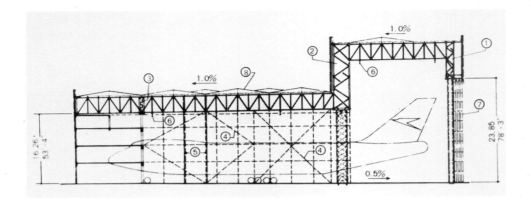

6.30

Cross section through a space frame for a hangar at London Airport which houses two 360-passenger jumbo planes, designed by Zygmunt Makowski and completed in 1973. The wingspan of these planes required a clear span of 138 m (453 ft). The careful control of temperature needed for maintenance made it desirable to limit the volume of air to a minimum, and this argued against the use of domes or suspension structures (Ref. 6.51, pp. 905–916).

6.7 SUSPENSION STRUCTURES

Suspension cables have a long history as a bridge form. Monkeys use vines to swing from one side of a gap to the other, and it may be assumed that humans did so in prehistoric times. They then secured the vine to a tree on the other side to form a single-cable suspension bridge not unlike those still employed by armies in jungle training. A second cable with cross connectors made the bridge more comfortable and less dangerous to use; bridges of this type made by primitive people in the Amazon jungles have been found by explorers. Suspension bridges of bamboo (Ref. 6.46) with spans of 60 m (200 ft) have been used in China. With the introduction of iron as a structural material, suspension bridges became longer in span. In 1826 the Menai suspension bridge (Section 6.1) was first to set a world record for span, and, except for an interval of forty years (Table 6.1), this record has been held by suspension bridges ever since. The history of the cable in architectural structures has been less successful.

In part this is due to the greater complexity of the three-dimensional surface structure compared with the plane bridge structure. Flexible suspension bridges, such as the Menai Bridge (Fig. 6.1), are statically determinate (Fig. 6.31). In a roof structure additional cables are required at right angles to support the roof sheeting, and for stability these cables must be interconnected to form a net. The resulting structure is statically indeterminate. A high-tensile steel cable spanning the relatively modest distance of 50 m (164 ft), tensioned to the relatively modest stress of 500 MPa (72.5

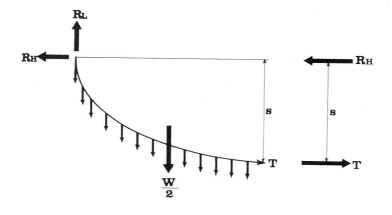

6.31

The cable provides a moment resisting the vertical forces because it sags. The resistance moment is formed by the horizontal reaction R_H, the cable tension T, and the sag s. Because the sag is far greater than the resistance arm a in the beam of Fig. 3.12, the cable can span further for the same tension T.

ksi), extends 125 mm (5 in.), which is not a small deformation. The conventional theory of elastic statically indeterminate structures applies only to structures whose deformations are so small that their effect on the geometry of the structure can be neglected; this is evidently not true of cable structures.

In addition it is necessary in practice to prestress at least those cables at right angles to the primary loadbearing cables. This is due to their sensitivity to thermal movement. A steel cable 50 m long expands 15 mm (0.6 in.) in a change of temperature of 25°C (45°F). Because flutter of loose roof cables would certainly damage the roof sheeting and might endanger the safety of the structure, it is necessary to ensure that the cables, fixed on a day of normal or perhaps cool temperature, will not loosen at the hottest time of the year.

Thus it is necessary to determine the geometry of a network of interconnected cables under the combined action of loads and prestressing forces.

The theory of prestressed cable networks was derived independently in two doctoral dissertations, one submitted by F. K. Schleyer to the Berlin Technical University in 1960 and the other by Avinadav Siev, to the Israel Institute of Technology in the following year (Ref. 6.47, p. 47). The earliest suspension roofs predate this theory, however. In 1895 V. G. Zhukov patented in Russia a method of roofing buildings with structural steel tents. In the following year four pavilions were roofed by this method at the All-Russian Exhibition in Nijni-Novgorod.There are contemporary etchings of these structures and their erection; some have been reproduced by I. G. Liudkovsky (Ref. 6.47, pp. 176–177), but the buildings have been demolished and forgotten.

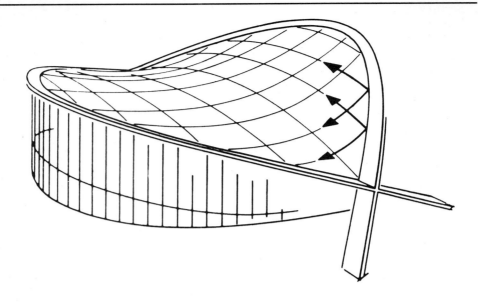

6.32

The Arena at the State Fair in Raleigh, North Carolina, the first modern architectural suspension structure, completed in 1953.

In 1950 Maciej Nowicki conceived the suspension structure for the Arena at the State Fair in Raleigh, North Carolina. He died in the same year and the structural design was completed by Fred N. Severud; the architect was William H. Deitrick. Built in 1952–1953 (Fig. 6.32), the structure consists of two intersecting arches, considered hinged at their supports and joints, from which hang suspension cables with a span of 325 ft (89 m). At right angles to these loadbearing cables are prestressed secondary cables. The corrugated galvanized iron sheets are supported on this network. Severud, describing the structure in 1956 (Ref. 6.22), likened it to two men who counteracted the pull on their arms by transferring the pressure to their feet; by comparison, a suspension bridge, whose cables span between towers and are anchored to the ground, require two additional dead men for anchorage (Fig. 6.33). The concrete arches are heavy, and the resultant of their weight and the pull of the cable must lie within the plane of the arches if the arches are to be in compression only.

The building produced some unforeseen problems that added greatly to the cost and defeated the original objective of an economical structure. In spite of their weight, the arches were, in fact, long and slender compression members that showed a tendency to buckle. Guy wires with strong earth anchorages had to be added to ensure their stability. Internal wires were fixed to the flatter portion of the roof to prevent buckling, and damping springs were inserted at the cable connections to avoid flutter of the roof structure.

In spite of these teething troubles, the Raleigh Arena initiated a decade of inventive and varied suspension roofs. The most notable collection was built at the Brussels

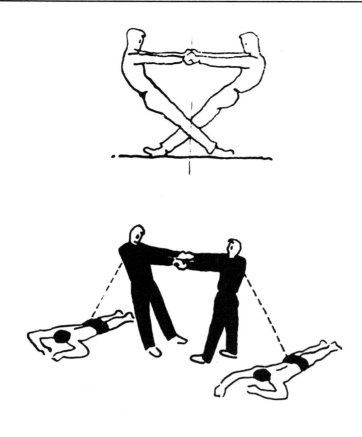

6.33

F. N. Severud's analogy for the structural principles of (top) the Raleigh Arena and (bottom) a suspension bridge.

International Exhibition of 1958 (Ref. 6.47, pp. 156–167), the costliest and most popular exhibition held until that time; it was visited by more than forty-one million people and was dominated by buildings that used cables in some manner. The two that attracted the greatest attention were designed by René Sarger, a Paris structural engineer. One was the French Pavilion whose roof was formed by a suspended spur and two symmetrical hyperbolic paraboloids, each 70 m (230 ft) square, with one side in common; it presented the appearance of a giant butterfly. The other was the Marie Thumas Restaurant, roofed by four conoids supported by eight inclined masts.

A less exotic but larger suspension structure roofed the American Pavilion (architect, Edward D. Stone; consulting engineer, W. Cornelius). It had a circular roof, 100 m (330 ft) in diameter, formed like a bicycle wheel, with an upper and a lower layer of cables that converged on a common outer compression ring. This concept was carried to its logical conclusion in the Utica Memorial Auditorium (Fig. 6.34), completed in 1959.

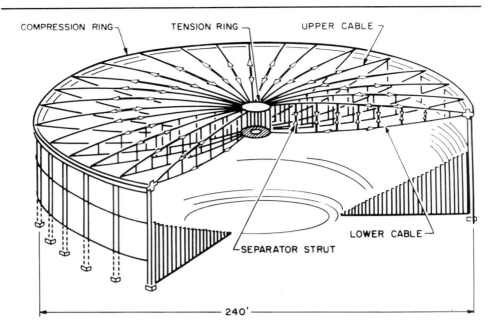

6.34

Circular prestressed cable roof for the Utica Memorial Auditorium in New York State, with a span of 240 ft (73 m). The upper and lower cables were separated by struts that caused them to act in conjunction. The upper and the lower cables were given different prestress tensions that resulted in different natural frequencies. Vibrations in one set of cables were thus out of phase with the other, and the opposing internal forces damped the vibration of the entire cable system. Architects: Gehron and Seltzer. Structural design: Lev Zetlin.

Another notable circular suspension structure of this decade was the Municipal Stadium in Montevideo, built in 1957 with a dishshaped roof, 308 ft (94 m) in diameter, and designed by the Preload Company. The roof was formed by a single layer of cables, covered with precast concrete slabs, which were loaded temporarily with building materials to prestress the cables. The joints were then grouted with cement mortar (Ref. 6.23). Because the circular brick walls provided ample strength for the outer compression ring, an economical roof structure was produced; the effect, however, was marred in the interior by a central drain pipe for rainwater.

The drainage problem was solved by Skidmore, Owings, and Merrill in their design for the Oakland-Alameda County Coliseum (Ref. 6.24), built in 1967 with a span of 420 ft (128 m). In this structure the rainwater is *pumped* off the roof; if more than 160,000 gallons collect on it, the excess is dumped on the arena floor to protect the structure against collapse. The cables are stiffened by concrete ribs and a cast-in-place gypsum roof.

6.35

Sidney Myer Music Bowl in Melbourne, Australia, a 40,000-sq ft (3700-m²) canopy of aluminum-bonded plywood supported on flexible galvanized wire ropes that spans 340 ft (140 m). It was completed in 1957. Architects: Yunken, Freeman Brothers, Griffiths and Simpson. Structural design: Irwin and Johnson.

Three notable suspension roofs with *parallel cables* also belong to the decade 1953–1963. One covers the indoor swimming pool at Wuppertal, West Germany, built in 1956, which is an inverted lightweight-concrete-covered vault with a span of 65 m (213 ft); the cables are secured to the inward-sloping concrete supports of the banked seats for the spectators. The building was designed by the Wuppertal municipal architect and the structural engineers were Fritz Leonhardt and W. Andrä (Ref. 6.25).

The roof of the terminal building for the Dulles International Airport in Washington DC completed in 1962, is also an inverted concrete-covered vault secured to inward-sloping concrete buttresses. The architect was Eero Saarinen and the structural engineers were Ammann and Whitney (Ref. 6.26, pp. 102–113).

The roof of the Ice Hockey Arena at Yale University, also designed by Saarinen, has a curved central spine from which parallel cables are anchored to ground on both sides. The timber-covered roof, completed in 1959, has a span of 185 ft (56 m). The structural engineer was Fred N. Severud (Ref. 6.26, pp. 60–65).

The most notable Australian suspension roof, the Myer Music Bowl (Fig. 6.35), belongs to the same decade.

A large proportion of suspension structures have concrete roofs. Among the nine described, four have flexible roof surfaces, four are covered with concrete, and one has a concrete-stiffened gypsum roof. The dividing line between a suspension structure and a prestressed concrete shell lies in the method of construction rather than in the structural behavior.

Dyckerhoff und Widmann (Section 6.3) have used prestressed edge members in cylindrical shell structures since 1936. The edge beams are prestressed after the concrete has hardened and thus flex the shell upward, lifting it off its formwork. This is a load-balancing action (Section 3.4) that increases the potential span; it also lifts the shell off the soffit, which makes it easier to strip the formwork.

Following the design of the Raleigh Arena, a cable structure was considered for the Schwarzwaldhalle, built in Karlsruhe, West Germany, in 1953 (Ref. 6.27, p. 159). The architect, E. Schelling, proposed an oval-shaped flat saddle roof with a major axis of 73 m (239 ft), which was the span of the loadbearing cables, and a sag of 4.5 m (14.8 ft), a ratio of only 1:16. It was estimated that a flexible cable roof would deflect 500mm (20 in.) due to snow and ± 300 mm (12 in.) due to wind. The design was therefore changed to a prestressed concrete saddle roof 58 mm (2¼ in.) thick. The deflection was reduced to 20 mm (¾ in.). The structure was designed and built by Dyckerhoff und Widmann.

The Schwarzwaldhalle was followed by other prestressed concrete saddle roofs, some in the form of hypar shells with straight edge beams (Section 6.4).

An unusual prestressed hypar shell was the Phillips Pavilion at the Brussels International Exhibition of 1958, designed by Y. Xenakis and Le Corbusier (Ref. 6.28). The span of this structure was quite small, but its complex shape, composed of numerous highly twisted hypar surfaces, made both analysis and construction difficult. Opinions differed on the aesthetic merit of this pavilion, but few denied that it was an engineering feat. The shell was constructed from precast concrete slabs, 50 mm (2 in.) thick, and assembled with prestressing wires placed externally. The structural analysis by C. G. J. Vreedenburgh was supplemented by two model tests (Section 5.2) carried out at the Delft Technical University by A. L. Bouma. A plaster model was used to test the stresses in the complete structure and a prestressed plywood model to test the method of construction. The pavilion was demolished after the exhibition.

The suspension structures considered so far were formed by interconnected cables. Most shapes familiar from shell construction can be designed as cable structures. Hypars, domes, and cylindrical vaults can be produced, although the last two must normally be inverted. The alternative and much older method is to use fabrics or skins. Tents made from animal skins have been used by nomads since prehistoric times. Canvas awnings are also of ancient origin. The ancient Romans are reported to have covered some of their open-air theatres with canvas awnings. In the case of the Colosseum in Rome this would have required a minimum span of 60 m (200 ft). In recent times circus shows have been produced in tents covering a large arena and many hundreds of spectators. This ancient tradition of tent building was examined systematically by Frei Otto in a doctoral dissertation presented in 1953 to the Berlin Technical University. In 1954 it was published in book form (Ref. 6.29). In 1955 Otto erected a number of small tentlike structures in Germany, which increased progressively in size and importance.

Frie Otto and Rolf Gutbrod won the first prize in a competition in 1965 for a design for the German Pavilion at the 1967 International Exhibition in Montreal (Ref. 6.30). They proposed a tentlike structure on eight supports, so complex in shape that it would be difficult to assign a clear span to it. It consisted of a net with a square mesh, covered with a translucent synthetic fabric. The tent was anchored at thirty-one perimeter points and covered 8000 m² (86,000 sq ft). The structural analysis by Fritz Leonhardt was supplemented by structural-model and wind-tunnel tests at the University of Stuttgart.

Otto designed a saddle-type net with translucent acrylic sheeting for the Athletic Arena at the 1972 Olympic Games in Munich (Ref. 6.31, pp. 112), with a maximum span of 135 m (443 ft). He also designed two structures of the same type with shorter spans for the Olympic Stadium and the Olympic Swimming Arena. The structural consultants were Leonhardt and Andrä. An important aspect of these designs was the determination of the shape that results from the stressing of the network. This was done by computer and the shape was plotted by a graphic output unit.

Although some delightful structures have been produced, suspension structures have not fulfilled the promise they appeared to hold in the late 1950s. Suspension cables and membranes have not become a common method of roofing, and the longest spanning roof structures remain domes of concrete or steel (Sections 6.3 and 6.6).

6.8 PNEUMATIC MEMBRANES

In 1917 F. W. Lanchester, who had built the first British motor car in 1896, registered British Patent No. 119,339 for an army field hospital which described his concept of air-supported structures:

The present invention has for its objective to provide a means of constructing and erecting a tent of large size without the use of poles or supports of any kind.

The present invention consists in brief of a construction of a tent in which balloon fabric or other material of low air permeability is employed and maintained in an erected state by air pressure and in which ingress and egress is provided for by one or more airlocks. . . .

The whole of the above have been securely staked and if necessary loaded by ballast the interior is inflated by moderate air pressure by a centrifugal fan, and the whole so inflated forms a tent of segmental form terminated by domelike ends. The marginal flap initially turned under in laying out the envelope now forms an air seal in contact with the ground, and where necessary is loaded by sandbags in order to maintain it in close contact and minimise air leakage. One or more doors in the form of an air lock, constructed as hereinafter described, are arranged at suitable points according to the use to which the tent is required to be put. The technique of construction in such matters as roping, netting, reinforcing, etc., may closely follow similar well-known methods at present practised in connection with dirigible balloons.

Twenty-one years later, in 1938, Lanchester was invited to give the Annual Lecture to the Manchester Association of Engineers. He chose to talk about span in general and air-supported structures in particular (Ref. 6.48). The old gentleman was listened to with the courtesy appropriate to a guest who had a distinguished record in the automobile and aircraft industries (I was in the audience), but few people took his remarks seriously. Lanchester outlined a proposal for a hemispherical dome, 1400 ft (427 m) in diameter, which is almost twice the size of the largest dome in existence in 1975. The sketch accompanying the paper showed a base decorated with coats of arms and classical columns.

Lanchester dismissed arguments that an air-supported building of this size would be dangerous because it would collapse if the air blower failed and because of the risk of a general conflagration if the fabric caught fire. In retrospect his assurance on the first point was correct. The excess air pressure required in the interior is quite small, and it would be feasible to provide an additional standby plant in case the main air compressor and its normal standby plant failed. Fireproof fabrics have also been produced, but the resistance of the roof covering to fire and puncture remain problems to be solved.

In 1946 Walter W. Bird, while head of the Aeronautic Laboratory of Cornell University, experimented with air-supported structures for radomes; that is, covers for the instruments of the American Defense Early Warning Line which ran through northern Canada and Alaska beyond the Arctic Circle. These domes had diameters up to 210 ft (64 m). In spite of winds exceeding 240 km/hr (150 mph), they performed satisfactorily with an excess air pressure of only 70 mm of water pressure (685 Pa or 14.3 psf), corresponding to 6.8×10^{-3} atmospheres, or an increase of internal pressure of 7% above external pressure. This figure agreed closely with that predicted by Lanchester (Ref. 6.49, p. 35).

In 1956 Bird and some of his colleagues formed a company solely for the purpose of manufacturing pneumatic structures, which successfully produced covers for

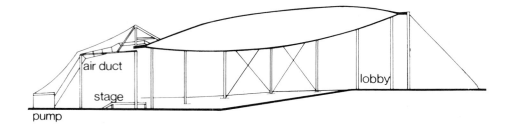

6.36

Structure of the roof for the Boston Arts Center Theater, Boston, Massachusetts. Lack of funds suggested a tent, but this did not afford a large enough area unobstructed by supports. A roof consisting of a 144-ft (44-m) diameter air-filled disk, 20 ft (6 m) deep at its center, was designed and built by Birdair in 1959 (Ref. 6.49, p. 40).

outdoor swimming pools to make them usable in winter, and air-supported tempo-rary assembly plants and warehouses. Several other companies were formed, both in the United States and Europe, to produce air-supported structures.

Later the method was found useful for temporary classrooms and hospital wards, which were used until more permanent accommodation could be provided; these air-supported structures were then dismantled and reused elsewhere.

In 1959 the architect Karl Koch designed a pneumatic roof, with Paul Weidlinger as structural consultant, for the Boston Arts Center Theater (Fig. 6.36).

The most persuasive proponent of pneumatic structures in the 1960s was Frei Otto. His book on this subject, published in German in 1962 and English in 1967 (Ref. 6.50, Vol. I), contained a wealth of new ideas but few details of structures actually built. Otto advocated pneumatic structures for large domes, for tanks holding liquids, for fountains, and for vertical features. Jens Pohl (Ref. 6.33), now Professor of Architecture at the California Polytechnic State University, San Luis Obispo, has suggested the design of multistory buildings held up by a small internal excess air pressure.

Pneumatic structures became the architectural theme of the Osaka International Exposition of 1970, as suspension structures had been in Brussels in 1958. A variety of ingenious and some entertaining pneumatic structures were built (Refs. 6.32 and 6.51), of which the longest spanning was the American Pavilion. This was sunk in the ground and covered with an oval-shaped, vinyl-treated fabric supported on cables 2¼ in. (57 mm) in diameter. The larger diameter of the oval was 460 ft (140 m), the smaller diameter, 262 ft (80 m). The building required an excess internal air pressure of 5 psf (240 Pa or 0.23% above external pressure). This is the longest spanning pneumatic structure so far.

A pneumatic structure is supported by a small excess internal pressure sufficient to carry the weight of the roof membrane; this part of the structural action could be self-balancing. In addition, however, more excess pressure is required to allow for the effect of wind and, in cold climates, snow. When there is no snow and the wind produces an uplift, the excess internal pressure produces tension in the roof mem-brane, which eventually limits the span. Dynamic effects on very large spans have not yet been fully explored. They may limit a span which at present seems potentially much greater than any that could usefully be required (Section 6.9).

Air-supported membranes have also found use in the construction field. For many years unsuccessful attempts were made to cast concrete domes on inflated mem-branes. The concept is attractive because the formwork is cheap and reusable. It is impossible, however, to cast a big concrete dome on a membrane inflated at low pressure or to spray it with a concrete gun without distorting its shape by the uneven distribution of the concrete during placement; the dome that results has an unac-ceptably distorted shape. In the 1960s Dante Bini, an Italian architect, solved this problem by casting the concrete on the ground on top of a carefully folded neoprene membrane and *then* inflating the membrane (with the wet concrete on top) with low pressure, using electric fans connected to pipes laid under the floor slab. The concrete was reinforced with springs that expanded as the balloon was inflated and controlled the shape of the dome (Fig. 6.37). The average thickness of the concrete was only 40 to 70 mm (1½ to 2¾ in.). More than forty domes of this type have been built singly or in combination for various purposes with spans up to 36 m (118 ft).

6.37

Dome for the Narrabeen North school library in Sydney, Australia. After the concrete for the shell had hardened sufficiently, the balloon was deflated and removed for future use. The necessary holes for doors and windows were cut into the concrete. An insulating layer was then sprayed on both the outside and inside of the concrete to give the dome the necessary insulation and stiffness. A waterproofing finish was sprayed over the insulation. Until the insulating layers have been placed on the dome it is sensitive to temperature gradients.

6.9   THE SOLUTION OF THE PROBLEM OF SPAN

Admiration for big objects is an ancient condition. The Seven Wonders of the Ancient World were all remarkable for their great size, not for their beauty (Ref. 1.1, Section 3.1). This fascination has continued to the present day. In the eighteenth century architects on the Grand Tour would travel long distances over atrocious roads to see a building noted for its great height or span. Young gentlemen who had no professional interest in the technicalities of long spans would do the same.

We discussed the limitations on the height of buildings in Section 4.13 and noted that the maximum is at present determined by the adequacy of transportation rather than the technology of the buildings themselves. This also applies to span.

In A.D. 123 the Romans built the Pantheon (Ref. 1.1, Section 3.9) with a span of 43 m (143 ft). This record held for 1700 years. During the nineteenth century this span was more than doubled by the use of structural steel. The Galeries de Machines (Section 3.3), built in 1889, had a span of 113 m (370 ft). In the twentieth century long-span structures were built in reinforced concrete (Sections 6.2 and 6.5), which was, in a sense, a return to masonry construction, but spans did not exceed 100 m until the 1950s, when there was a rapid increase. The CNIT Exhibition Hall, completed in 1958 (Fig. 6.17), has a span of 219 m (718 ft). No longer spanning building has been erected since then (i.e., 1974).

There has also been a great reduction in the amount of material required (Table 6.2). St. Paul's in London, generally accepted as the lightest of the great domes built by traditional methods, has an 18-in. brick dome topped by an 18-in. brick cone which carried the lantern and the timber roof, a total of 36 in. (900 mm). This corresponds to a span/thickness ratio of 37, three times better than the Pantheon.

In a modern reinforced concrete shell of comparable span the thickness can be reduced to 90 mm so that the ratio of span to thickness becomes less than that of an egg. In a prestressed concrete shell this can be further reduced to 60 mm and in a pneumatic dome it is only 2 to 3 mm.

It is difficult to compare the cost of an old and a modern building. The weight of the material used, however, may give some indication of cost if we compare ancient concrete or masonry with modern reinforced concrete. Evidently the reduction has been great.

For large spans we require a double shell; but double shells are more resistant to bending than single shells, and two concrete shells of 60 mm (2⅜ in.) could span farther than 219 m (718 ft). Similar considerations apply to metal and pneumatic structures. Spans in excess of 300 m (1000 ft) seem feasible with our present technology.

Buckminster Fuller (Ref. 6.34) has suggested an air-conditioned geodesic dome, two miles (10,560 ft or 3220 m) in diameter to cover midtown Manhattan. Without a detailed investigation it would be impossible to say whether this is practicable, but it is doubtful whether such a structure would serve a useful purpose commensurate to its cost. Most of the buildings are air-conditioned and so are most of the vehicles. Communication between the buildings could be effected more easily by underground passages or overhead bridges. This has already been done in the center of Montreal and in cities in Siberia, where winter temperatures can be extreme.

The advisability of erecting individual longer spanning buildings is also open to question. S. Sophia and the Blue Mosque in Istanbul, S. Pietro in Rome, and the Duomo in Florence all impress because of their great size, but they still bear a relation to the human scale. The CNIT Exhibition Hall when empty seems altogether too large for people, and a few columns would not interfere with the exhibits.

There is a better case for long-spanning sports arenas, but it is noteworthy that the Astrodome in Houston (Section 6.6) has closed-circuit television receivers among the seats so that people can see clearly what is happening. One asks whether it would not be preferable if some of the audience watched the game at home on a much larger color screen and the hall were smaller. To bring 60,000 people into one large building, all at the same time, requires a vast area for parking and access roads, and there are still no satisfactory solutions to the problems of planning such spaces and

Table 6.2

Year of Completion (A.D.)	Name, Place, Type of Roof, and References	Span (meters)	Average Thickness of Shell or Combined Thickness of Double Shell, (millimeters)	Ratio of Span to Thickness
—	Large hen's egg	0.04	0.3	130
123	Pantheon, Rome; solid concrete dome with relieving arches (Ref. 1.1, Fig. 3.20)	44	4000	11
1434	Duomo, Florence; double dome of masonry and brick (Ref. 1.1, Fig. 7.2)			
1710	St. Paul's London; brick dome surmounted by brick cone (Ref. 1.1, Fig. 7.4)	33	900	37
1924	Planetarium, Jena, East Germany; reinforced concrete shell (Section 6.3)	25	60	420
1927	Market Hall, Leipzig, East Germany; ribbed reinforced concrete shell (Section 6.3)	66	95	700
1953	Schwarzwaldhalle, Karlsruhe, West Germany; prestressed concrete saddle shell (Section 6.7)			
1958	CNIT Exhibition Hall, Paris; double reinforced concrete shell (Fig. 6.17)	219	120	1,800
1946	Radome for northern Canada and Alaska; fabric dome pneumatically supported (Section 6.8)	64	3	20,000

for transporting crowds of that magnitude. Perhaps the time has come when sports arenas should be reduced rather than increased in size.

We have reached a turning point. Curved structures were popular in the late 50s and early 60s. The modern style of architecture had become generally accepted, and most of the new buildings were rectangular boxes. Architects and their clients were looking for a change: they found it first in concrete shells and then in complex steel structures. Spans increased very quickly.

In the 1960s the great sense of achievement in creating new records for span no longer existed. The emphasis was shifting to the interior and exterior environment, and the new ambition was to create a building with a perfect interior, set in a perfect, or at least an improved, townscape.

CHAPTER SEVEN

Fire Protection,

Water Supply

and Sewage Disposal

If the father of our country, George Washington, was Tutankhamened tomorrow and, after being aroused from his tomb, was told that the American people today spend two billion dollars yearly on bathing material, he would say: "What got them so dirty?"

WILL ROGERS

This chapter deals with four interrelated problems that confronted the builders and administrators of medieval and Renaissance cities: the prevention of fires, the provision of an adequate water supply, the control of disease, and the disposal of sewage.

The seventeenth and eighteenth centuries made some progress toward their solution, but the most important advances were made in the second half of the nineteenth, commonly called the Victorian Age. It was an era of great discoveries in engineering and medicine.

The Victorian concepts of fireproof construction transformed structural design. The introduction of water pipes and drains made the bathroom and kitchen the most expensive parts of a house, and it changed the lifestyle of both the upper and the lower classes.

7.1 FIRES AND THEIR CAUSES

Fires do not claim so many lives at the present time as automobiles or accidents in the home caused by falls and poisons (Ref. 7.2, p. 293); they are, however, far more destructive than structural failures. Like other accidents, fires are caused mainly by people. Most are reported between 10 A.M. and 11 P.M., when people are normally awake and active (Ref. 7.1, p. 3). The majority are small and most break out in private homes; collectively they cause more damage and loss of life than the more widely reported conflagrations in public spaces.

Many more fires are reported in North America than in Europe. The incidence per head of population is about four times greater in the United States than in England (Ref. 7.1, p. 2); the energy consumption per head of population is also approximately four times greater (Ref. 7.1, p. 5). Thus the continuing high incidence of fires in modern times is at least partly due to the new sources of energy which pose additional fire hazards.

The damage and loss of life from major fires is also still a cause for concern. The burning of entire cities, frequently recorded by ancient and medieval historians, is no longer a danger, but the material damage and the number of deaths was in the past limited by the small size of the cities; ancient Rome was an exception. The total destruction of Lugdunum (now Lyon in France) in A.D. 59 caused less damage than the partial burning of Rome in A.D. 64 (Ref. 1.1, Section 3.8) because Rome was a much bigger city. The burning of most of London in 798, again in 982, and again in 1212, involved fewer buildings than the Great Fire that burned part of the city in 1666 (Ref. 1.1, Section 7.9) because London had grown much larger.

Loss of life has generally been light in conflagrations that had a small beginning and spread only gradually; thus the Great Fire of London caused the loss of only six lives. Many casualties have occurred when crowded theaters with insufficient exits were destroyed; about 750 people were killed in a fire at the Burg Theater in Vienna in 1881. Heavy loss of life has also resulted from fires started by natural disasters or by enemy action. About 700 people died in San Francisco on April 18, 1906, but most of them perished in the fire that followed the earthquake. The Tokyo earthquake on September 1, 1923, caused a firestorm; estimates of the number of people killed range from 74,000 to 143,000. Deaths from the firestorm following the 1945 air raid on Dresden were estimated at 35,000 and from the atomic bomb dropped on Hiroshima in the same year, at 75,000 (Ref. 7.7)

No reliable figures exist for the damage caused by ancient and medieval fires, but some modern fires have probably been more destructive. A fire in the covered bazaars of Istanbul on November 27, 1954, caused damage estimated at $178,000,000, and a fire in a shopping center in Ilford, a London suburb, on March 16, 1959, resulted in a property loss of about £15,000,000 (Ref. 7.7).

The fires that have influenced the design of buildings and the organization of fire-fighting services have not necessarily been those that produced the greatest destruction or loss of life. The Great Fire of Rome in A.D. 64 marked the beginning of modern rules for fire-resistant construction. Following the Great Fire of London in 1666, the first fire brigades were organized by insurance companies, first in London, then in other European cities.

In 1824 Edinburgh became the first British city to amalgamate its independent fire brigades (of which there were twelve at the time) into a single service; but before the new organization could establish itself, a fire broke out on November 15 of that year in a printer's shop in High Street. It spread to the tall buildings of the medieval city, some of which were twelve stories high. From there windblown firebrands spread the fire to Parliament Square and the Tron Kirk. The water supply failed and contradictory orders hampered fire fighting. On November 17 a torrential rain saved the city. This disaster led to clear rules for the conduct in the event of fire of firemen, police, magistrates, and property owners (Ref. 7.18).

In 1832 the insurance companies of London combined their fire brigades into a single service and appointed James Braidwood, the first fire master of the unified Edinburgh brigade, as their fire chief. In 1834 the old Palace of Westminster, the meeting place of the Houses of Parliament, caught fire, and the House of Commons and the House of Lords were destroyed. Braidwood was only able to save medieval Westminster Hall. Although this was a relatively small fire compared with others mentioned in this section, it focused attention on the need for properly organized and equipped fire brigades.

On December 16, 1835, a fire in New York destroyed 530 buildings in the business part of the city. Fire fighting had been greatly hampered by a shortage of water, and this led to the construction of the Old Croton Aqueduct. (Ref. 7.19), the first modern water supply comparable in magnitude to the Roman aqueducts.

In 1861 Cotton's Wharf and adjacent parts of London caught fire. The result was the passing of the Metropolitan Fire Brigades Act of 1865.

Two great fires started on October 8, 1871, in the Middle West. The more destructive of the two, at Peshtigo, Wisconsin, wiped out seventeen small towns and caused 1052 deaths. A fire the week before had burned the telegraph lines, and news of it reached the outside world only when a boat arrived at Green Bay two days later. By that time the newspapers were occupied with the Great Fire of Chicago, which started on the same day but killed only 300 people.

A smaller fire had broken out in a woodworking factory in Chicago on October 7; it was brought under control at 3.30 A.M. on the eighth after four city blocks had been almost completely burned out. The fire that started about 8.30 P.M. on the same day seemed at first a minor one by comparison. The subsequent enquiry established that the fire had broken out in a barn behind a laborer's shingled cottage on a property measuring 100 × 25 ft (30 × 7.5 m), surrounded by other timber cottages. Most accounts state that a cow kicked over a kerosene lamp, which set fire to some straw (Ref. 7.6). The fire was fought first by the O'Leary family, who owned the barn, and their neighbors, and no alarm was turned in until 9.40 P.M. The firemen were still tired from the earlier fire and the fire engines were slow to arrive. Although Chicago had several powerful steam-operated pumps, they were unable to extinguish the blaze, which was fanned by a high wind estimated by the U.S. Weather Signal Office at 60 mph (27 m/sec). Windblown firebrands spread the fire to a timber store and a gas works. By 1 A.M. on the ninth it was out of control, and even buildings considered fireproof started to burn. At 3 A.M. a flaming piece of timber, said to have been about 12 ft (3.6 m) long, pierced the roof of the waterworks, which supplied the mains of the entire city. It was built of cream-colored stone with battlements and turrets, and the shingle roof had recently been replaced with slate to render the building fireproof; yet the firebrand set the roof timbers alight and the waterworks was destroyed.

The water tower remained, the only building to survive (Fig. 4.22—the crenellated building in the foreground). On October 9 more fire engines arrived from Milwaukee, the wind changed direction, and at 11 P.M. rain started to fall. On the tenth the fire died down.

The Chicago fire had far-reaching consequences for architectural design because it was responsible for the fireproof reconstruction of the city buildings (Section 2.6).

The need for better protection of the iron and steel frames was demonstrated first by a fire in Baltimore in 1904, when eighty city blocks were destroyed, and again by the fire that followed the San Francisco earthquake of 1906 (Ref. 3.7).

Reference has already been made to the fire in Vienna's Burg Theater, in which about 750 people were killed and which resulted in the introduction of regulations for fire endurance and for egress from places of public entertainment. They became more stringent, however, when on December 30, 1903, at the Iroquois Theater in Chicago a fire caused 602 deaths because most of the exits were locked.

These regulations have not always been sufficiently enforced, particularly in small restaurants and night clubs. On November 14, 1942, a fire at a nightclub in Boston caused six deaths and led to an order that all similar premises be inspected. The Cocoanut Grove Club was inspected on November 20. The inspector stated that the conditions were satisfactory and the exits, adequate. At a subsequent inquiry he claimed that he had tried to ignite the imitation palm trees and the half-coconut light fittings and found them not flammable. The club was licensed for 500 people, but on the night of November 28 it held about 1000 (Ref. 7.4, pp. 26-30). One guest unscrewed a light bulb to dim the lighting. A waiter attempted to put it back, and because he could not see the fitting in the darkness he lit a match. A tree caught fire and the flames spread rapidly over the cloth-covered walls. One door to the street had been locked to prevent guests from leaving without paying their bills. A revolving door, which was unlocked, jammed with people inside it, and the pressure of others behind blocked it. Most of the 493 people who died in the fire were killed by noxious fumes in less than a quarter of an hour.

In spite of the publicity that this disaster received, similar fires, fortunately with less loss of life, have occurred since, the most recent (at the time of writing) in November 1970 at the Cinq-Sept dance hall in the small French town of St. Laurent-du-Pont.

The dance floor had no windows. The entrance had a turnstile and the main exit was controlled by a pedal operated by the cashier; four other exits were locked. There was no emergency lighting, no telephone, and the hydrant was not connected to the water supply. A chair caught fire, for reasons unknown; this ignited a foam-plastic cushion, and the fire spread to the walls which had been sprayed with foam plastic to imitate a grotto. Most of the 146 deaths occurred within five minutes.

Plastics were also a major cause of the fire in the Summerland leisure complex on the Isle of Man, off the west coast of England, which had a capacity of 5000. It was totally destroyed on August 2, 1973; most of the 3000 people inside escaped but fifty were killed (Ref. 7.5). The fire was started by three schoolboys who claimed that they had discarded a lighted cigarette, but the Commission of Enquiry thought that they might have been playing with matches. There was no suggestion of deliberate incendiarism. A fiberglass kiosk was ignited, and this set fire to the coating on the steel sheet that formed the wall behind. Because of the high thermal conductivity of the metal, fuel vaporized quickly from the coating and the fire spread in the cavity between the sheet metal and fiberboard lining. Within ten minutes substantial quantities of flaming vapors were ejected into the amusement arcade, and the fire then spread rapidly through the entire complex, which contained large quantities of acrylic plastic.

7.2 FIRE-RESISTANT CONSTRUCTION

Fire can damage buildings in two entirely separate ways. It may set alight some of the materials or finishes that can produce heat, smoke, and noxious fumes, and people may be killed by any of them; the burning building may then set other buildings alight. A fire may also weaken the structure of the building to such an extent that it will collapse; in the process it may cause casualties, particularly to fire fighters, and damage to other buildings (Fig. 7.1).

Some aspects of the construction of "fireproof" floors in the late eighteenth and nineteenth centuries have been mentioned in Sections 2.6 and 2.7 (Figs. 2.10-2.14) and in Ref. 1.1 (Section 8.6 and Figs. 8.11 and 8.12). One of the methods employed attempted to make the timber incombustible by painting it with a liquid that solidified on drying. This technique had already been employed in ancient Rome, where alum was used for fireproofing assault batteries to protect them from firebrands dropped by defending forces. A favorite composition of the eighteenth century was Wood's Liquid, which varied in composition but generally contained alum, borax, and ferrous sulfate.

In another technique the timber was protected by a coat of stucco, a mixture of lime, sand, and sometimes plaster of Paris, reinforced with hair or chopped hay. In addition, or as an alternative form of protection, the timber might be covered with iron fire plates.

A safer method replaced the timber planks with incombustible floors composed of hollow pots, brick jack arches, or concrete arches and protected the wooden beams with plaster or iron or both (Ref. 1.1, Fig. 8.11). In the next stage the timber beams were replaced with cast iron (Ref. 1.1, Fig. 8.12). This method of construction was incombustible, for none of the materials could burn or support combustion, but it was

7.1

Collapse of wall at the Buckingham Department Store in Sydney, Australia, in 1968. The demolishers lit fires to dispose of the rubbish, which set stacked timber alight. The heat was so great that the fire engines had to be withdrawn from the opposite side of the road, and three floors of a hotel on that side were severely burned. This type of collapse is a particular danger in buildings constructed with masonry walls and timber floors. The masonry walls are likely to fall when the restraint of the timber floors is removed by the fire (photograph courtesy of the Sydney Morning Herald).

not entirely fireproof. The iron girders failed if subjected to a sustained high temperature. Wrought iron and steel creep; that is, they deflect gradually to a dangerous extent at temperatures above 450°C (840°F). Cast-iron girders and columns can generally sustain temperatures of 550°C (1020°F) because the material is brittle (Section 4.10), and therefore it does not creep to the same extent; however, it is likely to crack suddenly, either because of the thermal expansion or contraction of a neighboring part of the structure or because it has suddenly been drenched with cold water from a fire hose.

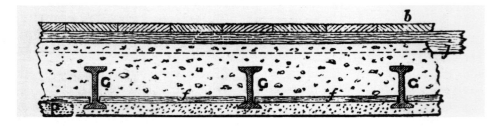

7.2

Fox and Barret's fireproof floor, formed by wrought-iron girders G, across whose bottom flanges were laid rough fillets f of timber. Concrete was then filled in between the joints. The underside was given a coating of plaster P (Ref. 7.8, p. 371).

In the nineteenth century structural fire protection was carried to another stage; the iron was protected from overheating by a cover of plaster, concrete, or hollow tile. Building fires may reach a temperature of 1200°C (2200°F) (Fig. 7.4), and the reason for the cover was to provide sufficient insulation to keep the temperature of the iron within the safe range.

The third edition of Rivingtons' *Building Construction* (Ref. 7.8), a popular British reference book published in 1887, still listed Sir W. Fairbairn's fireproof floors (Fig. 2.12); it gave several methods in which wrought-iron girders were protected by plaster (Figs. 7.2 and 7.3) but did not mention reinforced concrete.

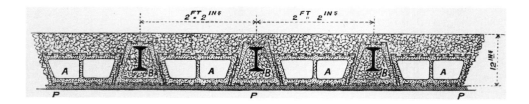

7.3

Hornblower's system, described in 1887 as a modern fireproof system recently introduced. A were large hollow fireclay tubes. B were smaller tubes of the same material containing iron girders and filled with portland cement concrete. Concrete was also packed between and over the tubes. P was the plaster ceiling below, the key for which was afforded by grooves on the lower side of the tubes (Ref. 7.8, p. 373).

In North America the development of tall buildings attracted particular attention to the protection of iron and, later, steel columns (Sections 2.6, 4.2, and 4.12). The early method of covering the columns with thin tiles, with air spaces between them and the column, proved unsatisfactory in the Baltimore fire of 1904 and the San Francisco fire of 1906 (Section 7.1 and Ref. 3.7): later designs provided a greater depth of cover, using hollow tiles closely fitted to the columns or gypsum on metal lath. American designers retained a preference for lightweight fireproofing, whereas European designers tended to prefer encasement in solid concrete that provided a cover of 2 in. (50 mm) over the metal to insulate the steel. In recent years structural steel, protected by sprayed coatings of vermiculite or similar lightweight insulating materials, has been used in North America. Elsewhere reinforced concrete has become the most common material for tall buildings (Section 4.12), at least in part because it does not require extra fire protection if the cover over the reinforcement is sufficiently thick (generally 40 mm or 1½ in.).

7.3   FIRE RESEARCH AND FIRE ENGINEERING

In 1893 William Merrill, a young engineer, was employed by several insurance companies to inspect a huge electric light exhibit erected at the World's Columbian Exposition in Chicago. Merrill became interested in the fire hazards associated with electrical equipment and persuaded the insurance companies to sponsor Underwriters' Laboratories Inc., which was established in Chicago in 1894. This organization tested electrical equipment, but soon built furnaces for determining the fire-resisting properties of various forms of construction. Although theories of fire resistance have been developed since then (Ref. 7.20), the only reliable method even now is to conduct an actual fire test, and this requires expensive installations.

The architect Edwin Sachs took the initiative in founding the British Fire Prevention Committee (BFPC), which established laboratories first near Regent's Park and then near Westbourne Park in London. Its *Publications,* later known as *Red Books,* became accepted as standards for the fire-resisting properties of various types of construction; altogether 257 were published. In 1903 it convened the first International Fire Prevention Congress in London. Sachs died in 1920 and the BFPC became inactive. A new testing station was built by the Fire Offices Committee at Borehamwood (Herts.) in 1938 (Ref. 7.21).

When a fire starts in a building containing enough combustible material, the temperature is initially that of the surrounding air. As the fire gains hold, the temperature rises rapidly, and within five minutes it may exceed 500°C (930°F). In 1916 Underwriters' Laboratories established a standard heating curve for testing columns. In 1918 it was adopted by the American Society for Testing Materials as the standard fire curve for testing floors and walls. Similar standard fire curves have since been adopted by other countries (Fig. 7.4). Unless a test is conducted in accordance with a standard fire curve, its result cannot be compared with that of another test, for the construction may perform quite differently at higher or lower temperatures.

A building element may merely need to be incombustible (i.e., not to catch fire) or it may be required to be fireproof; in the latter case the time required to produce a

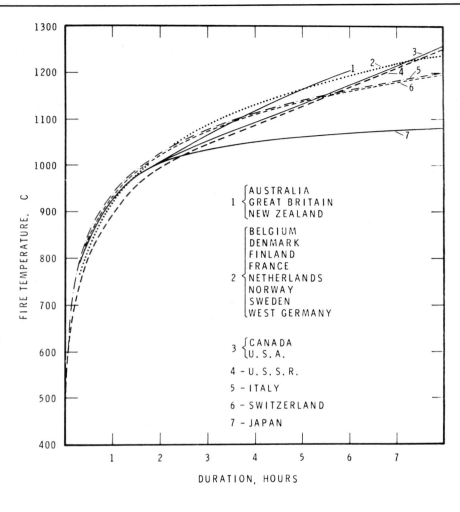

7.4

Standard temperature-time fire curves used in various countries for testing building components (Ref. 7.20, p. 35).

specified increase in temperature on the far side of the specimen, the development of a crack through which a flame could pass, or the collapse of the specimen is measured.

The concept of fire load also originated in the United States. The first model fire code, produced in 1905 by the National Board of Fire Underwriters in New York and subsequently adopted with some variations by many city building authorities, defined the duration of fire endurance required for various building types, ranging from four hours to half an hour. It was based on the observation that the type of occupancy affected the duration of the fire. Thus the contents of a warehouse full of combustible goods would burn longer than those of an office building.

In 1922 at the National Bureau of Standards S. H. Ingberg constructed a model office containing the furniture and other contents to be normally expected in such a room. He studied both the rate at which a fire in this room developed and its duration (Ref. 7.9, p. 5). Subsequent tests were carried out for several other types of occupancy. The duration of the fire (the fire endurance required, in hours) depends on the amount of combustible material per unit area (the fire load, which is sometimes stated by mass in pounds or kilograms or sometimes in thermal units—Btu, calories, or joules).

The twin concepts of fire endurance, as determined by fire tests, and fire load, depending on the type of occupancy and established by fire-load surveys, have become the basis of structural fire design.

The 1939–1945 war demonstrated the enormity of the damage that can be done by fire and the extent to which it can be reduced by appropriate design. Most European countries, as well as Australia, Canada, and Japan, have established new fire research organizations, and research on the fire endurance of various building elements has been greatly increased.

In addition, however, attention was given to the effect of smoke and gases, which frequently cause more casualties than the high temperatures or flames of a fire. There are particular problems in tall buildings. Such buildings cannot be evacuated by outside ladders, for fire ladders do not normally exceed 144 ft (44 m); this is one reason why many building regulations limited the height of buildings before the 1950s to 150 ft (50 m).

The use of stairs is restricted by the physical effort involved, but a more serious limitation is posed by smoke that can render stairwells unusable within five minutes (Ref. 7.10). Studies of the rate of movement of people on stairs made by the London Transport Board suggest that, accepting a limit of five minutes, stairs are not a practical means of egress for buildings taller than fifteen stories.

It is generally agreed that a floor on which a fire has broken out cannot be evacuated by elevator, because the heat of the fire might prevent the doors of the elevator shaft from closing, and the fire would then spread through the elevator shaft. Opinions differ, however, on whether elevators may be used to evacuate floors other than the fire floor (Ref. 4.63, Vol. 1b, pp. 526, 560, and 615; Ref. 7.22).

One solution for very tall buildings is the establishment of smokefree compartments and egress routes to them kept smokefree by excess air pressure. The most important aspect for the fire protection of these buildings is the early detection of fires and their automatic extinguishment.

7.4 FIRE FIGHTING

The ancient Romans had well-organized fire brigades and engines that could project water under pressure (Ref. 1.1, Section 3.8). Medieval fire fighting relied mainly on chains of people passing buckets of water. A London Ordnance of 1189 required each house in the city to have a ladder and a barrel full of water ready in case of fire. Parish accounts show occasional entries for repairing buckets and ladders.

Fire engines that pumped water through leather hoses and delivered a low-pressure jet were already in use at the time of the Great Fire of London in 1666.

7.5

American steam fire engine ("steamer") used in the 1870s. The engine cylinder had a diameter of 7½ in. (190 mm) and a stroke of 8 in. (200 mm). The tubular boiler provided a capacity for 500 strokes per minute, working at a pressure of about 90 psi (620 kPa). The engine weighed 8000 lb (3600 kg) and was drawn by two horses (Ref. 4.65, Vol. 9, p. 236).

Following the fire, insurance companies set up brigades (Ref. 1.1, Section 7.9) which were quick to reach the scene but had initially only simple equipment. The first fire brigade to own more than one engine was that of the Royal Exchange Insurance Company in 1722. In the 1720s Richard Newsham started building two-cylinder fire engines in London which could project a continuous jet of water at an even pressure; in 1731 the City of New York ordered one of these engines.

Hand-operated fire engines had limited power, and the men were likely to become tired after a prolonged outbreak. They also needed a prodigious number of men, for the largest had a crew of fifty. The first steam-operated fire engine was built in America in the early nineteenth century (Ref. 7.11); it provided a flow of 170 gal/min (12.8 liters/sec) to a height of 90 ft (27 m). Self-propelled steam-powered fire pumps followed soon. "Steamers" increased in power during the nineteenth century and remained in use until the 1930s. General A.P. Rockwell of Boston, who contributed the article on *Fire* to Vol. 9 of the ninth edition of the *Encyclopaedia Britannica*, published in 1879 (Ref. 4.65), illustrated a typical "steamer" of his time, pulled by horses (Fig. 7.5). The hoses were carried on a separate horse-drawn cart, as were the

7.6

American ladder-carriage with a capacity for 20 ladders and 12 men. Two ladders spliced together reached 70 ft (21 m) (Ref. 4.65, Vol. 9, p. 327).

ladders (Fig. 7.6). In 1876 (Ref. 4.65) New York had fifty-six steamers, London, twenty-six, and Paris, five; Berlin still relied on hand-operated pumps. Self-propelled motor pumps were first used in 1904.

Canvas chutes, buckets, and sectional ladders were all in use as fire escapes in the eighteenth century. In 1837 Abraham Wivell in England invented the first portable fire escape mounted on wheels. Rockwell described in his article on *Fire* (Ref. 4.65) an "aerial ladder" that when fully extended reached a height of 100 ft (30 m). It was self-supporting, could be moved after the ladder had been extended, and was used both for fire fighting and as a fire escape from multistory buildings. At present the longest extension ladder is generally 144 ft (44 m).

An automatic extinguishing device was patented in England by Ambrose Godfrey in 1723. It consisted of a cask of water to which was attached a cylinder of gunpowder connected to a fuse. The fire ignited the fuse, which exploded the gunpowder and released the water on the fire. It is reported that this dangerous contraption extinguished a fire in London on November 7, 1729 (Ref. 7.12, p. 167).

About 1806 John Carey, also in England, developed a system in which open nozzles were attached to empty pipes. The water valves were kept closed by weights hanging from combustible cords; the fire burned the cords and released the water (Ref. 7.13), which covered the entire area with a drizzle. This general distribution of water did much unnecessary damage if the fire remained small.

In 1874 Henry S. Parmalee invented a sprinkler kept closed by a spring, which was opened when the fire fused link *A* in Fig. 7.7. Frederick Grinell's sprinkler patent of 1882 used as the heat-sensitive element a liquid-filled glass bulb that burst at a designated temperature and released the spray. These are still the two main types of sprinkler head. In an automatic sprinkler system the spray is released only over the area affected by temperature to minimize water damage in minor fires. Frequently the sprinkler system is connected to an automatic alarm which is set off by the flow of water within the system caused by the opening of a sprinkler.

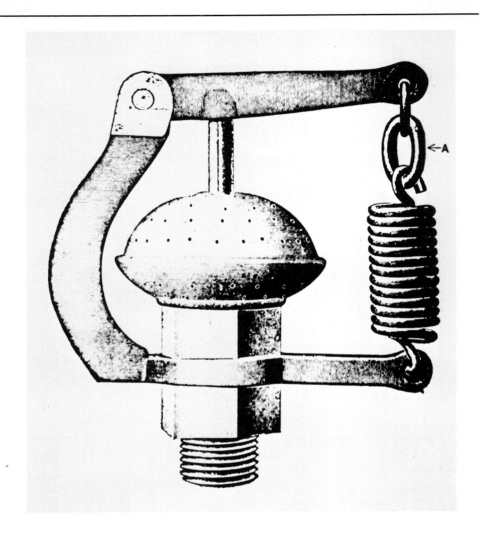

7.7

The Parmalee Sprinkler Mark I, patented in 1874. *A* is a fusible link.

Bells have been used at least since medieval times for fire alarms. Sometimes church bells were rung in a distinctive manner or a special bell was reserved for the alarm. In the second half of the nineteenth century electrically operated fire alarms were introduced. These were iron boxes, about a foot square, placed in conspicuous locations. The larger cities had more than a hundred and were usually kept locked and accessible only to policemen and fireguards. When a handle was pulled, an electric telegraph sounded the alarm at fire headquarters and indicated the number of the box from which the alarm had come. Glasgow had a particularly elaborate system of fireboxes; all were interconnected, and each could serve as a telegraph transmitting and receiving station. Thus the fire chief could, on arrival at the fire, use the nearest firebox to communicate with headquarters and with engines in other parts of the city by telegraph.

Automatic fire alarms were installed in important buildings about 1860. They were set off by a rise in temperature and were generally spaced at distances of about 25 ft (7.5 m). There were two types (Ref. 4.65, *Fire*): one was operated by the rise of a column of mercury which made an electric contact, the other by the expansion of a spiral spring which also closed an electric circuit. Most automatic alarms simply rang an electric bell, but some were connected to the nearest fire station, where another alarm was sounded and the location of the fire was automatically indicated.

Modern automatic detection systems are mainly of six types. In addition to those operated by the sprinkler system, there are heat detectors that utilize fusible links, bimetallic strips, thermocouples or thermistors; flame detectors that measure reflected radiation; smoke detectors that measure the loss of light from a beam directed at a photocell; and detectors of inflammable vapors operating on a catalytic principle. Ionization detectors, used since the 1950s, are small chambers containing a small quantity of a radioactive substance that ionizes the air within the chamber; combustion products reduce the current flowing through the ionized air, and these detectors are very sensitive. All modern detectors automatically report the location of the fire to the fire department.

7.5 WATER SUPPLY, HYGIENE, AND SEWERAGE

The improvement of the quantity of the water supply in the midnineteenth century greatly aided the fire fighters (Section 7.1), but from the point of view of the community as a whole the simultaneous improvement in the quality of the water was even more important. In the eighteenth century there were already a number of pumps that delivered water to taps in houses and public places (Ref. 1.1, Section 8.9). This process accelerated during the nineteenth century when steampower was used to drive the pumps. London's water supply was still in the hands of nine private companies, whose capacity and performance varied. The Grand Junction Water Company provided 8780 buildings with the most generous supply, 350 gallons (1590 liters) per day, with a maximum head of 152 ft (46 m) above the level of the Thames at an annual rate of £2.8.6 (Ref. 7.14, p. 198). The Lambeth Water Company supplied 16,682 buildings with an average of 124 gallons (564 liters) per day, with a maximum head of 185 ft (56 m) at an annual rate of £0.17.0. The water companies were in competition and in some streets their water mains ran in parallel.

The quality of the water supply varied greatly. That pumped from wells or the Thames was often polluted by sewage. John Simon, later Sir John Simon and the first medical officer to the General Board of Health, analyzed the deaths caused by the cholera epidemics of 1848–1849 and 1853–1854 in relation to the source of the water supplied. In 1848–1849 customers of the Lambeth Water Company had about the same ratio of deaths per thousand of population as those of the adjacent Southwark and Vauxhall Water Company. In 1853–1854 it was only $2/7$, the Lambeth Water Company having in the meantime moved its water intake upstream, *above* the first sewer outfall that entered the Thames from the London area.

The private Manchester and Salford Water Company, established in 1809, pumped its water from wells. The Manchester Town Council's decision to take it over and to build a reservoir and an aqueduct was not taken primarily for reasons of hygiene but because the wells did not provide sufficient water. The Manchester Waterworks, established in 1847 by an Act of Parliament, began building a reservoir in the Longdendale Valley in Derbyshire, whence the water was conveyed by aqueduct. In 1885 it went farther afield to Lake Thirlmere in the Lake District, a distance of 95 miles (153 km). The use of a clean source of water, conveyed to the city by aqueduct, was in fact a return to the Roman system of water supply.

Several years earlier, in 1837, work had been started on the Old Croton Aqueduct (Ref. 7.19) which, when completed in 1842, gave New York a clean water supply in excess of its immediate requirements; but fire fighting, not hygiene, was the principal reason for its construction (Section 7.1).

The cholera epidemic of 1848 stimulated several physicians to investigate the relation between the quality of the water supply and the spread of the disease. The first outbreak of cholera in the history of England was derived from an epidemic of the disease in Bengal (now Bangla Desh) in 1817; it spread slowly across Asia, reached European Russia in 1829, spread across western Europe, and was first reported in England at the port of Sunderland in 1831 (Ref. 7.15, p. 4). There were further epidemics in England in 1848–1849, 1853–1854, 1865–1866, and 1893. John Simon estimated (Ref. 7.14, p. 88) that between 1848 and 1859 cholera caused 237,500 deaths (out of a total population of 17 million).

This appalling death toll produced a number of investigations, of which Dr. John Snow's *On the Mode of Communication of Cholera,* published in 1849, was particularly influential. Snow, who was born in 1813 and qualified in medicine in 1838, continued his research during the 1853–1854 epidemic.

As mentioned above, Simon had already noted the effect on the incidence of cholera of the removal of the Lambeth Water Company's intake above the first London sewer outfall to the Thames. Toward the end of August 1853 Snow traced the cholera epidemic in the part of London known as Soho to a single pump in Broad Street, which worked a well close to a cesspool infected by a cholera victim. He recommended the removal of the pump handle, and the cholera outbreak subsided. On September 1 there had been 143 new cases, and the number remained high until September 8 when the pump handle was removed; by the tenth the number of new cases had fallen to five.

Snow produced the convincing medical evidence that cholera was caused by drinking water polluted by untreated sewage and thus provided the impetus for the necessary legislative action.

The provision of a pure water supply and of proper sewage disposal had been advocated earlier by several social reformers, of whom Edwin (later Sir Edwin) Chadwick and Lord Ashley (Anthony Ashley Cooper, later the seventh Earl of Shaftesbury) were the most successful. Chadwick was a barrister who had been appointed an *Assistant Commissioner of the Inquiry into the State of the Poor Laws* in 1832. In 1834 the Commission became a semipermanent department of state, and Chadwick became its secretary. The Public Health Act of 1848 changed its name to the General Board of Health.

Chadwick was the author of the Commission's *Report on the Sanitary Conditions of the Labouring Population of Great Britain,* published in 1842 in three volumes (Ref. 7.16). The report contrasted unfavorably the sanitary conditions of contemporary Britain with those of ancient Rome, and after much public debate it led to a number of reforms, including an act passed in 1875 that required local authorities in Britain to set up sewerage works.

One immediate result of Chadwick's report was an investigation in 1844 of Windsor Castle by Prince Albert. He found fifty-three overflowing cesspits and ordered that the commodes be replaced by water closets and drains; but the work was incomplete when he died in 1861 and Queen Victoria ordered that the Castle be left exactly as it was.

More important was the London Main Drainage Scheme, started in 1856, which within fifteen years provided London with a proper sewerage scheme. It was not the first modern system; Hamburg, then still an independent Free City, had started one in 1842. Brooklyn instituted a sewerage scheme in 1857 and Paris soon followed.

Once the sewage was removed by drains, purification of water drawn from wells or rivers became practicable. Sand filters had been used since the 1830s. The result, however, was still not entirely satisfactory in the late nineteenth century, except in places in which the water supply came from a distant clean source, such as a reservoir or lake in rural surroundings, carried to the city by an aqueduct. Writing in the late 1870s a Mr. A. B. MacDowall of London (Ref. 4.65, Vol. 9, *Filtration*) commented:

Though the arrangements for water supply for most of our large towns include filtering processes by means of which, as a rule, excellent drinking water is abundantly provided . . . it is generally thought a wise additional safeguard to employ one of these instruments (i.e. a domestic filter). In country places . . . their use is quite often imperative if the laws of health are to be respected.

The author noted that the first domestic filter was patented in England in 1790 and illustrated several different types (Fig. 7.8).

The Chadwick Report marked a turning point in the history of sanitation and as such it has received deserved praise. The annual reports of the Local Government Board, founded in 1871, showed a marked improvement by 1886. Sir John Clapham, Professor of Economic History at the University of Cambridge (Ref. 7.17, p. 466) commented on the 1886 report:

The cities of Britain were among the healthiest of the world, and were certainly the healthiest of the Old World. The death rate of New York was worse than that of Manchester; those of Paris and Berlin worse than that of Liverpool; and that of the whole city of St. Petersburg (now Leningrad) nearly 15% worse than that of a small, selected, black London area such as Clare Market.

In Manchester and other industrial cities the "by-law streets," built in accordance with the successively stricter legislation promoted by Chadwick, improved in quality as the century progressed. By 1900 they provided a standard of sanitation far ahead of working-class housing in the rest of Europe, and the houses are still sound, a testimony to the quality of the workmanship of nineteenth century bricklaying. This unfortunately discouraged their replacement when the sanitary provisions, consid-

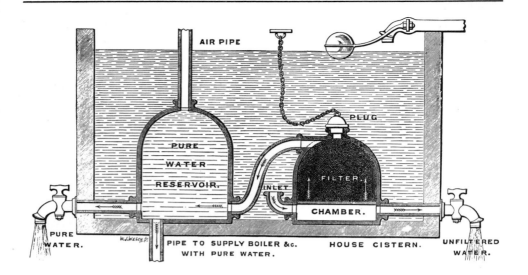

7.8

A water filter for installation in a water-storage cistern, illustrated in 1879 (Ref. 4.65, Vol. 9, p. 169). This was one of many types of filters used in the nineteenth century.

ered good in the nineteenth century, had become inadequate, so that many of the terraces have become a blight on the industrial cities of northern England. The present-day slums are not those described by Friedrich Engels in 1844 in *Condition of the Working Class in England,* but the sanitary by-law houses that replaced them.

7.6   TOILETS, BATHROOMS AND KITCHENS

Toilets flushed by running water have been used since prehistoric times and were a normal part of the equipment of buildings in Imperial Rome (Ref. 1.1, Section 4.3). They are still in use in some rural places where a brook can be polluted without causing trouble.

Toilets flushed with buckets of water have been used at least since medieval times in monasteries and the homes of the wealthy (Ref. 1.1, Section 5.9). The two principal methods of sewage disposal until the nineteenth century were by digging a hole in the ground (which was reasonably satisfactory under rural conditions but a great sanitary problem in urban areas) or by bucket; the bucket was usually carried away, but there are reports of nightsoil being regularly thrown out of windows in various European cities as late as the eighteenth century.

The invention of the water closet thus transformed the atmosphere of buildings as no other technical innovation has done. The fittings and drains, however, added to the cost of the building, and today the WC's, bathroom, and kitchen have become the

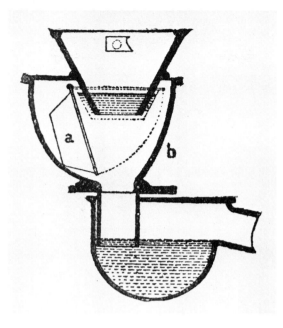

7.9

Pan toilet (Ref. 4.65, Vol. 21, p. 716).

most expensive part of the average house or apartment. Furthermore, the location of the drains has restricted the planning of buildings. Water under pressure can run horizontally and around fairly sharp corners; the drains, which operate under atmospheric pressure and carry suspended solids, must have an adequate fall and gentle bends.

It is not known when pan closets first appeared, but some were in use in the early eighteenth century. Apparently they were an English invention because in France they were described as *cabinets d'aisance à l'anglaise*.

Sir Alfred Ewing, better known for his work on magnetism and the strength of materials, contributed the article on *Sewerage* to the *Encyclopaedia Britannica* in 1886; he was then Professor of Engineering at University College, Dundee. In it he described the pan closet:

Water-closets used to be almost invariably of the "pan" type, but wherever sanitary reform has been preached to any purpose the pan closet is giving place to cleaner and wholesomer patterns. The evils of the pan closet will be evident from an inspection of fig. 14 [Fig. 7.9]. At each use of the closet the hinged pan a is tilted down so that it discharges its contents into the container b. The sides of the container are inaccessible for cleaning, and their upper portions are out of reach of the flushing action of the pan. They gradually become coated with a foul deposit. A gust of tainted air escapes at every use of the closet; and it rarely happens that the container is airtight, and that the filth it has gathered does not cause a smell even at intervals of disuse.

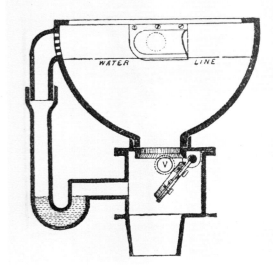

7.10

Bramah valve toilet (Ref. 4.65, Vol. 21, p. 717).

In 1775 Alexander Cummings patented the first valve closet, and in 1783 Joseph Bramah, one of the great engineering inventors of his time, best known for the hydraulic press, patented an improved version, greatly commended by Ewing (Fig. 7.10). Valve closets are still used today in railway trains.

The washout closet (Fig. 7.11) was developed in the 1870s by Thomas William Twyford, a potter in Hanley, England, who observed that the glazed earthenware parts of the valve closet were much cheaper than the metal. In addition, the one-piece closet without moving parts was easier to keep clean. By 1889 Twyford claimed to have sold 100,000 washout closets of their "National" brand. Washout closets are still in use. The washdown closet, developed in the late 1880s, is the type most commonly used at present. In it the bowl is reversed, so that the flush passes through unimpeded.

At about the same time the siphonic closet appeared in America. On pulling the handle, a fast flush of water which starts the siphonic action, is followed by a slower afterflush. It is widely used in the United States.

Some of the late nineteenth-century toilets had elaborate decorations. Armitage made a bowl supported on a lifelike earthenware lion. Doultons placed a skyline of the city of London in Delft blue on the inside of one of their bowls, with the waterline reaching to the springings of the dome of St. Paul's.

Water Closets required soil drains which, when installed in existing houses, were most easily fixed to the outside rear wall. The practice of attaching the drains to the outside of the house persisted, particularly in England and Australia, even when the

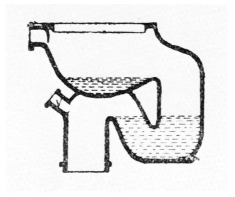

7.11

Washout toilet (Ref. 4.65, Vol. 21, p. 717).

drains were part of the original construction. It is a matter of opinion whether the slight saving in cost compensates for the untidy appearance of the back elevation (Fig. 7.12).

The installation of water closets did not itself solve the sanitary problem. The sewage was removed from the building and from the streets; but during the late eighteenth and the early nineteenth centuries most drains discharged into the nearest river. As long as sewage was not purified and the water supply was drawn from the same river, the danger of disease remained, as the London cholera epidemics showed (Section 7.5). The principal methods of sewage disposal in use today were known in the 1870s and were described in Ewing's article (Ref. 4.65, Vol. 21, p. 712) in 1889: discharge into the sea if conditions are suitable; chemical treatment and precipitation; filtration through artificial filters; and broad irrigation or downward filtration on suitable agricultural land. By the end of the nineteenth century most northern European cities had sewage disposal plants, but progress elsewhere was slower.

The installation of soil drains encouraged their use for other purposes, and baths and kitchens with running water and drains became common after water closets had been put in.

Portable baths of copper or tin, used by the upper and middle classes since the eighteenth century, were filled with buckets of hot water and eventually emptied by bucket. A bath shown at the Great Exhibition held in London in 1851 had a charcoal heater to ensure that the water would remain hot, but the bath was filled and emptied by hand. By 1870 baths with running water and drains had become common in Britain in the houses of the wealthy, but they did not make their appearance in working class houses until the twentieth century.

Running cold water was piped into some bedrooms in the 1870s, but hot water was brought in a jug from the kitchen as required.

Baths in the nineteenth century, even when fixed to a water supply and drains, were usually free-standing and supported on four legs; cleaning the floor underneath

7.12

Rear view of terrace houses in Bath, England, showing the drain pipes and their vent pipes.

Fire Protection, Water Supply, and Sewage Disposal

was difficult, and in addition carpets were not uncommon. Bath tubs built-in for easy cleaning and the hard vitreous enamel finishes now common on cast-iron tubs appeared in the early twentieth century.

England had been the leader in the development of bathrooms in the nineteenth century. However, one bath per house, or at least shared bathrooms, had been considered adequate. At the turn of the century well-to-do Americans began to insist on a bathroom for each bedroom, and some hotels provided private bathrooms. The American standards were soon adopted in other countries of the New World, but only gradually in Europe.

Running water and drains were generally installed in kitchens at the same time as in bathrooms, but the traditional fittings were modified only slowly. One-piece sink and drainboard combination fixtures with integral splashback were introduced about 1920, two-compartment sinks about 1930, and stainless steel sinks about 1950.

CHAPTER EIGHT

Environmental Design
and Building Services

Nature and Nature's Law lay hid in night;
God said "Let Newton be!" and all was light.

ALEXANDER POPE

This could not last; the devil howling "Ho!
Let Einstein be!", restored the status quo.

ANON.

In this chapter we consider the thermal, illuminating, and sonic environment in buildings and the means by which it has been improved. We also examine intrabuilding communications and transportation.

Twentieth-century building services, such as heating, ventilation, air conditioning, artificial lighting, and elevators, have led to a great increase in the consumption of electricity; the present energy crisis is due in part to the demands made by them.

Conflicts between the ideals of environmental efficiency and structural econoomy have not yet been fully resolved.

8.1 DESIGN OF THE INTERIOR ENVIRONMENT REPLACES STRUCTURAL ANALYSIS AS THE PRINCIPAL SCIENTIFIC PROBLEM OF ARCHITECTURE

Before the nineteenth century, except for about three centuries in ancient Rome (Ref. 1.1, Chapter 4), the design of the interior environment was confined to the decorative treatment of the interior surfaces. The physical environment was left largely to chance (Ref. 1.1, Sections 5.9, 7.10, and 8.9). Until the eighteenth century, except during the Roman Empire, artificial lighting levels remained low and heating was primitive. Artificial cooling was impracticable. Sound insulation received no specific attention, although street noise was noted as a nuisance by a number of writers. The acoustic quality of medieval and Renaissance churches is probably a secondary result of the structural design.

Nevertheless, the interior environment of many of the surviving medieval and Renaissance buildings is surprisingly satisfying. The thick masonry walls provide insulation to reduce the cold at night during winter and midday heat in summer, and they attenuate street noise.

The limitations of the interior environment were compensated by differences in life-style. People wore heavier clothes in the northern European winter, and stopped work in the afternoon heat of the southern summer for a siesta, as many do today.

Even the music may have adapted itself to the buildings: it is notable that medieval and Renaissance church music sounds better in the original buildings than in a modern concert hall.

During the nineteenth century heating improved as a by-product of the Industrial Revolution. This was not an unmixed blessing: the air pollution of the industrial towns of northern England, which grew steadily worse until the midtwentieth century, was due more to the smoke from thousands of domestic coal fires than to industrial fumes (Section 8.4). Lighting levels improved even faster during the same period. Although cooling was not a major problem in the countries of northwest Europe, where the Industrial Revolution started, artificial ventilation was used in factories and some public buildings. The building industry now required the services not merely of a structural expert, but also of a mechanical engineer. Electricity became an important building service in the early twentieth century.

The fabric of most buildings, however, continued along traditional lines until the midtwentieth century. Even when they were supported by a structural frame (Section 2.6), the walls were thick and provided thermal and acoustic insulation.

8.1

The Orangery at Longleat, in Wiltshire, built in 1814 by Jeffry Wyatt (later Wyatvile).

Orangeries (Fig. 8.1) with large windows had been erected since the early eighteenth century. Greenhouses, winter gardens, and shopping arcades had been constructed with roofs of glass since the early nineteenth (Section 9.3). Paxton, a noted builder of greenhouses, designed the Crystal Palace in 1851 (Section 10.1). It was immediately imitated for exhibition buildings and department stores which before the development of electric lighting required glass-roofed internal light wells (Section 8.9).

The Crystal Palace was acclaimed in its own time as a marvel of industrial construction and admired as such by contemporary critics. It remained outside the mainstream of architecture (Section 10.1), however, and its importance was recognized only in the 1930s when the theorists of the modern movement, notably Siegfried Giedion (Ref. 3.12), revised the history of architecture.

Apart from its industrialized construction (which we consider in Chapter 10), the Crystal Palace interested architects in the twentieth century because of its ingenious use of glass. Glass and natural crystals attracted many designers of the Modern Movement. Walter Gropius, in 1919, in the foreword to the first program of the Bauhaus, the new name of the school of design and architecture in Weimar, referred to the buildings his students would design as the crystal symbol (*kristallines Sinnbild*) of a new faith (*Programm des Staatlischen Bauhauses in Weimar*, April 1919, quoted by H. M. Wingler, Ref. 8.1, p. 31).

Gropius used large glass windows in the Fagus Works, a factory for making shoe lasts he designed in Alfeld, Germany, in 1911. In 1914 he encased the staircase of a model factory and office building for the exhibition of the Deutscher Werkbund (a society for the promotion of industrial design) entirely in glass and used glass extensively on the facade. In the building erected in Dessau in 1926, after his school of architecture had moved there from Weimar (both the school and the building were called *Bauhaus*), the glass wall extended without a break through three floors.

Mies van der Rohe, who succeeded Gropius as director of the Bauhaus, used full glass facades in projects for two glass skyscrapers that were never built but foreshadowed the glass curtain walls of the 1940s and 50s. Mies' use of plate glass extending from floor to ceiling without horizontal interruptions of any kind created an interrelation between the exterior and interior spaces. This was possibly a reflection of contemporary trends in painting which also emphasized the infinity of space. Many critics consider the house he designed for Dr. Edith Farnsworth in Fox River, Illinois, near Chicago, in 1946 (completed in 1950) as one of his masterpieces (see Section 8.2). Reyner Banham (Ref. 8.2, p. 56) described it as follows:

Space exists here between two given planes, the floor slab and the roof slab, and has no upward or downward extensions whatsoever, except that some sensitive spirits feel that it flows down the four little steps from the floor slab to the terrace.

These two limiting slabs are the only opaque surfaces on the exterior of the house, everything else is floor to ceiling glass, or nothing at all. As a result the interior space is in almost total communication with infinite space outside, visually speaking—so much in communication with infinity that some visitors feel a sense of risk from stepping off the edge of the floor slab.

Between the upper and lower slab there are no visible connections, except the six regularly spaced uprights of the structure. . . .

The use of glass enabled Mies to emphasize the lightness of his structural frame, in accordance with the dictum "Less is more" which he enunciated frequently during his directorship of the school of architecture at the Armour Institute (later the Illinois Institute of Technology).

Le Corbusier liked glass for a similar reason:

The history of architecture shows that down the centuries an unremitting battle has been joined on behalf of light against the obstacles imposed by the laws of gravity: the history of windows.

Another proponent of the use of glass in the architecture of the modern movement was Bruno Taut, who built a small pavilion for the glass industry at the already-mentioned exhibition of the *Deutscher Werkbund* in 1914. Paul Scheerbart, a Berlin novelist who had been an early advocate of glass in architecture, was invited by Taut to open the pavilion; he produced a rhyme as dedication:

> Das Glas bringt uns die neue Zeit
> Backsteinkultur tut uns nur Leid.

(Glass brings us into a new era, while living with brick walls does us only harm) (Ref. 8.3, p. 189).

In Section 2.6 we examined the invention of the skeleton frame, and in Chapter 4, the decrease in the weight of this frame with improvements in structural theory. In the late 1940s and 50s the advance in structural theory and the architectural concept of a light structural frame clad in glass combined to produce the American glass curtain wall. Whereas earlier buildings designed by architects of the modern movement had been relatively small, the new glass-walled towers, such as the United Nations Secretariat (1947–1950) in New York, the Lake Shore Drive Apartments (1951) in Chicago, the Lever Building (1952) and the Seagram Building (1956), both in New York, were too prominent to be ignored. They aroused hostility and inspired emulation, and there are few countries today without buildings with glass curtain walls supported by light structural frames and few without critics of such buildings.

The crystalline boxes with sealed glass facades would have been impracticable without the invention of air conditioning (Section 8.5) in the 1920s. The heat transmitted by ordinary glass, unshaded and uninsulated, was so great that air-conditioning systems could not always cope with the cooling loads in some of the earlier buildings. Moreover, as each floor was conditioned as a unit, it proved difficult to cool the sunny side of the buildings without overcooling the shady side. Clearly there was a conflict between structural lightness and environmental efficiency. On the one hand, advances in structural engineering made it possible to build light frames that had adequate strength when covered only with a transparent curtain wall. On the other hand, problems of the thermal environment, which had been solved reasonably satisfactorily by massive masonry walls, now required elaborate mechanical and electrical equipment.

Heat-reflecting glass and double glazing (Section 8.3) were used to reduce the cooling loads together with insulation in the nontransparent portions of the wall (Section 9.4); but in some countries, such as Australia, glass was soon replaced by materials with better insulating properties, such as concrete.

The growing appreciation of the problems posed by curtain walls coincided with a decline of interest in structural innovation. During the 1940s and 50s architects had used new structural concepts and new materials freely for aesthetic purposes (Sections 6.4 to 6.8). In the 1960s many felt that the theme had been exhausted, and in particular that there was limited potential for further increases in span (Section 6.9). Instead, attention shifted to the creation of a perfect environment. In North America the standard of air conditioning and vertical transportation improved greatly. In other countries in which air conditioning had been rare before 1960 it became normal practice for prestige buildings. The quality of the architecture came to be judged by the unobtrusive perfection of its interior environment. Because most of the environmental services required electricity, the increase in the use of electric power by buildings was large (Section 8.6). In New York the consumption of electricity in summer came to exceed that in winter because of the demand for air conditioning.

The late 1960s produced increasing disillusionment, particularly among students of architecture, with the affluent society and the buildings designed for it.

The ideal of a perfect interior environment, with its huge demand for energy, came under criticism from radical environmentalists, and doubts rose even in the minds of conservative people whether the increasing cost of energy made it possible to sustain any longer the concept of a glass box whose environment was maintained at its most perfect temperature and humidity by an unlimited supply of electric power.

8.2   CLIMATE, THERMAL COMFORT, AND THE DESIGN OF BUILDINGS

Vitruvius offered advice on the aspect of rooms for various purposes (Ref. 1.1, Section 4.5), and similar rules were common in the architectural books of the Renaissance. Most of these rules were based on sound experience and are still valid.

Climatology is, however, essentially a modern science. The ancient Greeks appreciated that climate depended on latitude, and the Alexandrian astronomer and geographer Ptolemy divided the earth's surface into parallel zones according to their inclination to the equator; indeed, the Greek word *klima* originally meant inclination. Ptolemy's concept of climatic zones assumed a motionless atmosphere. Alexander von Humboldt in 1817 discussed the origin of tropical storms and the effect of air movement on climate and devised isothermal lines, that is, lines of equal temperature at a certain time, to illustrate weather patterns. The first regular weather maps, drawn in the 1830s, used data sent in by mail. In the late 1850s Urbain J. J. Leverrier compiled maps from more up-to-date information supplied by the electric telegraph.

Climatological data are essential for calculating the heating and cooling loads for air-conditioning systems, and since the 1870s this information has been available for most major European and American cities. These data, however, apply to the reference point at which the observations are made. The microclimate on top of a ridge, in a hollow, in a confined public space, or on top of a building may differ appreciably. Data on microclimate are not generally available, although they can often be predicted from the general weather data by making a model of the area and observing the wind patterns in a wind tunnel.

Studies of the effect of air temperature on human comfort indoors go back to the eighteenth century. In 1775 C. Blagden observed (Ref. 8.51) that a man could survive a dry-bulb temperature of 260°F (127°C) for eight minutes. The predominant view, however, almost to the end of the nineteenth century was that the malaise experienced in crowded rooms was due to toxic gases produced by the people in the room rather than by the heat they generated. A. L. Lavoisier, who discovered the composition of the atmosphere, thought in the late eighteenth century that the toxic substance was carbon dioxide. Various compounds exhaled from the skin or the lungs in minute quantities were examined during the nineteenth century.

In 1905 J. S. Haldane established temperature, humidity, and air movement as the principal criteria of human comfort. Sir Leonard Hill finally disproved the anthropotoxin theory in 1914 (Ref. 8.4). He shut a group of students in a small room and observed the rise in temperature, humidity, and carbon dioxide content and the increasing distress of the people. Turning on a fan provided great relief. A control group of students outside made to breathe the same air, but at normal temperature and humidity, were not adversely affected and showed no distress.

The earliest investigations were not concerned with establishing the criteria for ideal comfort but with ameliorating the most adverse conditions that occur in factories, mines, and ships. One problem encountered in textile factories was the high humidity maintained artificially to reduce the breakage of threads. Heat was particularly distressing in deep mines and in the boiler rooms of ships in the tropics. Because Britain was still the leading industrial and naval power and a major operator of mines, most of the early work was done by British physicians such as Haldane and Hill. Two committees set up by the British Home Office on Ventilation in Factories

(1907) and on Humidity and Ventilation in Cotton Factories (1909) accepted the view that overheating was the chief problem raised by bad ventilation.

The effect of air movement on human comfort was noted by John Arbuthnot, a friend of Jonathan Swift. In 1733 he published *An essay concerning the effects of air on human bodies* (Ref. 8.52, p. 46) which explained that the wind causes chilling by dispersing the layer of warm moist air over the skin. D. Boswell Reid (Section 8.4) found in 1844 that the thermometer alone was not a reliable indicator of comfort and that air movement must be taken into account. We have already noted Hill's demonstration in 1914 of the usefulness of a fan in relieving high temperatures.

The effectiveness of radiant heat in improving comfort in cold weather was observed by Thomas Tredgold (see also Section 2.3) in 1824. In *Principles of warming and ventilating public buildings, dwelling houses, etc.*, he pointed out that people warmed by radiant heat from an open fire could be comfortable even when the air temperature was relatively low.

The earliest attempts to define precisely the criteria for human comfort were based on physical measurements. Devices for measuring temperature and humidity have been in use since the sixteenth century; indeed, Leonardo da Vinci had in the fifteenth century described an instrument for determining humidity, based on measuring the weight of a ball of wool, which increased in high humidity because it absorbed water vapor. The first mercury-in-glass thermometer appears to have been made by Gabriel Daniel Fahrenheit in 1714, the first hair hygrometer by Horace Bénédict de Saussure in 1783, and the first wet-and-dry bulb hygrometer by C. W. Boeckmann in 1802.

During the course of his investigations in 1914 Hill (Ref. 8.4) invented the katathermometer for measuring the effect of air movement on the human body. It was a thermometer with a large bulb that was heated a little above 100°F; the time required for a drop in temperature from 100 to 95°F, that is, at body temperature (98°F or 37°C), was observed. According to Hill, the dry kata gave a measure of the cooling power of the atmosphere by radiation and convection, the wet kata gave it by radiation, convection, and evaporation, and the difference between the two gave it by evaporation. If the air temperature is known, the dry katathermometer can measure the speed of the air and is still used for that purpose.

In 1930 H. M. Vernon devised the globe thermometer for measuring the effect of radiation. It consists of a hollow copper sphere coated with black paint containing an ordinary thermometer at its center. If the mean radiant temperature is higher than the air temperature, the temperature recorded by the globe thermometer is also higher and vice versa. If the air temperature and speed are known, the globe thermometer determines the mean radiant temperature of the surroundings.

Numerous instruments have been developed since then for the purpose of assessing comfort criteria, but all are based on the measurement of temperature, humidity, air movement, and radiation. An early attempt to simulate man by a physical instrument was the eupatheoscope built by A. F. Dufton at the British Building Research Station in 1932 (Ref. 8.6) for work on human comfort in cool climates. It was an electrically heated cylinder 22 in. (560 mm) high by 7½ in. (190 mm) in diameter. When a thermostat inside the cylinder reached a temperature of 78°F (25.5°C), the heating current was switched off, and after the temperature at the thermostat had dropped below 78°F it was automatically switched on. The instrument took account of temperature, air movement, and radiation, but not humidity.

The temperature of an environment, which required the same power input, Dufton called the *equivalent temperature,* and this was a temperature corrected for the effect of radiation and air movement at which the clothed human body, according to Dufton, was expected to be comfortable.

Although physical instruments give objective measurements, it is difficult to be certain that the various factors affecting human comfort have been correctly included in the design. The alternative is to rely on the subjective response of a group of observers, and this is the basis of the scale of *effective temperature*. It was developed at the Research Laboratory of the American Society of Heating and Ventilating Engineers (now called ASHRAE—American Society of Heating, Refrigeration and Air Conditioning Engineers) in Pittsburgh by F. C. Houghten and C. P. Yaglou (Ref. 8.7), mainly for warm-weather conditions. It thus takes account of temperature, humidity, and air movement, but not of radiation. The people taking part in the experiment were asked what conditions of temperature, humidity, and air movement gave the same sensation of warmth while they were engaged in light physical activity; the effective temperature is the temperature in that group of equal sensation of warmth at which the air is fully saturated with water vapor and the air movement is zero. The ASHRAE comfort chart until 1969 (Ref. 8.22a, p. 122) was still based on the concept of effective temperature (E.T.), and surveys conducted in the 1940s indicated that people are most likely to feel comfortable in an E.T. of 68°F (20°C) in winter and an E.T. of 71°F (22°C) in summer if the air movement is 15 to 25 ft/min (0.08 to 0.13 m/sec). As an indication of what this means in practical terms, an E.T. of 71°F (22°C) for summer corresponds to a dry-bulb temperature (i.e., the temperature indicated by an ordinary thermometer) of 75°F (24°C) and a relative humidity of 60% or a dry-bulb temperature of 78°F (26°C) and a relative humidity of 33%.

Evidently the comfort criteria are dependent to some extent on local customs. In the humid tropics people commonly wear very light clothing and are acclimatized to warm, humid weather so that higher effective temperatures are preferred. Surveys of people of Asian and European descent in Singapore (Ref. 8.8) indicated a preferred temperature of 81°F (27°C) at a relative humidity of 80% and an air speed of 80 ft/min (0.4 m/sec); this corresponds to an effective temperature of 77°F (25°C). For the same reason the preferred indoor winter temperature is about 9°F (5°C) higher in the United States than in England because Americans tend to wear lighter indoor winter clothing than the English.

In the late 1950s ASHRAE decided to reevaluate its effective temperature scale. The work was interrupted by the decision to close the ASHRAE laboratories in Cleveland in 1960. They were, however, reopened as the Institute of Environmental Research of Kansas State University, and several studies were undertaken as joint projects of KSU and ASHRAE, the largest by F. H. Rohles, a psychologist, and R. G. Nevins, an engineer, on 1600 students of college age (Ref. 8.22b, p. 138). These results led to a revision of the ASHRAE comfort chart (Ref. 8.22b, Fig. 17, p. 137).

After his return to the Technical University of Denmark P. O. Fanger, a Dane who worked at Kansas State University in 1966–1967, deduced from the KSU tests a series of comfort equations (Ref. 8.9, p. 41) which include the thermal resistance of the clothing worn and the percentage of the body covered, the metabolic rate and the surface area of the human body, the external mechanical efficiency of the body, the air velocity, the air temperature (or dry-bulb temperature), the mean radiant temperature, and the vapor pressure in the air (which determines the relative humidity). From these data he prepared by computer forty-three charts (Ref. 8.9, pp. 43–67), now known as the Fanger comfort diagrams, which show the comfort condition in terms of these variables. Some of these charts are reproduced in the current ASHRAE Handbook (Ref. 8.22b, pp. 140–142).

R. K. Macpherson inclines to the view that *Homo sapiens* developed in tropical forests (Ref. 8.11); although apes were to some extent protected by a fur-covered skin, the hairless human body was ill-suited to exposure to arid heat or cold without clothing. Like most animals, early man may have been a creature engaged in only a low level of physical and mental activity, for modern man retains that tendency in the tropics. Native people in the humid tropics are inclined to live off the land and not to work too hard. A European may deliberately go to a hot climate for his annual vacation in order to lie on the beach doing nothing.

The concept that different human races have different thermal comfort criteria is not supported by observations in the United States, England (Ref. 8.10), or Singapore (Ref. 8.8). It is, however, a concept of ancient origin of the greatest historical significance.

Vitruvius had already assumed a different racial response to climate (Ref. 1.1, Section 4.5), and racial prejudices were probably widespread during the voyages of discovery in the fifteenth and sixteenth centuries. The conquerors found in various parts of America, and on many tropical islands, native people who were disinclined to work on plantations, even under the threat of force, and they imported African negroes as slaves. Later (East) Indians were transported as indentured labor, more or less voluntarily, to the West Indies, Mauritius, and Fiji; and Chinese and Japanese to Hawaii. The concept that Europeans could not work physically in the tropics, whereas colored people could, was still widely held in the early twentieth century after the discovery of cures for most tropical diseases had reduced the risk of serious illness.

F. S. Markham argued in *Climate and the Energy of Nations* (Ref. 8.53) that a basic reason for the rise of nations in modern times has been their control over climatic conditions. He considered that acclimatization was achieved by gradually lowering one's efficiency to that appropriate to the climate. Colored labor was no better suited for work in the tropics than white labor. This has been borne out by experience in tropical Australia where political pressure prevented the importation of colored labor since the end of the nineteenth century, and the heavy work in the canefields was done entirely by people of European descent.

It follows that human efficiency in the tropics can be greatly increased by improvements in the indoor climate. Even if little can be done to improve the outdoor working conditions, ability to sleep comfortably at night in an air-conditioned room is beneficial.

In countries with a hot climate inhabited by people with high incomes, such as the southern United States and northern Australia, there is a clear case for air conditioning, although the thermal shock resulting from the repeated interchange between air-conditioned rooms and the open air should be reduced by adopting somewhat higher indoor temperatures than those ideally comfortable for continual indoor living. In developing countries the low average income and the lack of skill in maintaining equipment makes the widespread use of mechanically based air conditioning impracticable, but other low-cost measures can be taken to improve the indoor climate, such as better thermal inertia and ventilation.

More attention is now given to the environmental aspects of vernacular construction. Amos Rapoport (Ref. 8.12) has pointed out that although traditional house forms are the result of a response to social and cultural customs rather than to climate many are well adapted to their physical environment and some are more comfortable to live in than the modern buildings that replaced them.

Many native buildings in the hot-humid tropics are built from reeds, bamboo, or timber slats that allow the free passage of air and thus produce a desirable reduction in the "effective temperature." In Malaya and Indonesia houses can be found with curving gables which encourage the movement of air.

In the hot-arid tropics thick walls built of masonry, mudbrick, or mud reduce the peak temperature in the early afternoon by their high thermal inertia. At night such houses remain hotter than the outside air, but as the night is then pleasantly cool people can sleep on the roof or on the ground.

These building types are not always found in the right places. Traditional houses in the hot-humid region of Kumasi, Ghana, are built with mud blocks. According to tradition, the Ashanti arrived in the forest country, probably from a hot-arid area) in the sixteenth century, and it may be that these houses originated in response to a different climate. It has been impossible, however, to persuade the people to build wooden houses better adapted to the climate in spite of the plentiful supply of timber (see Ref. 1.1, Fig. 1.1 and 1.2).

Many architects of the Modern Movement, notably Le Corbusier, have looked to vernacular architecture for inspiration, but their interest was mainly in the visual and social aspects and their environmental design often ran counter to well-established principles. Although more traditional architects, including Frank Lloyd Wright (Sections 8.3 and 8.4) took effective advantage of advances in technology, there were spectacular errors in the thermal environment of some unconventional buildings. In 1933 Le Corbusier built a hostel for the Salvation Army, called the *Cité de Refuge*:

Six hundred poor creatures, men and women, live there. They were given the free and ineffable joy of full light and the sun. A sheet of glass more than a thousand square yards in size lights the rooms from floor to ceiling, and from wall to wall (Ref. 8.19, p. 18).

The building was a great success during the first Christmas, when it was pleasantly warm, but it was much too hot in the following summer, as Le Corbusier subsequently admitted. The southern facade was fitted with sun shades, and some of the sealed glass was replaced by windows that opened.

Mies van der Rohe never admitted that there was anything wrong with the Farnsworth House (Section 8.1). The client, Dr. Elizabeth Farnsworth, after completion described the house as unfit for human habitation and started legal proceedings against him. Although Dr. Farnsworth lost the case, she altered the house so extensively to improve its thermal environment and to screen the open terrace from insects that it ceased to be the visual masterpiece Mies had designed. The house has since been demolished to make room for a highway (Ref. 8.13, p. 19).

8.2

Room in the Red Fort at Delhi, built in the seventeenth century. The open screens are cut from slabs of solid marble.

8.3 SUN CONTROL

Rooms shaded from the sun which took advantage of any available air movement have been a central feature of the living quarters in many Muslim palaces (Fig. 8.2). In arid regions a pool of water was often provided whose evaporation reduced the temperature and increased the humidity (Fig. 8.3).

Overhanging roofs that protect the walls and the windows from the sun are common in warm climates. The Indian verandah was introduced into several British colonies, including Australia (Fig. 8.4). Although roof overhangs are easier to build with timber, they are also found in regions in which timber is scarce; brick arches are used instead (Fig. 8.5).

Summer in eastern and midwestern North America is hot, and many American architects have made use of large roof overhangs to provide shade; for example, most of the houses built by Frank Lloyd Wright in the Chicago region between 1889 and 1910 had carefully designed roof overhangs (Ref. 8.14). Some had flat roofs which showed that there is not necessarily an aesthetic conflict between their use and roof overhangs that shade the windows. Roof overhangs were an anathema to many architects of the modern movement, and this created environmental problems of the type described for the Farnsworth house (Section 8.2).

Le Corbusier appreciated the importance of shading after his experience with the *Cité de Refuge* in 1934. He then invented the sun screens that are an integral part of the building, still known by the French name *brise soleil*. He first used them in a building for the Ministry of Education in Rio de Janeiro, completed in 1937, for which he was consulting architect to a team led by Lúcio Costa. The *brise soleil* thereafter enjoyed a period of popularity, but it has several limitations. It radiates heat, interferes with the view, and dominates the facade. At the Ministry of Education its members were adjustable; but if the screens are adjusted differently the building is given a patchy appearance. Because of their large size and their exposure to the

8.3

Open audience balcony of the seventeenth-century imperial palace in Isfahan, Iran, overlooking an ornamental lake.

8.4

Deep verandah completely surrounding Experiment Farm Cottage in Parramatta, near Sydney, probably built in 1798, now a museum. It is one of the oldest surviving buildings in Australia, which was settled in 1788.

weather, movable members frequently get stuck and after a time adjustment becomes impossible. The fixed type of *brise soleil* has therefore seen more service in recent years.

The basic theory for calculating sunlight penetration was known to the astronomers of Alexandria. It is not easy, however, to evaluate the usefulness of different shading devices by mathematics alone because of the time taken to solve the equations. In 1932 Dufton developed the heliodon for experimenting with sun control devices (Ref. 8.15) and several variations of this machine have been built (Fig. 8.6). In the 1970s the development of computer graphics has provided an alternative and often a more convenient tool. Although the calculations are quite lengthy by slide rule because they involve spherical trigonometry and data for the movement of the sun, the problem is basically simple and easily handled by a short computer program. A graphic output unit can then draw the sunlight penetration on plan and elevation, and the piece of paper can be folded to give a three-dimensional model of the room or building (Fig. 8.7).

By calculating the effect of the movement of the sun more effective shading devices can be produced. The nineteenth-century verandah was generally much wider than the sunshading required (Fig. 8.4) because the extra width was used as living space. The colonnade of the Classic Revival houses in the southern United States also tended to provide more shade than was strictly necessary (Fig. 8.8) because the proportions were based on aesthetic rather than environmental considerations. In both cases the interiors were unnecessarily dark throughout the year.

In subtropical regions, in particular, the most effective sunshade is one that excludes the sun in summer but admits it in winter, and verandahs and balconies have been designed with this objective even if they are too narrow to provide useful living space; for example, a balcony placed above each window can be designed to provide a sunshade for the window below and a platform for cleaning the window above. Even more efficient as a shade is a horizontal screen of inclined slats of metal or wood which provides complete sun control without interfering with the view or air movement (Fig. 8.9).

External sunshading is less commonly used with glass curtain walls because it mars the visual continuity of the surface. Because of the ease with which glass transmits

8.5

Brick balconies built on squinches (short brick arches; see Ref. 1.1, Fig. 5.4a) at a religious college erected in the early eighteenth century in Isfahan. The rooms behind are fully shaded from the sun.

8.6

Shadows obtained on a heliodon with an architectural model of an office building in Belconnen, a suburb of Canberra. The table on which the model is placed is rotated to give the altitude and azimuth corresponding to the latitude and to the time of the day and year; these must be calculated. The sun is represented by a spotlight placed as far away as possible; in this instance 10 m (33 ft). A direct-reading instrument is also available in which the model is placed on a horizontal table, and the lamp is rotated at the end of a long arm. The latitude, the time of the day, and the time of the year are read directly on dials. However, as the length of the arm carrying the arm is limited by practical considerations, the machine is therefore less accurate. (*Architectural Science Laboratory, University of Sydney.*)

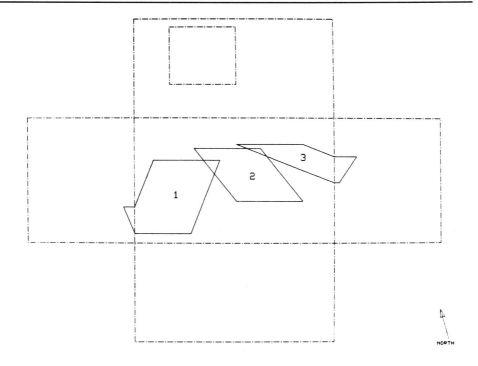

NORTH

8.7

Graphical output of sunlight penetration program for a rectangular room with an off-center window facing 15° east of north. The drawing shows the plan of the room and the elevations of the four walls. These can be cut out and folded along the dotted lines to produce a three-dimensional model. The polygons 1, 2, and 3 then show the sunlight penetration through the unshaded window at 10 A.M., 1 P.M. and 3 P.M. on April 22 in Sydney, Australia (latitude 33.92 degrees east; longitude 151.17 degrees south; time 10 hours later than GMT).

heat, this poses a problem in summer, even in temperate zones. The "greenhouse effect" is due to the ability of normal glass to transmit heat radiated at a high but not a low temperature. Solar heat but not reradiated interior heat is therefore transmitted. Increasing the thickness of glass has only a slight effect on its thermal resistance, but a double layer is much more effective than a single one. This has been known for a long time, and in 1831 a camellia house with a double glass roof was built in Vienna for the Archduke John. Hermetically sealed double glazing was commercially available in the 1930s, but its large-scale use in curtain walls dates only from the 1950s.

Although exterior sunshading devices are often visually unacceptable on the exterior of glass, they have limited use on the inside because they do not prevent

8.8

Classical colonnade shading the windows of a house in New Orleans.

solar radiation from passing through the glass. In the 1960s remotely controlled venetian blinds were placed inside hermetically sealed panels of double glazing, so that they stopped the sunlight before it reached the inner panel of glass without spoiling the appearance of the continuous glass wall.

Heat-absorbing glass was used in the Lever Building in New York in 1952. Heat-reflecting glass is, however, more effective in hot weather because more of the heat is reflected back to the outside air. The most efficient type is mirror glass, which also has a distinctive visual quality because it reflects it surroundings, but this can

8.9

Horizontal slats shading the windows of the facade of a building at the University of Sydney facing northwest, constructed in 1962. The device is unobtrusive and shades the entire facade below it except for a few thin lines of sunshine.

create a problem for buildings opposite for heat is reflected on them by the mirror wall. There have been cases in which windows fitted with correctly designed sun screens have borne unexpected heat loads reflected by mirror walls subsequently constructed.

8.4   HEATING AND VENTILATION

Heating improved rapidly from the end of the eighteenth century. Tile stoves, which made an appearance in Germany and Scandinavia during the seventeenth century, flourished during the nineteenth in central and eastern Europe. The stoves in palaces were sometimes designed by leading artists and especially made by porcelain factories. Because of their large volume and surface area, they generated a substantial quantity of heat at a comparatively low temperature, partly by radiation and partly by convection. Some palaces had special passages from which servants could maintain the fire in the stoves without disturbing their masters. In middle-class houses the stoves were generally plainer, but they were frequently shaped so that one could sit on a cushion on the warm tiles. Tiled stoves were still being built in the twentieth century and some are in use today.

In Britain the open fireplace was preferred during the nineteenth and the first half of the twentieth century. Coal production increased in the United Kingdom from about ten million tons in 1790 to about 250 million tons (254 million tonnes) in 1910, and toward the end of that period even poor people could afford it. In the houses of the wealthy and middle class nearly every room was fitted with an open fireplace by the end of the nineteenth century, and there were plenty of servants who were willing to tend them for low wages. The thermal efficiency of these open fires was low, and an appreciable proportion of the hydrocarbons was turned into smoke that caused serious pollution. The handsome sandstone buildings of Manchester and the surrounding industrial cities turned to a sepulchral black.

It is surprising that the pollution caused by domestic fires attracted little criticism during the nineteenth century. Environmental protests were directed more against the railways which today's ecological groups regard as an environmental asset. Although English people, on the whole, remained attached to the open fireplace, heating engineers have for almost a century endeavored to promote a cleaner and more efficient means (Fig. 8.10), but only in the 1950s did clean-air legislation stop the pollution.

Each open fire required its own chimney (Fig. 8.11), and the cleaning of these chimneys kept a large number of chimney sweeps occupied. In the eighteenth century and most of the nineteenth boys were employed who were small enough to climb up the chimneys. A society was formed in 1803 to stop the practice, but the necessary legislation failed to get sufficient votes in Parliament and it was passed only in 1875 through the efforts of Lord Shaftesbury (see also Section 7.5).

In working-class houses in particular the open fire in the kitchen was often used to heat water. The hot water rose to a storage tank in the bathroom or under the roof and the water heater was re-filled from a cold-water tank. This hot-water system was economical in fuel and, provided the kitchen fire was lighted each day, automatic. Usually the hot-water pipes passed through an "airing cupboard" where damp

8.10

"View from an upper window in city," from a book on heating and ventilating published in Manchester in 1927 (Ref. 8.23, p. 307).

laundry could be dried. During a frost, however, the cold water would burst its pipes, and plumbers in Britain had a busy time so long as the gravity water system remained in common use.

In North America the climate was more severe, servants were scarce and expensive, and the general standard of living during the nineteenth century overtook that of Europe. Domestic heating was generally supplied by a boiler (Fig. 8.12) which in time became more and more automatic. The actual heating was done by cast-iron

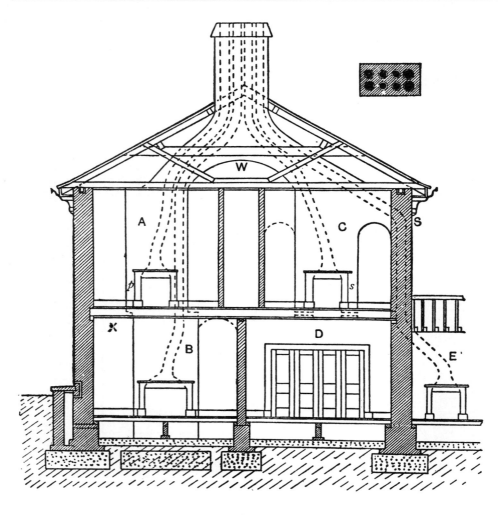

8.11

Collection of chimney flues into a central stack, from a book on building construction, first published in 1875 (Ref. 8.24, p. 142).

8.12

Boiler for heating houses and for domestic water supply made by the Richmond Stove Company, Norwich, Connecticut (from an advertisement in Ref. 8.16).

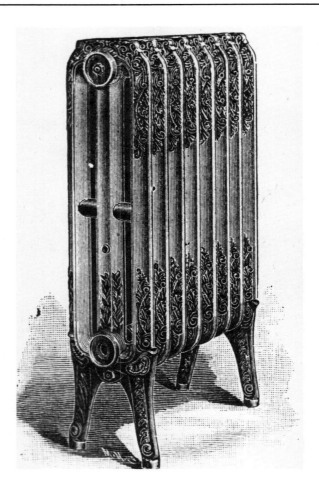

8.13

Radiator for domestic heating made by the Standard Radiator Company, New York, Boston, Buffalo, and Chicago (from an advertisement in Ref. 8.16).

hot-water or steam radiators (Fig. 8.13) which were frequently decorated with flower patterns or geometric curves. At that time radiators were far less common in Europe and usually less ornate.

American architects in the late nineteenth and early twentieth centuries were more receptive to innovations in heating than their European colleagues. In the Baker House, built by Frank Lloyd Wright at Wilmette, Illinois, in 1908, hot-water pipes were concealed in the wainscoting to supply radiators under the windowseats, which were slatted to permit the warmed air to circulate.

In 1914, on a trip to Japan, Wright described a visit to the house of Baron Okura in Tokyo, where he felt exceedingly cold in the dining room but pleasantly warm when they adjourned to the "Korean room" for coffee:

The Baron explained: The Korean room meant a room heated under the floor. The heat of a fire outside at one corner of the floor drawn back and forth underneath the floor in and between the tile ducts, the floor forming the top of the flues (or ducts) made by the partitions, the smoke and heat going up and out of a tall chimney at the corner opposite the corner where the fire was burning.
The indescribable comfort of being warmed from below was a discovery (Ref. 8.18, p. 99).

Although Wright did not say so, this construction, invented independently by the Chinese and Koreans, closely paralleled the Roman hypocaust (Ref. 1.1, Section 4.6).

I immediately arranged for electric heating elements beneath the bathrooms of the Imperial Hotel (in Tokyo)—dropping the ceilings of the bathrooms to create a space beneath each in which to generate the heat. The tile floor and built-in tile baths were thus always warm (ibid., p. 99).

Toward the end of the nineteenth century some of the wealthier American homes were both heated and ventilated by hot air, supplied to the rooms through plenum ducts terminating in grilles in the floor or the skirting. Because the air was fed into the rooms by pressure, there were no draughts.

By comparison an open fire produced a big flow of air up the chimney. Because there were, in general, insufficient grilles, or none at all, the air intake was balanced by an inflow of air under the doors and through any badly fitting windows. This created the draughts for which many of England's stately homes were famous.

There were fewer differences between American and European methods of heating and ventilating factories and public buildings. The British House of Commons was one of the first rooms in which a ventilating plant was installed (Ref. 1.1, Section 8.9). After the Palace of Westminster was destroyed by fire in 1834 (Section 7.1) D. Boswell Reid, a leading expert on ventilation, was engaged as a consultant, first for the temporary House of Commons and then for the House of Commons in the new Palace of Westminster. Although conflict between him and the architect, Sir Charles Barry, marred the result, it was one of the first attempts at a scientifically designed ventilation system, with some control over humidity. Reid used the concept that the air should be directed to keep the feet warm and the head cool. In a book published in 1844, *Illustrations of the theory and practice of ventilation* (Longmans, London; Ref. 8.52, p. 294), he described some of his experiences with the House of Commons. Apparently several members were vocal in their criticisms; one complained of suffocating warmth, whereas another at the same time said he was shivering with cold. Reid wrote that, after questioning, the first member said he preferred 52°F(11° C) and the other required 71°F (22°C). Even allowing for the fact that the English gentry wore heavier clothes in those days, 11°C is well below the 25°C now generally quoted as a desirable temperature (Section 8.2); however, it agrees with other statements from the more Spartan midnineteenth century; for example, Sir Douglas Galton, FRS, who contributed the article on *Heating* to the ninth edition of the *Encyclopaedia Britannica,* recommended in 1880 (Ref. 4.65 Vol. XI, p. 591) a temperature for living rooms of 54 to 68°F (12 to 20°C) and a temperature for bedrooms of "not less than 40°F (4°C)."

When Cuthbert Brodrick built the Leeds Town Hall (1853–1859), he made provision for ventilating the Council Chamber by incorporating extraction-duct outlets unobtrusively in the cornice moldings, an example that was followed in several other prestige buildings in Europe and America.

The technology of ventilation had been established for many decades by the use of fans in deep mines. The real problem was to transform the ventilation of buildings from a luxury to a sensible amenity. Hospitals and factory buildings led the way in this respect. In the latter the dust and heat resulting from industrial processes were a sufficient justification. Hospital ventilation was aided by a widespread belief that the malaise resulting from crowded rooms was due to toxic substances produced by the people in the room. Until the "anthropotoxin" theory was disproved in the early twentieth century (Section 8.2), there was a clear case for good ventilation in all hospital wards. The second half of the nineteenth century was an era of spectacular progress in public health, and the views of physicians commanded respect (see also Section 7.5). Manufacturers were often generous in their support of hospitals in their factory towns, and some of the buildings erected toward the end of the Victorian era are still in use. Bearing in mind the vast changes that have taken place in hospital design, it is a tribute to the quality of their construction that these ancient buildings can still be used, even with inconvenience, at the present time.

The Derbyshire General Infirmary, built at the beginning of the nineteenth century, was one of the first hospitals to be mechanically ventilated (Ref. 8.60). At the Johns Hopkins Hospital in Baltimore, built in 1873 with an endowment of three million dollars by Johns Hopkins, a full plenum underfloor duct system supplied air, heated as required by steam coils, and an air extraction system was installed in the ceiling (Ref. 8.16, pp. 144–163). This was designed to supply 1 ft³/sec (0.0283 m³/sec) of fresh air per hospital bed, with four times that supply in the isolation ward, but the system had greater capacity.* There are descriptions of a number of late-nineteenth-century American hospitals in *Engineering Record* which had similar systems (Ref. 8.16).

The British hospitals were, however, ahead in their control of air intake, perhaps because of the greater air pollution in the industrial cities. The Victoria Hospital in Glasgow had a wet air-screen in its ventilating system before 1895. The air, after being drawn in by propellers, was passed through a mesh of cords made of horsehair and hemp, closely wound over a top rail of wood and under a bottom rail, to form a close screen 16 × 12 ft (4.9 × 3.7 m). A constant stream of water, by which dust and soot particles were removed and carried away, trickled down the screen. By an automatic flush tank 20 gal (76 liters) of water were instantaneously discharged over the screen every hour (Ref. 8.16, p. 172).

---

*Modern recommendations range from 10 to 30 ft³/min per person; that is, patients and nurses (0.17 to 0.50 ft³ or 0.0047 to 0.0142 m³/sec). Sir Alfred Ewing (see also Section 7.5) in 1888 in the article on ventilation (2000 to 5000 ft³/hr per person, which equals 0.56 to 1.34 ft³ or 0.0157 to 0.0393 m³/sec per person.

In the Royal Victoria Hospital, Belfast, built in 1903, the air was washed in a manner similar to that described for Glasgow, heated, and circulated through brick ducts to air inlets below the ceilings of the wards.

In addition the humidity was measured daily with a wet- and dry-bulb hygrometer (Ref. 8.59, pp. 76-82). In the climate of Belfast the problem of humidity control was mainly that of adding water vapor when cold air was warmed, and the air-washing system, acting in conjunction with the heating plant, was able to increase the humidity to a more desirable level.

Thermal insulation is discussed in Section 9.4.

8.5 AIR CONDITIONING

Lowering the humidity in a warm-humid climate is much more difficult. Although summer is generally pleasant in northern Europe, in the American east and midwest it is frequently oppressively hot and humid, particularly in the southern states. In 1906 in Charlotte, North Carolina, Stuart W. Cramer used individual heads to spray chilled water to clean and cool the air and to control humidity. He called this method *air conditioning.*

In the same year Willis H. Carrier devised the dewpoint method of control in Buffalo. According to his own account (Ref. 8.17, p. 15), Carrier observed fog at the railway station in Pittsburgh:

Here air is approximately 100% saturated with moisture. The temperature is low so, even though saturated there is not much moisture. There could not be, at so low a temperature. Now, if I can saturate the air and control the temperature at saturation, I can get air with any amount of moisture I want in it. . . .

I can do it by drawing the air through a fine spray of water to create actual fog. By controlling the water temperature I can control the temperature at saturation. When very moist air is desired, I'll heat the water. When very dry air is desired, that is, air with a small amount of moisture, I'll use cold water to get low-temperature saturation. The cold-water spray will actually be the condensing surface. I certainly will get rid of the rusting difficulties that occur when using steel coils for condensing vapour in air. Water won't rust.

Many complaints had been aired about the uncomfortable working conditions in factories on hot days, and attempts were made during the nineteenth century to improve them. Refrigeration was needed to get really satisfactory results, but it did not become available in quantity at a reasonable cost until the 1920s.

Ice has been used for refrigeration since ancient times, but until the nineteenth century its use was limited by the difficulties of transport. Dr. Cargill Knott of the Imperial University of Tokyo, in the article on ice in the ninth edition of the *Encyclopaedia Britannica* (Ref. 4.65, Vol. 12) in 1881, referred to the profitable trade in American ice from Boston to Martinique, in the West Indies, and to Calcutta in India. At that time Norway exported annually 150,000 tons of natural ice to England. The same article described the construction of an ice house, essentially an underground structure insulated by a layer of earth above the roof, and called it "one of the desirable adjuncts of a country residence."

Ice during the nineteenth century was the main refrigerant for food stores and after 1851 for refrigerated railway wagons.

Yet mechanical refrigeration was already in use in the early nineteenth century. There were two main systems: one based on lowering the temperature by the rapid expansion of a compressed gas and the other on the latent heat required to volatilize a liquid. These methods are still used today. The choice between ice and a refrigerating machine was thus based mainly on cost and convenience. The first refrigerated steamship, the *Strathleven*, which carried meat from Australia to England in 1880, had mechanical refrigeration because the voyage took more than two months, mainly through the tropics.

The problem of applying refrigeration to buildings was complicated by the amount of cooling required if it was to serve a useful purpose. Ice and snow had been used by Roman emperors since the rule of Caligula for giving an impression of coolness in their palaces on a hot day. Considering the amount that would be required to achieve a physically significant result, it is unlikely that the effect was other than psychological. The same illusion was still employed in the 1920s in some American and Australian picture theaters, which in summer displayed large blocks of ice in front of fans in the foyer.

Air conditioning of buildings therefore depended on the more economical methods of mechanical refrigeration devised in the 1920s and on clients who were willing to pay for it.

Theater auditoria had always presented an air-conditioning problem because of the heat and humidity generated by the audience. Indeed, there are some nineteenth-century theaters in London which in spite of renovations become distressingly uncomfortable during a performance to a full house even in winter. Motion pictures had been a minor entertainment before 1914, but by 1920 they had acquired a large following and the industry fostered an image of smooth, luxurious comfort.

The first air-conditioned building was Graumann's Metropolitan Theater in Los Angeles, built in 1922 with a Carrier-made installation. The air was cooled by a refrigerating plant to a temperature whose dewpoint corresponded to the required humidity. It was then heated to the required temperature. The treated air was then fed into the auditorium at low velocity through diffusers in the ceiling and exhausted through grilles under the seats, thus providing the "cool head, warm feet" already stipulated by Reed (Section 8.4). In this respect Carrier's installation differed from most American ventilating systems that supplied the air through grilles in the floor and exhausted the foul air through the ceiling, often relying on natural ventilation to remove the warmer air (Ref. 8.16).

In 1928 the first fully air-conditioned office building was erected in San Antonio, Texas. The Milam building had twenty-one stories, the last few of which formed a faintly Gothic tower with pointed windows. George Willis, the building's architect, provided air supply ducts above the ceilings of the corridors, which were a little lower than those of the rooms, and the air intake was thus just below the ceilings of the rooms. Exhaust grilles were installed in the doors and the corridors acted as exhaust ducts.

The air-conditioning installation for Radio City, part of the Rockefeller Center complex, was completed in 1932, but conceived before the Depression. It remained the best installation until the 1940s and was much admired by Le Corbusier who referred to it in *When the Cathedrals Were White* (Ref. 8.19, p. 33) as "conditioned air throughout, pure, clean, and at a constant temperature." It was the technical realization of one of Le Corbusier's visions, yet which owed nothing to him. To this day most Americans consider the curtain-walled, air-conditioned building as a wholly American development, whereas many European architects view it as a European concept for which American engineers merely supplied the technical know-how (Ref. 8.20).

Although there was little new construction between 1935 and 1945, there were many advances in the technology of air conditioning. The gases used for refrigeration before 1930 had been more or less toxic and some had posed a fire hazard. This restricted the use of refrigeration plants for auditoria and office buildings. The freons, a group of chemicals based on fluorine, first manufactured in 1930 and first employed in refrigeration in 1934, are stable and physiologically harmless and are still commonly used for air conditioning (Ref. 8.21). World War II produced innovations in air conditioning machinery because of its usefulness in naval operations in tropical waters.

By 1945 the industry was ready for a major expansion, and air conditioning, which had been a luxury, became normal practice in office buildings all over the United States.

The acceptance of air conditioning outside America was slower. Since the late 1950s, it has become more common in the industrialized countries. In most cities air conditioning, once it had been introduced, spread rapidly for office buildings. Evidently businessmen considered air-conditioned accommodations either good for efficiency or good advertising and were willing to pay the premium.

In the early 1930s the American railways began the installation of small air-conditioning units in their trains, and by 1936 dining and sleeping cars on the important routes were almost entirely air-conditioned. The production of air-conditioning units small enough to use for a single room had been discussed in engineering journals at least since 1931, but the first practical units did not become commercially available until 1948. This brought air conditioning within the reach of homeowners and made it possible to air condition a single office or workroom without conditioning the entire building.

8.6 THE GROWING USE OF ELECTRICITY IN BUILDINGS AND ALTERNATIVE SOURCES OF ENERGY

Elektron was the Greek word for an alloy of one to four parts of gold to one part of silver. It was later used also for amber, a fossil resin of similar color, and this was the meaning of *electrum* in medieval Latin.

In 1600 William Gilbert, physician to Elizabeth I, published *De Magnete* in which he described a number of experiments on amber, in particular its ability to attract light objects when heated and rubbed. According to the Oxford English Dictionary, N. Carpenter in 1635 called substances with this property *electrical bodies*. In 1672 Otto von Guericke explained a primitive frictional-electrical machine in *Experimenta nova Magdeburgica* (New Experiments in Magdeburg, Germany).

Benjamin Franklin, American scientist and statesman, invented the lightning conductor in 1750 (Ref. 8.25), and in 1769 St. Paul's in London became the first cathedral to be so equipped.

The light produced by an electric discharge between two rods of carbon was noted independently by a number of observers in 1801. In 1808 Sir Humphry Davy, professor at the Royal Institution, gave a public display which demonstrated the potentialities of the carbon arc for illumination. In 1844 it was used in a performance at the Paris Opera, and in 1858 it was installed in the South Foreland Lighthouse on the English coast near Dover.

Michael Faraday, who succeeded Davy at the Royal Institution, demonstrated in 1821 the interrelation between the electric current and magnetic field by causing a magnetic pole to rotate around an electrical conductor; this is the principle of the electric motor. During the next ten years Faraday continued his experiments and in 1831 obtained a powerful direct electric current (dc) by rotating a copper plate in a short magnetic gap. The paper he read before the Royal Society in that year quickly led to the production of more practical electric generators, the first by Hyppolite Pixii in Paris in 1832, which used wire armature windings. At the 1851 International Exhibition in London (see Section 10.1) a variety of electrical generators was shown, but there were at that time few practical applications except to operate electromagnets; the largest on show could lift one ton (one tonne). When the South Foreland Lighthouse was fitted with its electric arc light in 1858, an electric generator was required to produce the current. After a trial supervised by Faraday, two dc generators were built of 2¾ hp (2-kW) capacity, each driven by a steam engine through a belt.

In 1875 arc lights were used to illuminate the Gare du Nord railway station in Paris, and in 1877 the Place de l'Opéra. During the next few years numerous arc lights were installed in lighthouses and public spaces.

Joseph Swan demonstrated his invention of the incandescent electric lamp on December 18, 1878, in Newcastle-on-Tyne; it was a much smaller light source, better suited for use inside buildings. Thomas Alva Edision, working independently on the same problem, solved it ten months later at Menlo Park, New Jersey. Both inventors registered their patents and considered litigation but decided instead to form the Edison and Swan United Electric Light Co. Ltd., which immediately set about building the first two public power stations, one at Holborn Viaduct in London and the other on Pearl Street in New York. The Holborn Viaduct power station opened first, in 1882; the electricity was generated by an Edison dc generator driven by a 125 hp (93 kW) steam engine and supplied a number of nearby buildings, including the Old Bailey, the General Post Office, and the City Temple. Electric lighting quickly became popular, and many more power stations were built before 1900 to supply the necessary current.

George Westinghouse demonstrated in 1886 the practicability of transmitting alternating current (ac), and in 1887 Nikola Tesla, a Croatian engineer who had settled in America three years before, invented the ac induction motor. By 1888 ac supply was becoming common in the United States, and in the following year Sebastian de Ferranti opened the first ac power station near London.

Thereafter the use of electricity in buildings increased rapidly (Refs. 8.25 and 8.26) and was responsible for the improved environmental services in modern buildings, although none ever depended entirely on electricity. Cooking and heating is done quite satisfactorily with solid fuel or gas and fans were driven by steam engines for a century. Bright artificial light was already produced by the incandescent gas lamp, and hydraulic elevators are still used in some cities. Communication systems, although limited in scope, existed before the telephone.

Except for heating, electricity has proved more efficient for all of these services. In addition, the technology of electric wiring imposes few restrictions on the design of a building.

The factories of the early nineteenth century, powered by a single water wheel or large steam engine, required a compact, near-cubical layout (Ref. 1.1, Section 8.6). Individual machines were driven by belts from overhead shafts, which, in turn, were driven by belts from the central power source. The longer the belts and shafts, the greater the loss of power. Factories were built multistory at a time when there were no passenger elevators in order to make the run of the belts and shafts as short as possible. When electric motors were substituted, about 1900, to drive each shaft, the flexibility of the layout was greatly increased. With the development of the fractional horsepower motor in the twentieth century, each machine was given its own power supply and the factory could be made independent of the power source. Convenience of materials handling soon encouraged one-story factories.

There are similar advantages of flexibility in the design of residential and office buildings. Electricity, unlike water, does not lose power when it flows upward through a few floors. Telephone wires, unlike speaking or pneumatic tubes, do not require nearly straight runs. Electric conduits barely increase in diameter as the quantity of electricity is increased; in this respect they differ from water and gas pipes.

Electric appliances, such as the electric washing machine, electric iron, and vacuum cleaner, are all more convenient to use than their nonelectric predecessors.

Pollution in buildings has been greatly reduced by electricity. Candles, oil lamps, and gaslight before the invention of the Welsbach mantle (Section 8.9) all produced soot, as did open wood and coal fires. In spring, when, according to the more austere standards of the day, heating was no longer required and the days became longer so that lighting was needed for a much shorter time, British nineteenth-century housewives traditionally carried out a "spring-cleaning" of blackened curtains and cushions. Apart from the visible pollution, the early gas burners probably did not burn the gas completely because complaints of headaches were common. The smoke produced by coal-burning domestic fires has already been mentioned (Fig. 8.10).

The ease with which electricity can be supplied to the various parts of a building and the convenience of electric machines and communication systems were all evident advantages. The huge increase in the consumption of electricity since 1945 has, however, created serious problems.

This is part of a wider trend. The growth in the use of electricity and in energy in general has been steady, and this is one of the characteristics that distinguishes living standards in Europe, North America, and Australia today from those of the early years of the century. It is also a characteristic distinction between the developed and the developing countries (Fig. 8.14).

The extraordinarily fast rate at which the consumption of electricity was increasing, particularly in the United States, and the prospect of a continued increase

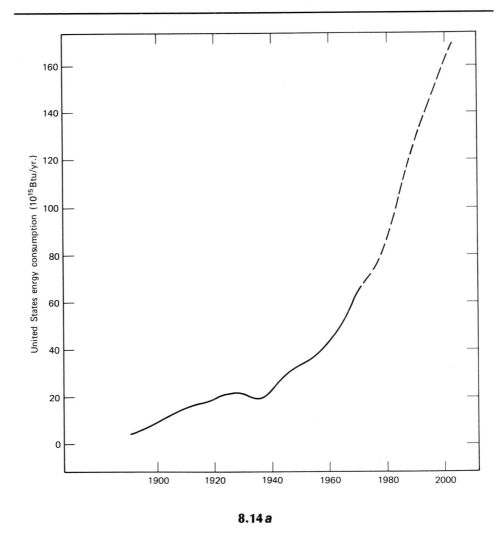

8.14 a

In the United States energy demand grew from 1880 to 1960 at an annual rate of approximately 3%. Between 1960 and 1970 this growth accelerated to approximately 4.2% annually. At 3% annual growth consumption doubled in 23 years; at 4%, in 17 years; at 5%, in 14 years (100 × 10^{15} Btu/yr equals 3,350,000 MW) (Ref. 8.5, p. 21).

became a cause of concern in the 1960s. Most of the sources of hydroelectric power near major population centers were already utilized; the production of power on a large scale from the tide and wind had so far failed to prove economical. Environmental protest groups opposed to the construction of traditional coal-burning and nuclear power stations created a serious shortfall of capacity in New York in the early 1970s which was, at least in part, due to the demands of elevators and air conditioning in the new buildings. The Arab-Israeli conflict produced a large increase in the

Environmental Design and Building Services

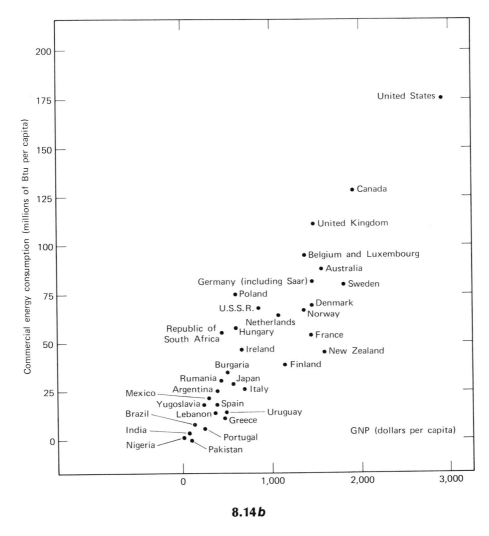

8.14b

Energy consumption per head of population of various countries in 1961, plotted against the gross national product (in U.S. dollars per head of population). The United States, which had 6% of the world's population accounted for 35% of the world's energy consumption (100,000,000 Btu. = 106,000 MJ) (Ref. 8.5, p. 23).

price of oil, which is still the fuel used in many heating installations and in some electric power plants.

Predictions of an eventual exhaustion of traditional energy sources, which had been ignored for several decades, received much publicity.

Solar energy emerged as a major topic of discussion in architectural conferences and publications. It is not always pointed out that most forms of energy in common use are derived from the sun. Plants (vegetables), stock (meat), and timber depend on solar radiation, but all of these are renewable resources. Coal, oil, and natural gas, derived from the decay of organic matter over millions of years, are being depleted in a few centuries, and this is the major cause of the present concern. Hydroelectricity is an indirect form of solar energy, for the sun evaporates the water that forms the clouds which eventually fall as rain to fill the reservoirs; its future is limited because we have already used the best sites. The present interest, however, is in a more direct utilization of this apparently limitless source of energy.

The fundamental principles were well known to physicists in the midnineteenth century. In 1872 a solar still was built in the Andes Mountains in Chile, which for about forty years supplied 6000 (U.S.) gal (22.7m³) of drinking water per day from brackish water for men and animals working a nitrate mine. Solar stills have been built many times since then in extremely arid regions remote from sources of drinking water (Ref. 8.27, p. 8).

In 1878 a solar reflector was set up in the gardens of the Tuileries in Paris, focused on a boiler that operated a steam engine; this ran a printing press. Slightly larger solar generators have been built since 1900 in Arizona, Egypt, and Soviet central Asia, but none has so far produced energy cheaper than that derived from conventional sources, except in desert regions remote from major centers of population. The capital cost of the equipment is high, and the cost of amortization, interest, and maintenance has so far not compensated for the saving in fuel.

Better economy has been achieved with lower temperatures which do not aim at the production of steam. Solar heaters produce hot water even in temperate climates. The water can be used for washing or heating. Except in permanently sunny climates, a solar heater must be supplemented by another power source, such as an electric water heater, to operate on days when the sunshine is insufficient, and this increases the capital cost. It seems likely that solar heaters will be used more in the future, as the rising cost of other sources of energy makes them more competitive. Their effect on the appearance of the house requires consideration, for they may dominate the design of the roof. There are few problems for one or two-story houses with sloping roofs (Ref. 8.28, pp. 65–107), but for buildings of several floors with flat roofs the problem is more difficult (Ref. 8.29). For tall buildings the necessary solar heaters would occupy too great an area to fit on a roof.

Multistory buildings are, however, a major cause of the rapid increase in the consumption of electricity. Although it seems unlikely at present that the problem can be solved by using solar energy or any other new power source, consumption could be reduced in many buildings. It may even be found that this will produce a better environment:

1. In a number of buildings with glass curtain walls the venetian blinds and curtains are kept drawn most of the time because of excessive sunlight penetration and artificial light is needed even during the day. This unnecessary cooling and lighting load could be avoided by suitable external sunshading.

2. The indoor temperature in air-conditioned buildings should be lower in winter than in summer. In some buildings it is actually the other way around. If people would generally wear heavier indoor clothes in winter and men dressed more in accordance with the climate in summer, the temperature could be adjusted so

that they would feel less discomfort on leaving an air-conditioned building and energy would be saved.

3. In most cities the outdoor climate is pleasant some of the time and in some it is pleasant most of the time. If windows could be opened, the air conditioning might be replaced with natural ventilation during that time. This would also help during a partial or full cut in the supply of electricity. Cuts are already common in some cities in which supply has not kept pace with demand (Section 8.1). Open windows do, however, admit noise and dirt (Section 8.10), and this is a serious problem on the lower floors of buildings close to heavy traffic.

4. In some buildings artificial light is unnecessarily substituted for daylight, and in others the lighting level far exceeds the necessary for the task performed in them (Section 8.9). Thermal radiation from electric lights is a major source of heat in most office buildings, and this increases the cooling load in summer (Ref. 8.49).

We may have progressed too fast in our search for the perfect environment, and a reduction of energy used in some services would actually produce more agreeable conditions.

8.7 VERTICAL TRANSPORTATION

Mining is an ancient craft and elevators have been used for many centuries to hoist materials and miners to the surface. Elisha Graves Otis's invention was not the elevator (or lift in British terminology) as such, but an automatic safety device that prevented it from falling if the hoisting gear failed or the hoisting rope broke. In 1854 he demonstrated it publicly at the Crystal Palace Exhibition in New York. He had himself hoisted to the ceiling and then ordered the rope to be cut; the safety catch thus released engaged ratchets in the guide rails.

The safety elevator was immediately accepted for goods, but only one passenger elevator was installed in the next thirteen years in the five-story Houghwout Store on Broadway in 1857. It was driven by a steam engine, traveled at a speed of 40 ft/min (0.20 m/sec) and cost $300. Three new office buildings were fitted with elevators in 1868, the New York Life, the Park Bank, and the Equitable Life. By 1872 more than 2000 were in service. The cost of the best had risen to $15,000 but they had several additional safety devices and their cabins were elaborately furnished.

The invention of the passenger elevator changed the established values. In the palazzos of the Italian Renaissance the kitchen, stores, and some servants' rooms were located on a service floor at or slightly below ground level. The *piano nobile,* which contained the reception rooms, was on the floor above, that is, a little above ground level, the bedrooms of the important members of the family on the floor above that, and so on. The less distinguished the person, the more stairs he had to climb. This continued as the general rule until the elevator made it easy to get to the higher levels. Hotels were among the first to realize that the rooms on the upper floors were "the most desirable in the house, whence the guest makes a transit in half a minute of repose and quiet and, arriving there, enjoys a purity and coolness of atmosphere and an exemption from noise, dust and exhalations" (Ref. 8.30). Blocks

8.15

Elevator hoisting gear, built about 1876. The vertical steam engine is on the right and its crankshaft is above. The cable drum at the left operated another winding drum on the top floor which hoisted the elevator cage; that is, it worked somewhat like a vertical cog railway (Ref. 8.30, p. 12).

of apartments, office buildings, and department stores found them equally useful (Figs. 8.15 and 8.16).

The first hydraulic elevator was installed by Otis at 155 Broadway, New York, in 1878, the second in the New York Stock Exchange in the same year. The steam engine worked a pump which originally operated a ram that pushed the elevator up. In later models the ram operated a cable, and these were capable of speeds of 700 ft/min (3.55 m/sec). The first electric elevator was installed in the Demarest Building

8.16

Steam-operated passenger elevator installed in Lord and Taylor's department store in New York in 1870 (Ref. 8.30, p. 8).

New York in 1889. It was a modification of the steam-driven drum type, the electric motor replacing the steam engine.

As elevator speeds increased, operators found the task more difficult; 700 ft/min equals 8 mph, at that time the speed limit for cyclists in New York streets. This even today is considered a normal speed for an elevator in a twenty-story building. Emphasis therefore shifted to better control mechanisms. The Ward-Leonard multivoltage control was developed in 1892 and first used for the elevator in Edison's Pearl Street Power Station (Section 8.6). The first push-button control was installed in the Vanderbilt residence in 1893, and gearless traction was first employed in the Beaver Building in New York in 1903 (Ref. 8.31, p. 8).

In the 1880s the lead in the construction of tall buildings passed from New York to Chicago. The first passenger elevator was installed in Chicago in 1870 (Section 2.6), but after the fire of 1871 all major city buildings were rebuilt as "elevator buildings"; that is, buildings designed around their elevators. The height of buildings increased rapidly, and the term skyscraper was coined to denote structures more than ten stories high. The first of these, the Montauk Building erected in 1882, had loadbearing walls so thick that there was insufficient space in the basement for the boiler, the engine, and the pumps of the two hydraulic elevators. This machinery had to be placed in an outdoor court behind the elevator shaft. Thereafter iron and steel were used more and more to give the buildings a skeleton frame that took up less space (Section 2.6).

The structure has probably never limited the height of a building (Section 4.13), but the space occupied by the elevators can be critical for its economic feasibility because it is space that does not produce income. Elevator speeds are important because the faster the elevators, the fewer needed. At present (Ref. 8.54, p. 64) 1800 ft/min (9.1 m/sec) is considered a suitable speed for buildings of more than sixty stories. It is then necessary to control acceleration and deceleration carefully to prevent the feeling of "seasickness"; this had not been perfectly achieved when the Empire State Building was completed in 1931.

Another factor that reduces the space occupied by elevators is the system of control. If elevators respond quickly to the call of passengers, fewer are needed. Signal control was developed in 1924; it relieved the operator of all duties except that of pressing buttons and closing the car doors. The maximum speed was attained automatically, and the elevator was leveled automatically to line up with the landing sill. The elimination of decision making by the human operator approximately doubled the speed of operation. This type of elevator was installed in the Empire State Building.

In 1946 the computers developed during World War II were utilized in elevator design. The "autotronic" system operated the cars automatically as an electronically coordinated group, which adjusted itself to changes in traffic indicated by the calls made and to cars being taken out of service. In 1950 the first self-service elevator system, entirely without attendants, was installed in the Atlantic Refining Company Building in Dallas.

In the World Trade Center buildings, completed in New York in 1973, sky lobbies were introduced (Fig. 8.17). Each building was subdivided into three tiers, the lowest served by an elevator lobby at ground level, and the other two by elevator lobbies "in the sky." These sky-lobbies are connected to the ground-floor lobbies by nonstop

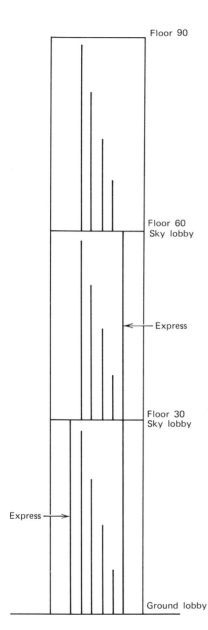

Floor 90

Floor 60
Sky lobby

Express

Floor 30
Sky lobby

Express

Ground lobby

8.17

Sky lobbies for a 90-story building divided into three tiers. The three sets of elevators use the same shafts which saves space. Each sky lobby is connected to the ground lobby by an express elevator. This system was used for the first time in the 110-story World Trade Center in New York.

elevators. Because three sets of elevators work in the same shafts, there is considerable saving of space, and, assuming that anybody wished to attempt it, an increase in height should be possible. However, the limit is set at present by horizontal rather than vertical transportation (Section 4.13).

As the foregoing account shows, the developments in elevator design have occurred mainly in the United States. There have, however, been some notable European lifts; for example those installed in the Eiffel Tower in 1888 carried passengers from the second to the third level, a distance of 150 m (493 ft) without stopping. There were two of them, tied together so that each acted as a counterweight to the other. Each elevator had two levels, a concept hailed during the last ten years as an important innovation, so that each lift could carry 100 people. They were operated by electric motors of 30 hp (22 kW) installed in each of the carriages (Ref. 8.32).

Moving platforms for pedestrians have been proposed since the midnineteenth century, but the first was actually constructed at the World's Columbian Exposition in Chicago in 1893 (Ref. 8.33, p. 33), designed by architect J. L. Silsbee and engineer Max E. Schmidt. In 1896 another moving platform was built at the Berlin Industrial Exhibition, and in 1900 the first moving stair, or *escalator,* was built by the Otis Company at the Paris Exposition. Since then escalators have become particularly popular in department stores where they take customers through intermediate floors past displays of merchandise on their way to their destinations.

8.8   INTERNAL COMMUNICATIONS

The earliest method of communication was the messenger who played such a prominent part in stage plays from Greek times to Shakespeare. Messengers are still used to convey files from one office to another.

Bells operated by wires were used for centuries to summon servants but were replaced by electric bells in the late nineteenth century. Before the installation of telephones the rooms in the better hotels had an array of bells for different services.

Speaking tubes used, for example, to transmit messages from the dining room to the kitchen consisted of ordinary tubes that confined and reflected the sound. Cartridges propelled along overhead wires by a spring were used in stores until the 1940s to take money to a central cashier and the need to give these wires a straight, level run imposed a restriction on the layout. The reason for this complex arrangement was probably the difficulty of checking on sales assistants before the development of recording cash registers.

Pneumatic tubes were used by businesses and post offices for delivering messages. Since the late nineteenth century many post office counters have been connected by tube to the telegraph and delivery rooms so that telegrams and express letters could be handled expeditiously. Some of these installations are still functioning. Compressed air from a rotary air blower is used at the central point to propel the carriers in the tubes and rarefied air then draws them back.

Most of these methods have, however, been superseded by the electric telephone, invented by Alexander Graham Bell in Salem, Mass., in 1875. The first telephone exchange was opened in New Haven in 1878, and the first automatic exchanges were built in the early twentieth century. By 1920 the clarity of speech had improved sufficiently to make the telephone a useful means of interoffice communication. It became simpler to pick up a telephone than to walk next door.

In the last twenty years the telephone has been supplemented by closed-circuit television for short-range communications. Banks and business firms have used it for checking signatures and for transmitting drawings and other documents. Hospitals have employed television monitors in intensive care units to enable a single nurse to observe at a central nursing station, for example, the heartbeats of a number of patients.

A teleprinter or visual display unit linked to a computer system has during the 1970s largely supplanted the telephone as a rapid method of transmitting data. This is the result of a conjunction of the electric telegraph, invented in the 1830s, the electronic digital computer, invented in the 1950s, and recent data processing methods.

It is possible that improved communications will result in a reversal of the old-established policies that financial houses should be within walking distance of the stock exchange, that business firms should be near the financial houses, and that architects and engineers should be close to their clients. If there were no longer a need for a central business district, because business could be conducted anywhere with the aid of the telephone and television, there would also be no longer any need for tall buildings near to one another to accommodate the offices traditionally associated with the city center. The implications would be far-reaching, indeed.

8.9 ILLUMINATION

Artificial lighting improved during the eighteenth century, partly because candle-making became mechanized and partly because the new spermwhale fisheries produced fuel to supplement wax and tallow (Ref. 1.1, Section 8.9).

In the nineteenth century the supply was further augmented by petroleum which had been used occasionally for lighting in antiquity. Some archaelogists think that lamps found in the Royal Tombs of Ur, dating from about 3000 B.C., used mineral oil, and Pliny described "Sicilian oil" which was of mineral origin and used directly in lamps. Most petroleum is too heavy, however, to be used directly.

The existence of petroleum has been reported from many parts of the world at various times. Marco Polo was particularly impressed by the oil springs at Baku, near the Caspian Sea. A French missionary first observed oil in North America in 1629, and in 1829 oil was struck in Kentucky by a group of people drilling for brine. A small quantity was used for medicinal purposes, but the rest went to waste, for it had no use.

In 1847 E. W. Binney, a Manchester chemist, distilled petroleum found at Alfreton in Derbyshire, and produced oil suitable for lighting. In 1851 manufacture of illuminating oil started in the United States, and after the discovery of the Pennsylvania oil fields in 1859 oil became plentiful. Oil lamps were improved by giving the wicks a spurred wheel adjustment and adding glass chimneys so that the flame was properly aerated. Oil lamps remained the normal form of lighting in remote districts until small electric generators became common.

Paraffin produced as a by-product from oil distillation provided a cheap and plentiful material for making candles, and their price dropped to a few cents per pound. The foot-candle became the first standard of illumination.

In the meantime, gas had become a serious competitor. A Mr. Spedding, who managed Lord Lonsdale's coal mines near Whitehaven on the edge of the English Lake District, had piped gas into his office in 1765, where he used it for illumination. The supply was so plentiful that he offered to light the streets of Whitehaven with it, but the magistrates refused permission (Ref. 8.55).

In 1805 Samuel Clegg installed gas lighting in the cotton mill of Henry Lodge in Halifax, Yorkshire, and several more factories were fitted with gas light in the next few years. The Gas Light and Coke Company was formed in London in 1810, but the early gas lighting was no better than oil lighting. The first burners were simply pipes with holes. Improvements were made by shaping the holes to spread the gas flame; for example, a few of these gas lights are still in existence in the Great Hall of the University of Sydney.

In 1855 the German chemist Robert Wilhelm Bunsen invented the very hot, smokeless burner still known by his name, and in 1885 Auer von Welsbach hung over a Bunsen burner a mantle of silk fabric impregnated with a mixture of 99% thorium oxide and 1% cerium oxide which produced an incandescent light far brighter than any then in use except the electric arc. It was the first type of artificial domestic lighting comparable in quality to that used today. The Welsbach mantle was shortly after also adapted to the oil lamp and greatly increased its brightness.

We noted in Section 8.6 that the carbon arc lamp had been used since 1844. It was, however, too bright a light and could not be subdivided. Thus arc lights have always remained restricted to streets, theaters, factories, and the *laterna magica,* the predecessor of the cinema and the modern slide projector. Arc lights remained in use until the 1930s for applications requiring a powerful, very bright light, when the new discharge lamps started to replace them.

We also noted in Section 8.6 that Swan and Edison invented the incandescent electric lamp in 1878 and 1879, respectively. Both lamps consisted of filaments made of carbonized threads placed in a glass tube from which the air was evacuated. The lamps were an immediate success, and although the demand for electric power had been insufficient to warrant the construction of a public power station in 1882 Edison built two of them, one in London and the other in New York (Section 8.6). Many more were built, and by 1900 it was estimated that in London alone 2½ million electric lamps were in use.

Complaints of headaches caused by gaslighting had been quite common before the invention of the Welsbach mantle, and Edision was not slow to point out the advantages of his invention. A poster distributed in 1882 stated:

This room is equipped with Edison Electric Light. Do not attempt to light with match. Simply turn key on wall by the door. The use of electricity for lighting is in no way harmful to health, nor does it affect the soundness of sleep (Ref. 8.34).

By 1890, however, Welsbach mantles had come into general use for gaslighting, which was again competing with electricity. In 1897 W. H. Nernst, one of Welsbach's assistants, made an electric lamp with a thorium oxide and cerium oxide filament, which became conducting when heated to incandescence by a heating filament inside. In 1905 Werner von Bolton, at the works of Siemens and Halske in Berlin, discovered the tantalum filament; this was replaced in 1908 by the tungsten filament, developed by D. Coolidge of the General Electric Company in Schenectady and still used today. The high melting point of tantalum and the even higher melting point of tungsten enabled the lamp to be run at a higher temperature that increased its efficiency. At these high temperatures, however, the filament evaporated in a vacuum and blackened the glass. In 1913 Irving Langmuir, also of the General Electric Company, suggested replacing the vacuum with the inert gases nitrogen or argon. This solved one problem but created another through the loss of heat by convection in the gas. Langmuir then coiled the filament around a central wire which was later dissolved in acid, leaving a coiled tungsten filament from which the loss by convection was much less. In the same year he suggested the coiled coil, in which the coiled wire was again coiled on itself, but suitable manufacturing techniques were not developed until 1934 (Ref. 8.55, p. 15). There have been no further major innovations in the design of incandescent electric lamps.

Since the 1940s fluorescent tubes have been used more and more for lighting in buildings. It had been observed since the early eighteenth century that a glow could be produced by an electric discharge through a rarefied gas in a glass tube. From time to time demonstrations of this phenomenon were given before meetings of scientific societies, particularly in London and Paris. The color of the light depends on the gas used: neon produces a red glow, mercury vapor gives a greenish light, sodium vapor a bright yellow light, and carbon dioxide, a light resembling daylight. The earliest practical lamps were made with carbon dioxide by Daniel McFarlan Moore in the United States in 1895. Peter Cooper-Hewitt, also in the United States, marketed a mercury lamp in 1901. In 1910 Georges Claude in France used neon and created the neon sign industry. In the 1930s sodium vapor lamps and high-pressure mercury vapor lamps were first employed for street lighting; they displaced the electric arc lights previously used where high brightness was required.

Some discharge tubes produced considerable quantities of ultraviolet radiation which was invisible as such but could be turned into visible light by coating the tube with a fluorescent powder. The visible light produced by these lamps is mainly due to the excitation of the fluorescent coating by ultraviolet radiation from the mercury, but the mercury vapor itself makes a small contribution.

Fluorescent tubes containing mercury vapor were first made commercially in the United States in 1938 and were used to light most of the New York World's Fair in the same year. After this success fluorescent tubes were immediately accepted for lighting American offices and factories because their consumption of electricity for the same level of illumination was only about a quarter of that of incandescent lamps; this more than compensated for the higher capital cost. They were less suitable for homes because the tubes take a little time to light up and most of the early ones did so with a disconcerting flicker. The fluorescent lamp is still operated in conjunction with control gear which heats the tungsten filaments, provides a sufficient voltage to start discharge between them, and limits the current through the lamp.

During the 1950s fluorescent tubes largely displaced other forms of lighting in office buildings and factories throughout the world, and the length of the standard fluorescent tube has become in many buildings the module for the design of the ceiling; this often means that it is also the module for the plan of the entire building (see Section 10.6).

Artificial lighting progressed rapidly during the twentieth century because the electrical industry evolved into a multimillion dollar business that was willing moreover to spend a significant proportion of its profits on research. Daylight is free, and the only sponsorship for daylight research has come from public bodies and glass manufacturers.

A great deal of thought had been given in the past to the effect of daylight on the appearance of buildings. The splendid interaction of light and shade in many Greek temples is probably not accidental. Stained glass was one of the important elements in the architectural quality of the Gothic cathedrals, and in Baroque architecture daylight was often employed with striking effect (Ref. 1.1, Section 7.10).

Although the quality of the daylight was evidently studied with care, there is no evidence that much consideration was given to quantity. This may not have been a serious problem before the nineteenth century because most buildings except for palaces and churches were small. During the nineteenth century, however, new types of building were erected, and in factories, offices, and department stores the quantity of light could have been critical before artificial light was perfected. Many were built with interior courts to admit additional daylight, but these were frequently reduced to the bare minimum required. Sometimes they were given a glass roof to allow the light in but keep the rain out. Toward the end of the century light wells frequently became open parts of the building under a glass roof.

Light wells have gradually disappeared with improvements in artificial lighting. This has not been an unmixed blessing. As rooms became deeper, daylight became insufficient to light the back of the room and artificial light was installed for this purpose. As artificial lighting improved and its cost decreased, the entire room was often lighted by it. Sometimes the curtains or blinds were kept drawn all day to avoid the glare from sunlight, although this could have been prevented by shades that did not exclude daylight. The result was the creation of an artificial luminous environment that consumed a great deal of electricity and in the process produced heat that had to be removed by additional air conditioning. A solution suggested for this problem (Ref. 8.38, pp. 458–484) is permanent supplementary artificial lighting at the back of the room, away from the windows and operated by a separate switch, which is in use all day long, plus artificial lighting for the rest of the room which is used only when the level of daylight near the windows ceases to be adequate. Provision must also be made for night-time lighting distributed evenly over the whole of the interior.

Actual measurements of daylight do not appear to have been made before the twentieth century. In the United States daylight studies have since 1916 been based on the quantity of light (Ref. 8.35). This gives the information in a form that can be related to studies of artificial light, but there is no universal answer, for natural light varies from day to day, throughout the day, and even momentarily due to clouds. In England daylight studies have been based on the daylight factor (Ref. 8.38, pp. 68–73), which is defined as the ratio of the illumination* at any point in a room to the illumination at the same instant on a horizontal plane exposed to the unobstructed sky. Its principal component depends on the geometry of the building and

* The technical term used by lighting engineers is *illuminance*.

8.18

Artificial sky for daylight studies. A model of the building or room is placed on a table at the level of the horizon plane and interior lighting levels are measured with photoelectric cells. The smallest size cell determines the size of the model, hence the diameter of the sky, generally 6 to 7 m (20 to 23 ft). Only the indirect lighting, not the artificial sun, is shown in this illustration (Architectural Science Laboratory, University of Sydney).

any surroundings obstructing the daylight, and this can be calculated. The actual quantity of daylight at any point in a room can be deduced from it if the illumination outside is known or can be taken from a standard table.

This difference of approach is at least partly due to differences between English and American law. English law has recognized "ancient lights" since medieval times. If a window has been used without interruption for a long time (defined in 1832 as twenty years), the owner of the building acquires the right to prevent the owner of adjoining land from obstructing the light received through this opening. In 1922 a judge adopted a daylight factor of 0.2 as the border line between adequacy and inadequacy of daylight in a room. Determination of the daylight factor is therefore a necessary exercise in many design problems in Britain. American courts rejected the doctrine of ancient lights at an early date, arguing that there was sufficient land available so that the first person to build did not have to locate his building on the boundary line.

Design by either method can be facilitated by tables and charts (Refs. 8.36 and 37). For more complex problems, however, model studies have been used since the 1930s (Ref. 8.38, pp. 377–394). These can be done in the open air under the real sky or under controlled conditions and an artificial sky. The most common type of artificial sky is a hemisphere (Fig. 8.18) lighted in accordance with a standard. In 1942 the *Commission Internationale de l'Éclairage (CIE)* adopted a proposal by Parry Moon and Domina Eberle Spencer (Ref. 8.39) for the overcast condition in the temperate zone whereby the luminance at the horizon is one-third of the luminance at the zenith and the luminance at intermediate points is interpolated by a cosine formula. This is still the international standard for comparing data. In 1968 the Indian Standards Institution (Ref. 8.40) adopted another formula for interpolating the luminance for a clear tropical sky, which has also been used in other tropical and subtropical countries. For a clear sky direct sunlight must be considered in addition to the light reflected by the sky, and an artificial sun is needed in addition to the illuminated hemisphere.

There is a comfort problem in relation to the luminous environment just as there is a comfort problem in relation to the thermal environment, although discomfort in relation to lighting is easier to correct. Inadequate light was a major problem before the twentieth century, and occasional difficulties are still due to bad design (Refs. 8.41 and 8.42). The problem of glare remains even when there is enough light. Glare can be caused by daylight or by artificial light; it does not depend on the absolute luminance of the light source as much as on the contrast in luminance within the field of view (Ref. 8.42, pp. 80–104).

8.10   ACOUSTICS AND NOISE

Greek and Roman theaters had steeply raked seats. The seats of the Teatro Olimpico were designed in 1580 in accordance with Vitruvius' rules for auditoria (Ref. 1.1, Sections 4.4 and 7.10). During the next century the seating gradually changed, and in Burnacini's Imperial Theater in Vienna, opened in 1690, some of the seats were on a gently sloping floor and others were arranged in multistory horseshoe galleries. This remained the pattern in the opera houses of the eighteenth and nineteenth centuries: the Teatro alla Scala in Milan (1778), the Royal Opera House at Covent Garden in London (1858), the Staatsoper in Vienna (1869), the Théâtre National de l'Opéra in Paris (1875), and the Metropolitan Opera House in New York (1883).

During the eighteenth century concerts were performed in the ballrooms of the aristocracy; these provided the pattern for the assembly rooms in which the upper middle classes gathered for social functions.

In the midnineteenth century the new centers of population created by the Industrial Revolution erected public buildings befitting their new status. Thus St. George's Hall was built in Liverpool as a concert hall in 1850–1854, and the Free Trade Hall in

Manchester followed in 1853–1854. Many of the new town halls included a large room which could be used as an auditorium, some of which were capable of holding more than 3000 people.

Halls were also built specifically for concerts in the older centers of population. Europe, the United States, and Australia acquired a large number between 1850 and 1914, where there had been few before. Most were enlarged versions of the assembly rooms: a rectangular room with a gallery along three sides. They include some of the best concert halls in existence: the Grosser Musikvereinsaal in Vienna (1870), St. Andrew's Hall in Glasgow (1877), the Concertgebouw in Amsterdam (1888), and the Grosser Tonhallesaal in Zurich (1895). The excellence of so many of the surviving nineteenth-century concert halls is to a large extent due to laws of natural selection: most unsatisfactory halls were demolished long ago unless they formed an integral part of an important building like a town hall.

The music of past ages was composed to fit the place in which it was to be performed, rather than vice versa (Ref. 1.1, Sections 5.9 and 8.9); but in the nineteenth century a composer might have his music performed in a number of different concert halls or theaters which, even if they had the same general shape, had different reverberation times and reflective surfaces in different locations.

In addition, a concert program in the late nineteenth century might contain music that required widely different concert-hall acoustics, as for example a Mozart piano concerto and a Brahms symphony. This had not been a problem in the eighteenth century when all the music likely to be performed at a concert required the same acoustic qualities.

The nineteenth century was a great age for music. The existing instruments were perfected and some new ones were invented (Ref. 8.56). Music played an important part in the entertainment of the middle classes, and playing the piano had by the turn of the century become as much part of a young lady's education as reading and writing. Famous orchestral music could be performed in the home as a *Klavierauszug*, a piano transcription, which filled roughly the same function as the modern phonograph record.

The architects of the late nineteenth-century concert halls were therefore dealing with an informed public. A great deal was written on architectural acoustics, but it was mostly guesswork. Charles Garnier found it of little help in the design of one of the most prestigious and successful opera houses. Leo Beranek (Ref. 8.57, p. 237) said that the Paris Opéra has "very good acoustics—not the finest, but very good". Garnier wrote a book on its design in which he stated:

The credit is not mine. I merely wear the marks of honour . . . . It is not my fault that acoustics and I can never come to an understanding. I gave myself great pains to master this bizarre science, but after 15 years of labour, I found myself hardly in advance of where I stood the first day . . . . I had read diligently in my books, and conferred industriously with philosophers— nowhere did I find a positive rule of action to guide me; on the contrary, nothing but contradictory statement. For long months, I studied, questioned everything, but after this travail, finally I made this discovery. A room to have good acoustics must either be long or broad, high or low, of wood or stone, round or square, and so forth (Ref. 8.57, p. 237).

There is no evidence that acoustics played a significant part in the design of any building in the eighteenth or nineteenth century, with the exception of Boston

Symphony Hall, completed in 1900 and discussed below, and the Festspielhaus in Bayreuth, completed in 1876. The architect of the Festpielhaus was Otto Brückwald, but Richard Wagner laid down the guidelines. It seems that, like many designers subsequently, he modeled it on an existing building. Wagner had conducted in a theater in Riga which he found very satisfactory. Thus the Festspielhaus, like the Riga theater, has steeply banked seats and a sunken orchestra. The problem was a relatively simple one because the Festpielhaus has only 1800 seats and it was not intended to be used for the performance of any music except that of Richard Wagner (Ref. 8.57, pp. 243–250; Ref. 8.43, pp. 616–620).

The science of *physical* acoustics made a slow start in the eighteenth century, but during the nineteenth a number of eminent physicists worked out the basic theory. Lord Rayleigh's *The Theory of Sound,* published in Cambridge in two volumes in 1877 and 1878, is still considered a standard work.

The foundation of the science of *architectural* acoustics was laid by Wallace Clement Sabine (Ref. 8.58), who was Professor of Mathematics and Natural Philosophy at Harvard University until his death in 1919. His association with the subject began in 1895, shortly after he had been appointed an assistant professor, when the president of Harvard asked him to "do something" about the lecture room in the newly completed Fogg Museum of Art. This led to an invitation in 1898 to give acoustical advice on the design of the new Boston Music Hall, completed in 1900 and now known as Symphony Hall. The Hall was immediately recognized as a masterpiece. In the 1950s Beranek asked twenty-three of the world's most famous conductors to rate fifty-four of the world's best-known auditoria in order of preference. With one exception, all rated Boston Symphony Hall as America's best and one of the three best in the world.

Thereafter Sabine's views commanded respect. He changed architectural acoustics from a qualitative to a quantitative subject. Acoustic design had consisted mainly of a determination of sight lines and first reflections. Sabine demonstrated the effect of reflecting surfaces and gave a formula, which still bears his name, for predicting reverberation time.

Relatively few concert halls and opera houses were built between 1914 and 1945 because many cities had already acquired satisfactory auditoria in the late nineteenth century. Beranek (Ref. 8.57) included only nine auditoria built during that period on his list of fifty-four and rated only two of them as outstanding, namely, the San Francisco Opera House, completed in 1932 and subsequently the meeting place of the first session of the United Nations, and the Konserthus in Gothenburg, opened in 1935.

The 1920s and 30s were, however, a great age for cinema building. The best provided seats more comfortable than those customary in live theaters and concert halls built during the same period, perfect sight lines, and excellent acoustics. The acoustic problem was much simpler for the sound was in any case recorded and could be adjusted electronically both during recording and reproduction. Moreover, the designer could place his loudspeakers so that everybody could hear well and then adjust auditorium characteristics so that there would be no significant echoes.

The radio and the phonograph also developed during this period, and architects began to install sound-reinforcing systems in conference rooms and lecture halls. Their morality was at first questioned, for the use of microphones and loudspeakers might interfere with the quality of a parliamentary debate or a public lecture. The main problem in the 1930s, however, was the imperfection of the equipment, and an experienced speaker could often make himself heard better without it.

Electronics made great advances during the 1939–1945 war, and sound amplification has been used increasingly since then. Young ladies now collected phonograph records instead of playing the piano and vocalists with popular bands used microphones.

Electronic corrections were even made in concert halls. The Royal Festival Hall in London was completed in 1951; Sir Robert Matthew was the architect and Hope Bagenal, the principal acoustic consultant. Conductors and critics disagreed sharply about the quality of the hall. Herbert von Karajan called it "the best of the modern halls"; Leopold Stokowski said that the sound was "metallic and tinny; there is no warmth" (Ref. 8.57, p. 330). There was general agreement that the hall was very good for hearing speech and chamber music and for orchestras of the size popular in Mozart's time, poor for the music of Wagner and the late Romantic composers.

In the 1960s it was decided that the auditorium, although excellent for some music, did not meet the requirements of a general concert hall, and its resonance was in 1964 "assisted" by installing eighty-nine microphone-amplifier-loudspeaker channels so positioned in space and adjusted in amplitude that each one prolonged the reverberation time at one frequency in the range of 70 to 340 hertz (Ref. 8.44).

Many concert halls have been built since 1945. Some of the best auditoria in central Europe were destroyed during World War II, and others no longer met current standards of fire safety. In America and Australia new orchestras were formed which required a home.

Acoustic consultants now had not merely a theory that could be used for quantitative design but also precise electronic equipment for acoustic tests which had not existed in Sabine's time. Some of the new halls have been praised for their excellence, but there are still some notable failures to show that the design of a concert hall or opera house, as distinct from a lecture room or conference hall, is still not an exact science. In 1957 Beranek, one of the most experienced acoustic engineers, was appointed consultant for the new Philharmonic Hall at Lincoln Center for the Performing Arts in New York. He has described the preliminary studies and the design (Ref. 8.57, pp. 511–540). When the hall was completed in 1962, a week was devoted to "tuning" it. The New York Philharmonic Orchestra played to an audience simulated with fiberglass mats, and the result was judged with instruments and by a panel of experts. Adjustments were then made, mainly by adding or removing absorptive or reflecting material, and the tests were repeated. In spite of the theory and experimentation used in the design of this hall, it was not a success and had to be extensively altered (Ref. 8.50).

The other aspect of acoustics is noise control. Contemporary writers in Rome in the days of the Empire and in London from the seventeenth century to the present time mentioned noise in the streets, and from time to time laws were passed to restrict certain noise-producing activity. Traffic noise is possibly no louder today than in Rome in the second century A.D., in Paris in the sixteenth century, or in London in the nineteenth. The nature of the noise made by automobiles and aircraft differs, however, from that made by horses and carts on stone pavements, and by street criers.

The problem has been aggravated by the current use of lightweight structures and materials and by the changing pattern of traffic noise in residential areas. Insulation against airborne noise depends mainly on the mass of the materials, and traditional construction therefore provided good insulation if the windows were kept closed. Interior airborne noise in earlier times was mainly due to talking and shouting, which is still a reason for complaints about noisy parties. In the last few decades the invention of radio, television, record players, tape recorders, and household "labor-saving" appliances has added a new source of domestic noise.

Impact noise caused by people walking about on the floor above is perhaps causing less trouble than previously; it can be stopped by floating the floor or by a thick carpet, and the necessary materials are now more readily obtainable at a reasonable cost.

Interior noise in offices is caused by people talking and by telephones, typewriters, and other office machines. In addition to the problem of excess noise, however, there is the possibility that people may be disturbed by involuntarily listening to other conversations when the room is too quiet. A more serious problem is that a confidential conversation may be overheard; for example, what is being said on the psychiatrist's couch may be heard by people in the waiting room. The terms "acoustical deodorant" or "noise perfume" have been used to describe masking noises. The ventilating or air-conditioning plant often serves this purpose. The ducts can act like an old-fashioned speaking tube in transmitting fan and regenerated noise from the plant room, but this is absorbed by attenuators or by lining the duct with a sound-absorbent material; yet some sound is always heard. Alternatively a background masking noise or music can be broadcast over a loudspeaker system, a method adopted in some offices without walls or partitions.

This is an example in which a noise problem can be solved by the use of electricity, but control of excess noise can be achieved only by the judicious use of sound-absorbing or sound-insulating materials.

Sabine built the first testing room about 1910 (Ref. 8.58, p. 245), but progress was slow until the Acoustical Materials Association was founded in the United States in 1933. In 1934 it started issuing annual bulletins of sound-absorption coefficients (Ref. 8.47, p. 1). The mass law, which provides a linear relation between the sound insulation in decibels and the logarithm of the weight of a partition per unit area, was proved after the construction of the first accurate acoustic transmission chamber in Geneva, Illinois, now the Riverbank Acoustical Laboratories of the Illinois Institute of Technology. Since then a vast amount of data has been accumulated in the testing laboratories of many countries on the sound insulation of various materials and forms of construction and are published from time to time in the form of a reference book (e.g., Ref. 8.45) or in proprietary literature.

Like air conditioning, noise control usually depends on keeping the windows closed; thus natural ventilation is practicable only in areas of low traffic noise or sometimes on the upper floors of multistory buildings (Section 8.6). Sound insulation can also be used to control interior noise originating in another room, but this is effective only if the noise cannot bypass the insulation. It is almost impossible to stop the noise from a neighbor's television set by insulating floors or walls if both parties keep their windows wide open.

Noise originating in the same room is attenuated by absorbing material, such as carpet, curtains, or acoustically absorbing surfaces, particularly if they are placed near the source of the sound; for example, a noisy machine. In 1941 three Dutch acousticians first related the absorptive capacity to the structure, thickness, and porosity of the material (Ref. 8.48).

Very loud noise poses special problems. The difficulty of hearing speech in excessively noisy areas was not seriously considered before World War II, when S. S. Stevens at the Harvard psychoacoustic laboratory investigated the psychological aspects of noise control and speech communication posed by tanks and military aircraft. This work has since been extended to factories. In very noisy areas that cannot be isolated people must be provided with personal ear protection.

It is possible today to build a conference room over-looking a main road or a hotel near an airport runway with quite satisfactory sound insulation. Because noise control has only a brief history this is a significant achievement. Soundproof construction is expensive and because the building envelope must be sealed air conditioning becomes essential; however, it is not at present an economical solution for private houses near airports.

The acoustical environment therefore poses a more difficult problem than the thermal or luminous environment. By means of electricity or some other form of energy we can heat or cool a building, raise or lower its humidity, and light it. We can use loudspeakers for masking sounds and improving audibility in lecture and conference rooms and night clubs, but electrical solutions are not at present fully acceptable for concert halls and do not exist as a means of noise control.

In a squatter settlement near an airport in Asia or South America people are unlikely to complain about aircraft noise: they have other environmental priorities. In developed countries higher environmental standards are expected. The noise from automobiles, aircraft, and electrical equipment is at present increasing (Ref. 8.46), and the strength of popular feeling is indicated by the use of emotive words like "noise pollution" and "Lärmbekämpfung." Many countries have legislated since 1950 to control noise and to require minimum standards of sound insulation in buildings, but these regulations are not easily enforced because of the cost involved.

CHAPTER NINE

Building Materials

I said it in Hebrew, I said it in Dutch,
I said it in German and Greek;
But I wholly forgot (and it vexes me much)
That English is what you speak.

LEWIS CARROLL.

This chapter considers some of the problems associated with building materials, principally natural stone, brick, tile, porcelain enamel, glass, lightweight materials made specifically for thermal insulation, precast concrete, galvanized iron, weathering steel, stainless steel, copper, lead, aluminum, thermoplastics, thermosetting resins, timber, plywood, and hardboard. The raw materials from which these materials are made are mostly plentiful, but the energy consumed in processing them could become a problem. Durability and the cost of maintenance must be considered in choosing a material. The choice can be made easier by tests of the properties of the materials.

9.1 NATURAL STONE

Natural stone was the most important of the durable structural materials from the Middle Ages to the nineteenth century and the form of the structure was to a large extent determined by the fact that it had high compressive but low tensile strength (Ref. 1.1, Chapters 3, 6, and 7). After iron was produced in sufficiently large quantities in the eighteenth century, cast iron, steel, and reinforced concrete, in turn, replaced natural stone for long spans. This was a gradual process, however, and two of the world's eight largest masonry domes were built in the midnineteenth century (see Ref. 1.1, Section 7.2).

As a building material, natural stone retained its importance throughout the nineteenth and early twentieth centuries. Although wages gradually rose, they were countered by more efficient methods of quarrying and of dressing the stone. Metallurgy and toolmaking made great advances particularly in methods of hardening tool steel. About the middle of the nineteenth century the diamond drill was invented, and steam engines were used to operate drills and saws.

In 1867 Alfred Nobel produced the first safe high explosive, dynamite, by combining unstable nitroglycerin with the absorbent inert mineral kieselguhr. Dynamite was simpler and more effective for use in quarrying and excavation than the traditional gunpowder.

The nineteenth century was not a great age for the art of architecture in the sense that it produced few original ideas. It was, however, a great age for architectural construction, and in virtually every city in the world the natural-stone buildings constructed since 1800 greatly outnumber those erected before that time. The quality of nineteenth-century stone construction is mostly excellent because the masons still had the old skills, better tools, and better mortar. Thus an English industrial city like Sheffield, which has few buildings of consequence predating 1800, acquired many handsome private houses and some good public buildings of natural stone. Hobart, the capital of Tasmania, founded in 1804, has public buildings, private houses, and warehouses (Fig. 9.1) of sandstone that would merit attention in a much older city.

After 1918 new solid stone buildings became rarer and their decoration less elaborate; after 1945 they disappeared almost completely. Repairs on old stone buildings continue to keep some stonemasons occupied. Their wages are much higher than those paid to masons before 1900 and more money may have been spent on the restoration of some medieval cathedrals in the twentieth century than they cost to build originally.

9.1

Sandstone warehouses and offices built in the midnineteenth century opposite the docks in Hobart, Tasmania.

In the last twenty years stone has made a comeback in curtain walls. Durable stones, such as marble and travertine, cut thin and highly polished, cost more but otherwise compare favorably with most curtain-wall materials in regard to appearance, thermal insulation, maintenance, and durability. An unusual example is the Rare Book Library at Yale University (Fig. 9.2).

Vitruvius recommended that stone be selected from a known quarry or quarried a long time before use and exposed to the weather. If the stone survived without damage, it was satisfactory (Ref. 1.1, Section 4.1). This was still the basic method used in the early nineteenth century. Thereafter the new sciences of geology and chemistry made it possible to predict the deterioration of stone in many cases from its chemical composition and stratification (Ref. 9.2).

9.2   CERAMICS

Both natural stone and well-burnt bricks were freely available in ancient Rome, and brick was substituted extensively for stone; presumably it was the cheaper material.

9.2

Beinecke Rare Book Library at Yale University, which has walls with "windows" formed by thin slabs of translucent marble. The rare books and manuscripts within are preserved in a central glass-walled stack, separately air-conditioned. The architect in charge of the design was Gordon Bunshaft, of Skidmore, Owings and Merrill (Ref. 9.1, p. 207). This curtain wall is, in a sense, a revival of an ancient technique used before glass was perfected (Section 9.3). San Miniato al Monte in Florence is a surviving late (eleventh century) example of the use of translucent marble windows (Ref. 9.27, p. 473).

During the Middle Ages the use of brick in Europe* was mainly confined to regions, such as northern Germany, the Netherlands, and eastern England, where natural stone was not locally obtainable. From the fifteenth century on brick manufacture improved and was used increasingly in London which had in the past imported natural stone for important buildings. Brick became gradually less costly than natural stone even in regions with a plentiful supply of good natural stone. In eighteenth-century London brick was frequently plastered and painted to look like Bath stone (Ref. 1.1, Sections 3.7, 5.7, and 8.8).

*During the same period some of the most artistic and structurally intricate brickwork was built by the Selijuks in Turkey, Iran, and Russian central Asia (see Ref. 1.1, Fig. 5.4a).

During the nineteenth century brickmaking became mechanized, and because of the great increase in population after the Industrial Revolution many new houses were needed. In England they were built of brick which was the cheapest form of fireproof construction. The surviving nineteenth-century brick terraces built for the working classes in cities like Manchester are one of the depressing heritages of the Victorian era (see also Section 7.5), but the fact that these houses still exist is a tribute to the excellence of the brickwork.

Toward the end of the nineteenth century brick became increasingly appropriate for the houses of the wealthy. Face bricks carefully selected for their color and texture were accepted as equal, perhaps even superior, to natural stone.

In England and Australia face brick now exists side by side with brick plastered and painted (Fig. 9.3). In Germany, however, plastered brick is almost exclusively found in the south (e.g., Frankfurt and Munich) where natural stone was used in medieval times; face brick is used mainly in the north (e.g., Hamburg) and in the Netherlands, which was already brick country in the Middle Ages. In one case a plasterer is employed to cover the roughly finished brickwork; in the other the bricks are carefully laid with regular joints. The necessary skilled labor is not readily available outside its traditional region.

From the technical point of view bricks have fewer problems of durability than natural stone. As they are manufactured, their performance is more predictable than that of stone taken from a quarry. If they are well-burned, they are unlikely to disintegrate due to climatic influences.

It is not known who invented cavity walls. S. B. Hamilton, who studied the subject (Ref. 9.3), found no references to them before 1821 when Thomas Dearn mentioned them in *Hints on An Improved Method of Building*, published in London. The wall described by Dearn was 11 in. (280 mm) thick, made up of two leaves 4½ in. (114 mm) thick, separated by a 2 in. (50-mm) air space, and tied together with iron ties. Most cavity walls are still built with these dimensions. Cavity walls provide thermal insulation superior to a solid wall that uses the same amount of material (Section 9.4), and any rain passing through the outer leaf is drained off from the cavity. Cavity walls had become popular in Britain by the midnineteenth century, and British architects exported them to North America, Australia, and Africa. One notable Canadian structure with cavity walls is the Parliament Building in Ottawa, completed in 1867 and designed by English-trained architects (Ref. 9.4).

Cavity walls have no disadvantages compared with solid brick walls in temperate and warm climates, but in countries with cold winters ice may form inside the cavity after the condensation of water vapor when the temperature drops at night. This is probably the reason why cavity walls, after a period of use in the midnineteenth century, are now rare in North America and why they have been little used on the European Continent.

Dampproof courses to stop moisture from the ground rising in porous natural-stone or brick walls have occasionally been used in important buildings in the past, the most common form being a layer of hard impervious stone. Toward the end of the nineteenth century dampproof courses were used even in low-cost houses. There were three principal types (Ref. 9.5, p. 214): a course of impervious stone, slate, or glazed tile laid without mortar in the vertical joints to prevent moisture from rising through them, a sheet of lead, or a $\frac{3}{8}$-in. (10-mm) layer of asphalt.

9.3

Plastered and painted brick house in Woollahra, an inner suburb of Sydney, Australia, built during the late 19th century.

Hollow ceramic blocks were also in demand at that time, particularly after presses were invented to mechanize the process. They had already been used by the Romans and were revived in the eighteenth century in Paris (see Ref. 1.1, Section 8.6). Sir John Soane used them in 1793 in the new building for the Bank of England, William Strutt used them in 1803 in a mill at Belper, and John Nash used them in 1824 in Buckingham Palace. Their main application at first was in fireproof floors and in fireproofing for steel columns (Section 7.2); later they were employed in interior walls. Hollow blocks consist largely of air spaces so that they are light but provide good thermal insulation (Section 9.4). This type of construction has been particularly popular in North America and Continental Europe.

Reinforced brickwork is actually older than reinforced concrete. It was first used in the circular shafts for the Thames Tunnel in 1825 by Sir Marc Isambard Brunel (Ref. 9.6) and has been used from time to time since then. In the 1933 earthquake in Long Beach, California, reinforced masonry showed appreciable extra resilience, and by 1940 most building codes in California required that all masonry construction be reinforced (Ref. 9.8, p. 1058). Reinforcement has also been found especially useful in tall masonry buildings.

We noted in Section 2.6 that walls became thicker as buildings grew in height and that this was the principal reason for the replacement of loadbearing walls with skeleton frames. The tallest building erected in Chicago in the 1890s with loadbearing walls was the sixteen-story Monadnock Building (215 ft or 65 m high) whose walls were 72 in. (1.83 m) thick at the base. In retrospect this was unnecessary. Rules for the thickness of tall masonry structures were partly based on extrapolation from the thickness used in lower buildings and partly on precedent. Most surviving tall masonry buildings have thick walls at the base; for example, La Garisenda Tower (Ref. 1.1, Section 5.5) has a base thickness of 7 ft 9 in. or 2.35 m.

Until the 1930s the design of rigid frames generally ignored the effect of brick infills in frames. We now consider that these walls act as shear panels by resisting the distortion of the frame in their own plane. Evidently this still holds true when there is no frame and only a loadbearing brick wall. In the 1950s a new method of designing loadbearing brick walls came into use (Ref. 9.7, pp. 314–326) by which the wind loads were resisted not only by the windward and leeward walls but also by the cross walls at right angles to them. Thus the length of the wall rather than its thickness becomes the critical depth for wind load. By this method wall thicknesses have been greatly reduced. At the time of writing the tallest masonry buildings without skeleton frames are the twenty-one story twin Liberty Park East Towers in Pittsburgh (Ref. 9.9, p. 569); however, their walls of brick and concrete masonry are only 15 in. (0.38 m) thick, which is 21% of those of the sixteen-story Monadnock Building.

Calcium silicate bricks, developed during the early twentieth century, are made of sand and lime and have a uniform white color which is attractive for face bricks. Their chemical composition and some of their physical properties differ from those of clay bricks.

Concrete masonry was already employed in the midnineteenth century (Section 9.5), often under the name *artificial stone*; both solid bricks and hollow bricks were in use. In the twentieth century hollow concrete blocks became particularly popular in the United States. The principles outlined above apply equally to loadbearing walls in concrete and calcium-silicate masonry.

Decorated clay products have an ancient history. Glazed *bricks* with superb colors were used in the Ishtar Gate in Babylon in the seventh century B.C.; it is now in the Pergamon Museum in East Berlin. Glazed *tiles* reached their peak of perfection in Muslim architecture (e.g., Fig. 8.5, and Ref. 1.1: Figs. 5.4 and 5.5). Terracotta, that is, decorated slabs of clay burned to a surface harder than brick, were used in ancient Greece. Encaustic tiles, which are hard-burned shapes made of varicolored clays to produce a pattern, were used in Gothic churches. All these materials were revived during the period of eclectic architecture in the nineteenth century and from a technical point of view were equal to any made in earlier times. In industrial cities the glazed surfaces and hard-burned terracotta showed a resistance to pollution from soot that most natural stones lacked, but opinions differ on the aesthetic merit of Victorian decorated ceramics.

Victorian bathrooms were often carpeted, and the use of glazed tiles established itself only gradually. Tiled bathrooms are essentially a twentieth-century development.

9.3 GLASS

Porcelain enamel is a glasslike nonporous material fused to iron castings and steel sheet. In many respects it resembles the glazing applied to ceramic tiles, and the surface is extremely durable and easily cleaned. The material has long been used as a finish for cast-iron bathtubs.

A technique developed for external use in building construction, by which alumina-borosilicate glass was fused to steel sheet at a temperature of 1400 to 1600° F (760–876° C), was first employed in the early 1930s. Since the 1950s it has been used extensively in conjunction with glass curtain walls (Section 8.1). Many different colors are obtainable but not all are permanent.

The cylinder glass and crown glass processes (Ref. 1.1, Sections 5.8 and 7.8) were still in use in the nineteenth century. In 1832 Robert Lucas Chance introduced from France a modification of the cylinder process. The cylinders were blown larger and allowed to cool. Instead of being cut hot with iron shears, they were cut cold with a diamond cutter, reheated, and flattened out on a bed of smooth glass instead of an iron plate covered with sand. The glass thus had a better surface, and larger sheets could be produced. Chance's Birmingham works supplied 956,195 ft² (88,800 m²) of glass for the Crystal Palace (Fig. 10.3) in 1851. The sheets measured 49 × 10 in. and were $\frac{1}{13}$ in. thick (1.24 × 0.25 m × 2 mm).

Processes for making sheet glass by drawing it directly from a pool of molten glass were patented by the Belgian Emile Fourcault in 1904 and by the American Libbey-Owens Glass Company in 1905. An improved process was patented by the Pittsburgh Plate Glass Company, also in the United States, in 1927.

Some distortion and lack of flatness must be accepted in sheet glass. Plate glass, defined as a perfectly flat material, has been made by casting and rolling since the seventeenth century and was originally used for mirrors. The glass was polished first with sand and then with rouge and water. In 1890 E. F. Chance introduced figured plate glass, that is, glass with a patterned surface, made by using indented rollers. In 1925 Pilkington Brothers at St. Helens, England, developed a continuous grinding and polishing machine for plate glass and in 1952 invented the float glass process by which the glass was floated on molten metal and both sides were fire-finished, thus avoiding the need for polishing and grinding.

In medieval times only small pieces of glass were available and these were assembled into windows with strips of lead, called cames. By the eighteenth century glass sheets had become large enough to be fitted directly into wooden frames with putty made of powdered chalk and linseed oil. This method was used in the Crystal Palace. This putty becomes brittle and falls out after some years, and for large buildings "patent" glazing was introduced in the 1870s to avoid the need for renewal. In patent glazing the window is not necessarily sealed; the glass is held in position by clips and the water is drained off in grooves. Neither method is suitable for the curtain walls of air-conditioned buildings, which must be sealed with a durable material. A number of synthetic materials have been especially developed for this purpose.

Greenhouses and orangeries were a feature of the nineteenth century. Owners of most of the great country houses (Fig. 8.1) and some suburban homes built them, and experiments in the breeding of tropical plants were carried out in botanical gardens in the temperate zone. Curved glass roofs (Fig. 9.4) for admitting as much sun as possible were first proposed in 1815 (Ref. 9.23, p. 210). The Great Conservatory built in the gardens of Chatsworth House in Derbyshire by Joseph Paxton for the Duke of Devonshire was the first of the really large greenhouses with a curved glass roof; it measured 277 × 132 ft and was 67 ft high at its highest point (84 × 40 × 20 m). It was assembled from 4-ft (1.2-m) lengths of cylinder glass made by Chance. In 1932 it was demolished because the Duke could no longer afford its upkeep. The Palm House at the Royal Botanical Gardens in Kew, near London, was started by Decimus Burton and Richard Turner in 1842 and is still standing. It has the same height as the Chatsworth conservatory and is 362 ft (110 m) long.

Equally remarkable glass-roofed structures were erected in France. The Galerie d'Orleans was built near the Palais-Royal in 1829. It was a wide lane with small shops on each side, covered by a vaulted roof made entirely of glass in wrought-iron frames. It was demolished in 1935 (Ref. 9.23, p. 240). The Winter Garden and Assembly Room was built in 1847 (Ref. 3.12, p. 242) near the present Avenue Marboeuf. It contained a ballroom, a café, a reading room, an art gallery, fountains and trees. It was even by present standards a large building: its greatest length and width were 300 and 180 ft, and its greatest height was 60 ft (91 × 55 × 18 m).

These buildings provided the pattern for Paxton's Crystal Palace in 1851. It is particularly noteworthy for its industrialized construction, and we shall consider it therefore in Section 10.1. The success of the 1851 Exhibition produced a large number of new glass-roofed buildings. Shopping arcdes were built in most cities and some still survive (Fig. 9.5). Two are on a grand scale: the Galleria Vittorio Emanuele II, in Milan, designed in 1861 by Giuseppe Mengoni, and the Galleria Umberto I in Naples designed in 1887 by Ernesto di Mauro. In both the glass vaulting covers a great pedestrian thoroughfare flanked by three-story buildings, and both have central glass domes (Ref. 9.10, pp. 290–295). Glass domes were included in many new buildings in London and Paris and were also added to some existing buildings; for example, Sir Charles Barry in 1880 built a glass dome for Sir William Tite's Royal Exchange (London's stock exchange); it was replaced in 1936.

9.4

Interior of the Conservatory at the Horticultural Gardens in Chiswick, 1841. (From *London*, edited by Charles Knight and published by Knight and Co., London 1842.)

Glass roofs also played an important part in the growth of department stores in the midnineteenth century. The concept of bringing together under one roof a number of departments offering different goods depended on the ability to provide a proper display of this large array of merchandise. Before the perfection of artificial light this could best be achieved by having a glass roof. Lord and Taylor opened the first store described as a department store in New York in 1826 (Fig. 8.16 shows the elevator in a later building). The *Magasin au Bon Marché*, about 1855, was the first in Europe. In 1876 new premises were designed for Bon Marché by the architect L. A. Boileau. The store had a number of glass-roofed courts which provided daylight for the ground floor and for iron galleries supported on iron columns, all designed by the engineer Gustave Eiffel (Ref. 3.12, pp. 238–241).

Complete glass walls were first constructed in the second half of the nineteenth century. The glass wall separating the stack from the reading room in Henri Labrouste's Bibliotheque Nationale in Paris (1858–1868) is an early example.

The first glass curtain wall was built in 1918 by the architect Willis Jefferson Polk in the Hallidie Building in San Francisco. The glass panels (which could be opened for ventilation) were carried by cantilevered floor slabs 3 ft (1 m) in front of the reinforced concrete columns. Although the glass formed a continuous facade, the decorative iron in front of it gave it a distinctive nineteenth-century appearance.

In 1918 San Francisco was still too remote from Europe to influence its design, and the precursor of the glass curtain walls of the *Bauhaus*, built in 1926 (Section 8.1), and the *Cité de Refuge*, built in 1933 (Section 8.2), are to be found in earlier European developments, such as Gropius' 1911 Fagus Works, Mies' unrealized glass skyscraper project of 1920 (Section 8.1), and the ideas Le Corbusier expressed in 1923 in *Vers une Architecture* (Ref. 3.11). The development of the American glass curtain wall in the 1940s and 50s has already been discussed in Section 8.1, and the effect of glass walls on thermal comfort and energy consumption has been considered in Sections 8.1, 8.2, and 8.6.

Glass blocks or lenses for casting into concrete were patented in Germany in 1907 by Friedrich Keppler, who founded the *Deutsche Luxfer Prismen-Gesellschaft* to exploit his patent, and by M. Joachim, who took out a patent in France in 1908 and called his method *béton translucide* (translucent concrete). Both systems were a development of the pavement lights set in iron, used since the midnineteenth century to light basement rooms. Glass has high compressive strength and the glass blocks can therefore replace some of the concrete in a reinforced concrete structure. Joachim et Cie. built complete ceilings in *béton translucide* in the early twentieth century. In some cases there was a courtyard above, in others it was the ground floor or the roof, and the levels of daylight obtained were generally quite satisfactory. The method is still used in northern and central Europe by a number of different companies. In countries with hot summers the heat admitted through the glass lenses poses a problem.

Glass blocks have also been used in walls in reinforced concrete buildings. Auguste Perret (Section 3.1) used glass lenses in the staircase of one of his early buildings, No. 25b Rue Franklin, Paris, in 1903. Le Corbusier employed similar lenses in the ground floor of the *Cité de Refuge* (Section 8.2) in 1933. Perret was particularly skillful in using the interaction of solid concrete and translucent glass, as, for example, in the church of Notre Dame at Le Raincy, a suburb of Paris, built in 1922. However, glass blocks have more often been used simply as a means of admitting daylight without transparency or without interrupting the loadbearing function of a wall.

Glass lenses used in pairs with an airspace between provide better heat and sound insulation. The Luxfer company made lenses designed to fit in pairs, but a great improvement in the thermal insulation was achieved when in 1935 Corning Steuben in the United States began making hollow glass blocks from two halves fused

9.5

Strand Arcade, Sydney, built in 1891 by John B. Spencer. It is one of the few remaining glass-roofed arcades in Australia, of which there were once many (Ref. 9.22).

together. A partial vacuum was set up as the block cooled and the air inside it contracted. The finished block was 4 in. thick and measured 11¼ × 11¼ in. (102 × 286 × 286 mm).

Cement is used in conjunction with colored glass in a technique called *dalle de verre* developed by A. Labouret and Pierre Chaudière in the 1930s. The rough surfaces of the glass diffuse the light and the strips of cement are wider than the lead cames in traditional stained glass. These windows blend well with modern concrete architecture.

The medieval art of painting stained glass windows probably started in the tenth century and flourished until the fourteenth. It then declined and had virtually disappeared by the midnineteenth when the Gothic Revival created a demand for suitable glass. In 1847 James Hartley patented "cathedral glass," a rolled unpolished plate glass made of tinted melt. It, however, did not produce the same effect as genuine medieval glass which was full of air bubbles (Ref. 1.1, Section 5.8), a technical imperfection. Medieval glass makers would probably have preferred to produce a more transparent glass with less enclosed air. The bubbles, however, gave the colors a brilliance that Hartley's cathedral glass lacked. Toward the end of the nineteenth century several attempts, including a revival of the medieval method of making cylinder glass, were made to put the bubbles back.

A considerable amount of stained glass presently in existence was in fact painted in the nineteenth and twentieth centuries. If the artists did not always succeed in recapturing the religious feeling of the Middle Ages, they had at their disposal a far greater range of colors, notably better shades of green and yellow. The windows in St. Vitus' Cathedral, Prague, which date from the 1920s and 30s, blend well with Peter Parler's fourteenth century stone fabric and have a color composition that was not within the reach of medieval glass painters. The same applied to purely modern designs, such as Marc Chagall's twelve windows at the Hadassah Hospital in Jerusalem, painted about 1960.

9.4   THERMAL INSULATING MATERIALS

Glass also makes a good insulating material. Glass fibers were first produced in 1908 in a German laboratory. In 1938 the Owens-Illinois Glass Company and the Corning Glass Works jointly developed the Owens-Corning glass wool process by which molten glass passed first through orifices in a platinum bushing and then between two converging jets of high-pressure air which cooled the material and turned it into a fleecy mass of glass staples. More glass wool is now made by the crown process which employs a rotating dish with hundreds of small openings.

Rock wool or mineral wool is a material made from certain rocks such as diabase or limestone. Slag wool is made similarly from blast furnace slag, a waste product of steel manufacture.

Vermiculite is the name given to a group of hydrated aluminum-iron-magnesium silicates, which, like mica, are laminar in appearance. If they are subjected suddenly to a high temperature, the flakes exfoliate to many times their original volume. Exfoliated vermiculite thus consists of accordionlike granules with a multitude of minute air layers.

Perlite is a natural glass of volcanic origin that can be expanded by heating. It is used as an aggregate for lightweight concrete, particularly as a backing for curtain walls.

Expanded blast furnace slag is made by watering the slag when it comes from the furnace and then roasting it; this causes the volume to increase many times.

Aerated concrete is made by mixing aluminum powder with cement and water to produce a concrete containing hydrogen bubbles which floats on water. Alternatively a foaming agent, similar to a household detergent, can be used to produce air bubbles in the concrete. Wood composition board is made from wood chips, wood wool, flax, or straw (Section 9.8). Various plastics such as expanded polystyrene or polyurethane foam, can be expanded to many times their original volume (Section 9.7).

All these materials have one thing in common: they are very light because they include a great deal of air or some other gas. That is why they are good thermal insulating materials (see also Section 9.10).

We noted in Section 8.1 that traditional thick masonry construction provided both good heat and sound insulation. We also noted in Section 8.10 that insulation against airborne sound depends on the mass of the material. If we substitute a 100-mm masonry wall for one that is 1 m thick we greatly reduce the sound insulation. If we substitute a 1-m wall of glass wool for a 1-m masonry wall, we also greatly reduce the sound insulation. That is why sound insulation presents difficult problems in modern lightweight construction.

Thermal insulation, on the other hand, is improved by light weight. If we substitute a 100-mm masonry wall for one that is 1 m thick, we greatly reduce the thermal insulation. If we substitute a 1-m wall of glass wool for a 1-m wall of masonry, we greatly increase the thermal insulation. A 1-m masonry wall has approximately the same thermal insulation as a 35-mm glass-wool wall, if both are kept dry, or approximately 3 ft of masonry versus 1¼ in. of glass wool (Ref. 9.11).

Thus it is possible to back a steel or aluminum curtain wall (which has negligible thermal insulation) with an insulating layer ranging from 13 to 100 mm (½ to 4 in.) in thickness, and obtain a result comparable to a traditional masonry wall (ranging from comparatively thin to quite thick).

The different materials available have different applications in practice. Aerated or perlite concrete may be cast into a tray of steel or aluminum to form the external skin of a building. Urethane foam can be injected into a cavity. The dry materials are easily placed as insulation on a thin ceiling of the type used in domestic roofs.

Lightweight insulating materials solve the thermal problems of buildings only in regard to thermal conduction: that is, the heat lost from a warm building to the cold air around it, or gained by the building from warm air. Another problem is due to radiant heat. Radiation from the sun is a particular problem in a hot, sunny climate, but a building also loses heat by radiation during a cold, clear night.

Absorption of radiant heat can be countered by a surface that reradiates the heat in the direction whence it came. Thus in hot-arid regions the buildings are traditionally painted white to reflect radiant heat; dark paint would absorb the heat. Galvanized iron roofs, now in use for more than a century, are sometimes painted green to blend with the landscape, or red to look like clay tile. They are better insulators if left unpainted, particularly while the galvanizing is still bright.

Aluminum foil has been used for reflective insulation. It remains bright for many years and thus acts like a mirror reflecting the radiation. Placed under the roof, it can act against absorption of undesirable radiant heat and against radiant cold.

Thermal insulation received relatively little attention before 1920, even though traditional houses with thick masonry walls could have benefited from insulation in the roof space to prevent entry or escape of heat through the ceiling and roof covering. Since the 1940s, however, more attention has been paid to the fuel that can be saved in cold winters and to air conditioning in hot summers, by providing adequate insulation. Many countries now have legislation that requires thermal insulation, partly for consumer protection and partly as a measure to conserve energy.

9.5 PRECAST CONCRETE

We examined the properties of concrete in Section 2.8 and the surface finish of concrete as a site-cast material in Section 3.1. After World War II concrete became the dominant structural material, and in addition was widely used for facing. This change was partly due to the great destruction wrought by the war and by the interruption to construction for civilian use, partly to the increase in labor cost, and partly to a lack of labor skilled in the traditional building trades. In eastern Europe the last argument was cited in particular as a reason for replacing loadbearing brick walls with factory-made loadbearing concrete panels.

Precast concrete had been manufactured since 1893 (Section 10.3), but in the late 1940s its use increased in both western and eastern Europe. In the middle 1950s the government of the USSR decided to make precast concrete the main material in its housing program, and other eastern European governments did the same (Section 10.4).

We considered the development of metal-and-glass curtain walls in Sections 8.1 to 8.6 and their insulation in Section 9.4. Most countries that adopted the curtain wall used glass and metal at first, but the high initial and operating costs of the large air conditioning plant required caused concern, particularly in the countries of southern Europe and Australasia. Nonloadbearing precast concrete panels became popular in the 1960s, even for the curtain walls of steel-framed buildings.

In many cities concrete became the material most in evidence, and concrete surfaces altered the character of the central business district and high-rise residential suburbs.

It had been realized since the earliest days of concrete construction that its surface finish presented a problem. In 1912 T. Elson Hardy wrote (Ref. 9.12, p. 241):

The reason offered by architects why reinforced concrete is nearly always covered up is its alleged lack of inherent beauty as a building material. Fortunately, the general adoption, some time in the future, of more artistic methods of finishing will remove the reproach.

He proceeded to explain the methods then available: the untouched surface with all the board marks showing; brush finish, by which a pattern is produced with an ordinary scrubbing brush, either while the concrete is still "green" or after it has been treated with muriatic (i.e., hydrochloric) acid; carborundum finish, by which successively finer material is used to rub off the surface; sand-blast finish, which requires compressed air; bush-hammered finish; glazed finish, achieved by using enameled steel forms; pebble dashing, in which the form work is lined with clay into which small pebbles, pieces of glass, or marble chips have been lightly pressed—the clay is washed off after the concrete has set, exposing the aggregate; and facing the concrete in the form, in which the concrete is lined as the work proceeds with a material containing an aggregate of superior quality that is subsequently exposed by scrubbing. This last method required a "granolithic plate" to allow the two different concretes to be placed together. As Hardy pointed out, plastering the concrete after it had set did not give satisfactory results. In addition he described twenty-two mineral pigments for coloring the cement.

In 1945 these were still the principal surface finishes available; but for precast concrete the technology of exposing the aggregate had greatly improved, and it was much easier to achieve a good finish when the slabs were cast horizontally. The methods called by Hardy "facing the concrete in the form" and "pebble dashing" were now called face-down and face-up. In the first the special aggregate to be exposed was placed at the bottom of the mold and the cement was removed subsequently. In the face-up process the special aggregate was rolled into the surface after the concrete had been placed. Almost any color and shape of aggregate could be produced, from pink to green and from rounded to angular. The resulting panels were relatively cheap and the best looked attractive at close quarters. When seen from a distance, however, the effect was often blurred. Thus the top of a tall building faced with a mixture of black and white aggregate became a dirty grey, even though the material looked splendid at groundfloor level. In the early 1950s J. G. Wilson, an architect working at a research establishment financed by the British cement industry, investigated the "readability" of exposed aggregate (Ref. 9.13), and found that the ⅜-in. (10-mm) stones normally used for exposed aggregate could not be distinguished beyond a distance of 75 ft (23 m). For aggregate to be "readable" at a distance of 350 ft (100 m) it needed to be retained on a 1½-in. (38-mm) sieve. Aggregate of this size looks "rustic" at a short distance. Thus a person viewing a tall building faced with panels of variegated exposed aggregate will either find the upper panels blurred or an aggregate on the lower panels that looks too large. In the 1960s, outside eastern Europe, exposed aggregate was made with white cement and white aggregate, such as marble or silica chips.

Basically there are two approaches to the problem of producing a good surface finish on concrete. One is to remove the cement that is the cause of the blemishes and expose the aggregate. The other is to superimpose a pattern or profile that draws attention from the blemishes. Late nineteenth-century profiles tended to imitate natural stone. Thus concrete blocks were made with a "rock-faced" finish which from a distance resembled carved, rusticated stone but nearby was an obvious cast imitation. In the 1920s geometric patterns were introduced. In the 1960s broken surfaces became popular, particularly in Australia where the strong sunlight produced irregular shadows on these surfaces. These are economical finishes that can be made by casting the concrete against ribbed timber forms and breaking the surface off the ribs with an ordinary hammer, by casting ropes into the concrete near the surface and pulling them out, leaving rope-texture grooves between broken surfaces,

9.6

Surface finish produced by splitting panels. The panels were made in pairs with bars cast through the middle, and then split along the line of the bars (Commonwealth Bank, Annandale, N.S.W., built 1970).

or by splitting precast panels and using the interior break as the exposed surface (Fig. 9.6).

9.6 OLD AND NEW METALS

In 1880 Charles Alder Wright wrote:

One of the greatest inconveniences in connexion with the use of iron and steel for constructive and general purpose is the tendency of the metal to oxidize and rust in the air under ordinary atmospheric conditions, i.e., in the presence of free oxygen together with moisture and small quantities of carbon dioxide (and in the case of the air near seaside places of saline spray) . . . . Protective coatings of paints of various kinds are generally applied to the exterior of large iron constructions . . .; the function of these is more mechanical than chemical, the coating simply preventing the metal from coming into contact with the oxidizing medium; but in some of the paints used the basic character of certain of the materials also diminishes the tendency to oxidation. In certain cases the corrosion of iron can be diminished by placing a more active metal in contact with it (e.g., zinc), so that by a galvanic action the oxidation is largely limited to the zinc; by causing the surface of the iron to be closely adherent to the protecting zinc coating (by dipping the brightened metal in fused zinc), a sort of permanent metallic paint coating is obtained, which acts as a preservative in the threefold manner of mechanically preventing contact with air, of galvanically confining the oxidation of the zinc, and of chemically causing the iron to be coated with a basic film of zinc oxide (when the zinc has become slightly oxidized) (Ref. 4.65, Vol. XIII, p. 357).

Corrugated iron was fabricated from wrought iron in the 1820s. An advertisement pointed out that "a sheet which will not bear its own weight, will bear after this process 700 pounds." Commercial processes for galvanizing were patented both in France and Britain in 1837.

Galvanized corrugated iron was already used in the midnineteenth century for entire prefabricated buildings (Section 10.2). Corrugated sheets were exported to the colonies where material for the walls and timber for the roof structure was generally obtainable but hygienic roofing material presented a problem. In many tropical countries galvanized iron has been around for so long that it has virtually become a vernacular material.

Galvanizing is not a practical means of protecting steel that needs to be joined by riveting or welding because the heat destroys the galvanizing, and it is impossible to dip large members into a zinc bath. Painting, however, is expensive on a long-term basis because it has to be renewed every few years.

Weathering steels were introduced in the United States in the early 1960s to overcome this problem. These are high-strength, low-alloy steels that contain small quantities of nickel and chromium and which under appropriate atmospheric conditions develop a durable, tightly adherent oxide coating. This changes over a period of several years from a light rusty orange to a dark purple-brown but remains water-soluble for many years. The building and sidewalks surrounding it must be designed to prevent rusty rain water, which drips from the steel, from staining other materials such as concrete, brick, or stone.

The two other methods of preventing corrosion in steel are too expensive to be used except for curtain walls and other applications involving a limited amount of material. One is porcelain enamel (Section 9.3) developed in the 1930s and first applied to curtain walls in the 1950s.

The other is stainless steel, which was invented independently by Harry Brealey in England, F. M. Becket in the United States, and Benno Strauss and Eduard Maurer in Germany. There are different types of stainless steel, but all contain at least 10% of chromium (usually about 18%) and many also some nickel. In 1929 the tower of the Chrysler Building in New York was sheathed in stainless steel, and in 1932 it was used for the column and mullion covers of the Empire State Building. The first stainless steel curtain wall was built for the General Electric Company's four-story Turbine Building in Schenectady of twenty-gauge (0.93 mm) corrugated sheet. In 1951 the entire surface of the twenty-four story Gateway Center in Pittsburgh was covered with flat twenty-two-gauge (0.78 mm) sheet backed by an insulating layer

Building Materials

of lightweight perlite concrete (Section 9.4). This thin sheet showed considerable waviness which detracted from the appearance of the building (Ref. 9.14, p. 38), and in most subsequent stainless steel curtain walls the sheet has been corrugated or embossed with a pattern.

Since the 1950s stainless steel has replaced glazed pottery in kitchen sinks and urinals.

Copper and bronze (copper-tin alloy) were the first metals employed by man, and in the oldest civilizations the beginning of the bronze age preceded the beginning of the iron age by more than a thousand years. The use of brass, the alloy of copper and zinc, dates from Roman times. Before iron became common, tools, nails, and occasionally girders were made of copper (e.g., the Pantheon, (Ref. 1.1-Section 4.2). Copper has been a common roof covering at least since Roman times and is still used for this purpose. Copper acquires a green patina on exposure to the weather; this is basic copper carbonate, which forms a hard surface that prevents further corrosion. The color is characteristic of copper roofs. Because of their good electrical conductivity, copper and brass are most commonly employed for electrical wiring.

Copper alloys have been used for curtain walls (Fig. 9.7); they are expensive, however, and a similar effect can be achieved with bronze-colored anodized aluminum.

Lead is a soft metal that shows elastic characteristics at a sufficiently low temperature. This metal was used extravagantly in ancient Rome and Byzantium (Ref. 1.1, Sections 4.2 and 5.2), and until the nineteenth century it was a favorite material for water pipes and cisterns which were often elaborately decorated. In the Middle Ages and Renaissance it was a common material for roofs, particularly in England, which was one of the main producers, and was still used extensively in the nineteenth century (Fig. 9.8). Lead was once a cheap material, but it has now become expensive, and many roofs covered with it have been reroofed with other materials if only to prevent the lead from being stripped off at night by thieves.

All the metals discussed so far were known to the ancient world. The word aluminum is derived from alum, a crystalline potassium aluminum sulfate, the *alumen* of the Romans, which had been used since the fifth century B.C. as a mordant in the dyeing trade. Sir Humphrey Davy (Section 8.6) in 1809 produced an aluminum-iron alloy by electrolysis from alumina (which can be prepared from alum): he proposed the name *aluminum*. Hans Christian Oersted was the first to produce metallic aluminum in 1825 by a chemical reaction. In 1845 Friedrich Wöhler made enough of the metal to determine its physical properties, such as density, color, and ductility. The modern electrolytic method of producing it was

9.7

Office building for Seagram and Sons, New York 1958. Designed by Mies van der Rohe and Philip Johnson, this was the first major use of copper alloys for curtain walls. The window frames are of bronze and the panels of Muntz metal, a bronze-colored brass (Ref. 9.15, p. 166).

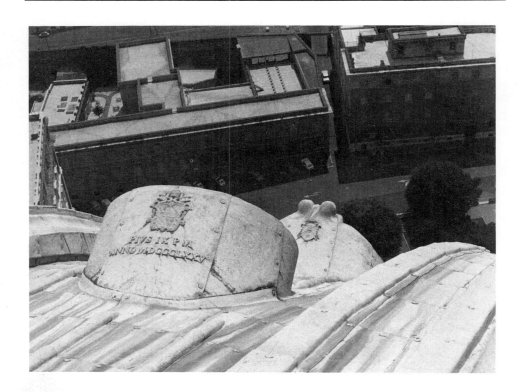

9.8

Lead covering over the dome of S. Pietro in Rome, installed in 1875. The horizontal projections protect windows lighting the space between the inner and the outer shell of the dome.

discovered almost simultaneously in 1886 by Paul Héroult in France and Charles Martin Hall in the United States. The *Encyclopaedia Britannica* in 1875 (Ref. 4.65, Vol. 1, p. 647) gave its physical and chemical properties and described it as a metal used in jewelry and, because of its light weight, in balance beams. In 1893 aluminum was used for the statue of Eros in Piccadilly Circus, and this statue, still in good condition, testifies to the durability of the metal. One of the earliest architectural uses was as an ornamental sheet-metal cornice on the Canada Life Insurance Building in Montreal in 1896. The first corrugated aluminum roof was laid on the Chief Secretary's Office in Macquarie Street, Sydney, in 1900. The sheets were fixed with copper nails, a mistake frequently made for many years thereafter, although the electrochemical series of metals was already well known at that time; these nails made holes in the aluminum sheets by electrolytic corrosion, but the sheets were otherwise in excellent condition when replaced in 1937.

9.9

The Alcoa Building in Pittsburgh, which houses the office of the Aluminum Company of America, was designed by Wallace K. Harrison in 1953. The curtain walls are formed by stamped aluminum panels, 0.125-in. (4.4-mm) thick, backed with 4 in. (100 mm) of perlite concrete (Ref. 9.15, p. 408).

About 1860 Napoleon III had a dinner service made of aluminum, but by the beginning of the twentieth century it was cheap enough for kitchen utensils.

World War I gave great impetus to aluminum production, for its lightness made it suitable for aircraft frames, but because of its high chemical reactivity the durability conferred by the rapid formation of a hard oxide skin was not at first appreciated, and the high price limited its use in buildings until the 1930s.

In 1932 cast aluminum spandrels totaling 1660 tons were used in Rockefeller Center, New York, and in 1936 the first aluminum windows were made for the Cambridge University Library, designed by Sir Giles Scott.

During World War II aluminum production almost quadrupled (Ref. 9.24, p. 30), and the surplus at the end when aircraft production was drastically curtailed was enormous. Aluminum was suggested for long-span structures, but it was not so economical as structural steel for general application; one notable aluminum structure was the Dome of Discovery (Fig. 6.26); it was also used to face the plywood of the Myer Music Bowl (Fig. 6.35). Aluminum was popular for prefabricated buildings for several years (Section 10.4).

Aluminum roof sheeting was more successful in competing with steel. Aluminum windows also proved competitive because complicated shapes could be produced by extruding the hot metal through a die; they are attractive in appearance, their weight makes them easy to handle, and they do not require painting.

Aluminum has become the most widely used metal for curtain walls since Wallace K. Harrison chose it for the trim of the United Nations Secretariat in 1950. In 1953 the same architect faced the Alcoa Building in Pittsburgh with stamped aluminum panels (Fig. 9.0).

For external applications aluminum is usually anodized; that is it is electrochemically processed to increase the thickness of the aluminum oxide on its surface between 0.0006 and 0.0008 in. (0.015 and 0.02 mm), which improves the weathering of the metal, particularly in an industrial environment. It is possible to color the oxide skin during the anodizing process. In the 1950s a number of organic colors which subsequently faded were used for curtain walls. The colors for external use are now always inorganic and permanent.

Aluminum has been used for a variety of fittings, partitions, and reflective insulation (Section 9.4). It has also been covered with an asphalt coating as a dampproof course, replacing the now expensive lead.

9.7 PLASTICS

The present century has seen a rapid development in the technology of plastics and they now serve as an alternative to metals and glass for certain purposes. They are used for dampproof courses, eaves, gutters and rainwater pipes, water pipes, ventilating ducts and electrical conduits, and unbreakable glazing.

Some natural materials can be formed plastically at a moderate temperature, much below that required for softening metals and glass. One is asphalt, known to the ancient Assyrians, and since employed in various parts of the world in which there are natural deposits. Another is shellac, a natural resin made from incrustations formed on certain trees by insects and used in the early phonograph records.

Even the synthetic plastics have been known for more than a century. In 1833 the Swedish chemist J. J. Berzelius introduced the term polymer, and in the same year Henri Braconnot at the University of Nancy prepared the first nitrocellulose by treating cotton with nitric acid. With a high nitrogen content this material became explosive *guncotton,* and with a low nitrogen content it became photosensitive *photocotton,* used under the name *collodion* in the early days of photography for sensitizing plates.

Collodion was the basis of the first commercially successful plastic. The game of billiards became popular in the early nineteenth century, and "billiard problems" figured prominently in later books on mechanics. The perfect billiard ball imparted the entire energy lost by impact to the ball with which it collided, and only ivory balls had the necessary elasticity. The demand for billiard balls threatened the extinction of the wild elephant herds, and in 1863 the American firm of Phelan and Collander, as a matter of enlightened self-interest, offered a prize of $10,000 to anyone who might develop a substitute for natural ivory. This prize was won by the brothers Isaiah and John Wesley Hyatt who made a satisfactory material by dissolving nitrocellulose in camphor. It was soon found to have a variety of other applications, and in 1871 J.W. Hyatt founded the Celluloid Manufacturing Company; among its products were dental plates and celluloid collars which were worn in place of starch-stiffened collars.

Celluloid is a thermoplastic material; that is, it can be formed into different shapes by the application of heat and pressure. By contrast, thermosetting materials become rigid on heating due to a chemical reaction, and they cannot thereafter be softened significantly. Leo H. Baekeland made the first commercially important thermosetting resin by treating phenol with formaldehyde. He called the material *Bakelite.* A similar material was invented in England by Sir James Swinbourne, whose patent application was lodged one day after Baekeland's. The two inventors eventually reached agreement, and Swinbourne became chairman of Bakelite Ltd. in England. Bakelite arrived just in time to supply the need for an insulating material that could be molded into electrical switches and other fittings (Section 8.6). These had been made of wood, which was liable to split, or of brass, which was inherently dangerous because of its conductivity.

Phenol formaldehyde is also the basis of hard-wearing laminates, such as *Formica.*

Bakelite was a functional and economical material, but it could be made only in dark colors. The urea formaldehydes were first investigated in the early 1920s, and commercial manufacture started in the United States in 1929; they can be made in delicate and brilliant colors. Melamine formaldehydes, which are harder, stronger, and more resistant to water, were not produced commercially until the 1940s.

Although celluloid was in most respects an excellent transparent thermoplastic, it was highly flammable. In 1901 Otto Röhm began work in Germany on deriving a safer product from acrylic acid which had been known since 1843. In 1927 he polymerized methyl methacrylate and produced *polymethyl methacrylate,* now marketed under the names *Perspex* and *Plexiglas.* Production expanded rapidly during the late 1930s because of the adaptability of the material to aircraft bodies.

The third important clear plastic was first made by Imperial Chemical Industries in England in 1938 by subjecting ethylene to pressures of about 1000 atmospheres (100 MPa or 15,000 psi). Large-scale production of polyethylene, often called *polythene,* started in 1940. This material is strong and cheap and can be made very thin. It has been used in the construction industry as a waterproof membrane, a water-vapor barrier, and as protection for surfaces of building components during construction.

During World War I Germany was cut off from supplies of natural rubber, and this stimulated research into the manufacture of macromolecules that might serve as a substitute. In the late 1920s and the 30s the chemical industry in Britain, France, and the United States also became active in the search for new materials. Before 1939 polyvinyl chloride (PVC), polyvinyl acetate, and polystyrene had been produced in commercial quantities; however, they had found little use in building.

During World War II both sides were cut off from their normal supplies of raw materials for long periods, and new ones were required for aircraft and other war equipment. After the new manufacturing capacity had become available for civilian production, plastics began to replace many traditional materials for a variety of purposes, from toilet seats in place of those made of wood to kitchen utensils in place of those made of glass. Plastics were particularly useful as thermal insulating materials (Section 9.4), suspended ceilings, floor surfaces, and work surfaces in kitchens and offices. They were employed in the 1950s by Walter Bird for pneumatic structures (Section 6.8) and by Buckminster Fuller for small geodesic domes (*radomes*) built to cover instruments of the American Defense Early Warning Line (Ref. 9.16, p. 74), but plans repeatedly produced since the 1940s to manufacture complete plastic houses have not proved economical.

The rapidity of the change can be seen by comparing the three editions of a popular book on plastics published by two industrial chemists (Ref. 9.25) in 1941, 1956, and 1968.

The chemical industry has also transformed the manufacture of paints and adhesives. Organic vehicles and natural pigments have been largely replaced by synthetics (Ref. 9.26). Traditional glues made of bones, fish waste, casein, soya beans, and cassava have been largely superseded by synthetic resins, particularly for exterior and other waterproof applications.

9.8 TIMBER

The development of new waterproof binders and adhesives has greatly extended the usefulness of timber. Particle board is related to timber as concrete is related to natural stone. The wood chips are to a large extent produced from forest residues (trees that are culled to thin a stand of timber), from pieces left over when the round trees are converted into rectangular sections, and from industrial offcuts. The chips are cemented with a suitable and sufficiently cheap thermosetting resin, such as urea formaldehyde. These boards can be made by hot pressing or extrusion.

Basically there are three types of wood composition board: thermal insulating board (Section 9.4), which weighs less than 25 lb/ft³ (400 kg/m³); particle board which has a density in the range of 25 to 50 lb/ft³ (400 to 800 kg/m³) and is used for partitioning, flooring, and general cabinet and joinery applications; and hardboard which has a density in the range of 50 to 60 lb/ft³ (800 to 960 kg/m³) and is used as a building board for walls and ceilings. These boards have been made since the 1930s, but their use has greatly expanded since the 1950s.

Like particle board, plywood manufacture is largely dependent on the plastics industry. Chairs were made from laminated timber in the Russian Baltic about 1875, with traditional carpenter's glue made of bones. In 1905 Otto Hetzer patented a process in Germany for making structural members from timber laminations joined with casein glue. Phenol formaldehyde was the first synthetic resin employed to produce a waterproof plywood for exterior use, but the acid hardener attacks some types of timber. In the 1930s urea formaldehyde was used for waterproof plywood, and this is still the most common type. Since World War II melamine formaldehyde and resorcinal formaldehyde have been employed; they are more expensive but also more resistant to heat and moisture.

Plywood overcomes the main limitations of traditional timber structures: the lack of strength across the grain and the occasional imperfections of timber resulting from knots, pockets of resin, or sloping grain. By using several laminations with the grains at right angles to one another, the effect of grain direction is eliminated and a defect is confined to one lamination. Although timber beams and planks behave like an assembly of structural members spanning in one direction (Section 4.3), plywood behaves like a two-way slab or wall. Plywood diaphragms prevent the distortion of timber frames under wind load and give additional strength to one-family houses in areas exposed to hurricane-force winds (Section 4.7).

Decorative plywood is made with a thin outer veneer from a timber especially selected for its grain or color. The veneer may be rotary cut (i.e., peeled from the log) which presents a flat grain or sliced at some angle through the log to produce more interesting patterns of the grain.

Timber structures can be glued together from individual laminations (Ref. 9.17). Large trees are rare and take hundreds of years to grow, but there is no limit to the depth of beam that can be produced by glued laminations. Curved structures can also be made by bending the laminations before gluing them (Fig. 9.10). Long-span structures made from glued-laminated timber have been built since the 1940s, particularly in North America.

Timber is still the most widely used material for one-family houses in many countries such as the United States, Canada, USSR, and New Zealand. It is well adapted to prefabrication (Section 10.2-5) and it has found some new uses; for example, in Australia loadbearing brick walls have been largely replaced by brick-veneer construction, which consists of a timber frame lined inside with a wood composition board or plaster board and outside with a single leaf of brickwork. The building looks like a brick structure but has a timber frame. The outer brick leaf protects the timber frame in the event of a bush fire.

The traditional timber house is a complex space frame. Considerable savings have been achieved since the 1950s by theoretical analysis and experimental investigation of the strength of full-size frames.

Higher timber stresses have also become possible by more careful stress grading of the timber. Predicting the strength of timber is relatively simple if all of it comes from the same species and has a uniform grain, as in Douglas fir grown on the west coast of North America. It is much more difficult when it is cut from virgin forest containing many different species, and if some of the timber has imperfections. Stress grading can be performed visually by forming a judgment of the strength group of the timber from its appearance. Mechanical stress-grading machines are based on a rough

9.10

Eagle Aviary in Bristol, England. Curved structural members can easily be produced in laminated timber even if the shape does not conform to any regular curve.

empirical law that the deflection of timber is inversely proportional to its strength. This is because the modulus of elasticity of timber is higher for the stronger species and because serious imperfections, such as loose knots and pockets of resin, greatly increase deflection. Each piece is passed through the machine, which applies force to the timber and squirts a spot of dye on it according to the deflection; the pieces are then sorted according to the predominant color of the spots of dye.

9.9   CONSERVATION OF RESOURCES AND ENERGY REQUIREMENTS OF BUILDING MATERIALS

Since the late nineteenth century a number of people have written about a future when an increasingly affluent society would have exhausted the earth's natural resources. World War II was a period when some raw materials were in short supply, and soon after the colonial territories that produced many of them obtained their political independence. As a result a number of detailed analyses of future trends were carried out in the late 1940s. The 1970s brought an energy crisis and another period of soul searching.

The building industry in terms of sheer volume is one of the largest consumers of raw materials, and it is therefore worth asking whether the supply might be restricted in the future.

Brick and tiles are made of clay or shale. Cement is made of limestone and either clay or shale. Concrete is made of cement, sand, and gravel or stone chips. Glass and porcelain enamel are made mainly of sand. These are virtually inexhaustible materials, but as traditional sources near major population centers are used up we must go farther afield; this will increase transport cost and fuel consumption.

We may exhaust some sources of natural stone, but it is worth noting that the Pentelic quarry, which supplied the marble for the Parthenon, and the Carrara quarry, which supplied Michelangelo's best blocks, are both still being worked.

Some of the traditional metals may eventually be in short supply, but only steel is essential for building. It is the indispensable component of steel structures, reinforced concrete structures, and prestressed concrete structures. Forecasts on the available supplies of iron ore vary; although their exhaustion is certainly not imminent, it poses a long-term problem. Aluminum is one of the most abundant elements, for clay is aluminum silicate; at present it is made from bauxite (hydrous aluminum oxide), of which there are large deposits.

Magnesium, which is a possible substitute for aluminum, is made from seawater. Many of the raw materials used in the plastics industry come from waste products of food processing and from by-products of petroleum refining. Although some of them will cease to exist if we run out of oil, it may be expected, in view of the past inventiveness of industrial chemists, that alternatives can be developed.

Timber is the major building material that is organically grown, and, provided we plant new forests, it is inexhaustible. The harvesting of the shrinking supply of virgin forests and the reafforestation of land with timber of a single species, planted in regular rows, has met with some opposition. It is worth noting that the forests of Britain, Germany, and France, which existed in Roman times, disappeared long ago and have been replaced by a man-made landscape. The landscape artfully contrived on great tracts of England in the eighteenth century still commands admiration. Other examples of planting have been done with less taste and skill. It seems inevitable, however, that in countries in which the native species do not yield softwood suitable or sufficiently fast growing for building there will have to be some planting of exotics; for example Monterey pine (*pinus radiata,* a native of California) in New Zealand and more recently in Australia. The only alternative is to substitute more steel and aluminum for timber.

Although there seems to be no real danger of a shortage of raw materials for building, the energy consumption of some building materials gives cause for concern. In this respect timber is the best, if one ignores the solar energy absorbed by the growing tree, and aluminum is the worst because of the electrolytic process in its manufacture; but because a tonne of aluminum goes a long way and there is little aluminum in most buildings steel is actually the biggest consumer of energy among the basic unfabricated building materials (Ref. 9.18). In addition, fuel is used by the machines that process these materials, by the vehicles that transport them to the building site, and by the cranes that hoist them in position. Thus buildings with concrete floors carried on loadbearing brick walls or on closely spaced columns and clad with brick walls containing timber windows are low consumers of energy. Steel frames with aluminum curtain walls and structures made of precast concrete components are high consumers of energy (Ref. 9.19).

In buildings occupied by a lot of people the energy required for their operation is far more important than the energy consumed in their construction. Heating and cooling are big users of energy (Section 8.6) and insulation (Section 9.4) is therefore important (Ref. 9.18). Lighting is a big consumer of energy in office buildings and hot water, in residential buildings.

9.10 MAINTENANCE, DURABILITY, AND TESTING OF MATERIALS

We considered quality control of structural materials in Section 4.6. Testing the strength of samples of a material is relatively simple, and according to probability theory we can predict the minimum strength of, say, 95% of the material in a structure (Fig. 4.15).

Testing durability and predicting the maintenance that will have to be done on a particular material during the lifetime of the building is much harder. Ideally, if a building is designed for a useful life of a hundred years, all the material should last a hundred years and one day and no longer. In practice, one assumes that paint and carpets, for example, will have to be renewed during the life of a building and that cleaning will have to be done at regular intervals. An attempt can be made to predict the time of replacement of a nondurable material and the annual cost of cleaning. It is then possible to work out the capitalized cost of the replacement and maintenance of various materials and finishes and to select the one that is most economical. This is often neither the cheapest to buy nor one that requires no maintenance at all.

There are some imponderables, such as changes in wages, inflation rates, and interest charges. One private homeowner may do his own painting and regard it as pleasant recreation, so that only the cost of the paint is an expenditure; another may hire a contractor. The amount that a business may write off the value of carpets and other deteriorating assets for the purpose of assessing tax varies greatly from one place to another, and this can make an otherwise expensive material cheaper to the owner (although not necessarily to the nation).

Building economics is, in principle, an ancient subject. Records of expenditure on buildings have been found in ancient Egypt, and arguments between kings and their parliaments or their councils of ministers on the extravagance of various buildings are a matter of historical record. Calculations of the cost of alternative methods of construction were already performed in the seventeenth century, but they did not include maintenance. Maintenance was considered qualitatively by Victorian architects and their clients. Since the 1950s the calculation of the capitalized maintenance cost of large buildings has been considered good practice, and the development of suitable computer programs in the 1960s has facilitated the choice between various alternatives.

To determine the useful life of a material or finish we can take a look at existing buildings to see how they have performed (Ref. 9.20). This is the most reliable method, but it is useful only if the same materials continue to be used over long periods, a state of affairs that applied to a large extent until the early twentieth century. Today's materials are constantly changing. Assuming that we want to examine the durability of a material claimed to have a useful life of twenty-five years, we might have difficulty in finding an appropriate sample. Except for some basic materials, such as concrete, we would probably find that the manufacturer no longer makes the brand or claims that the present product is much improved over that made twenty-five years ago, a claim that is quite often justified.

We can expose a material on a rack in the open air and give it a more severe exposure, than it would ordinarily get from the weather, for example, by watering it frequently. We can place a floor covering in a passage in which there is unusually heavy traffic and count the number of people passing over it by the interruptions to a lightbeam falling on a photoelectric cell. However, these experiments also take a long time.

We can measure the durability on a weatherometer, which speeds up the time scale. Depending on the material, the weatherometer exposes the sample to alternate wetting and drying, to alternate heat and cold, to intermittent ultraviolet light (which is a serious cause of deterioration in many plastics), or to any combination thereof. By having cycles much faster than they occur in real life we can obtain in a few weeks a measure of the durability of a material over a period of years. The speeded-up cycles do not, however, have the same effect as exposure at the normal rate. It is possible to compare two types of linoleum or two types of chlorinated rubber paint with a weatherometer, but it is not possible to predict the durability absolutely.

Abrasion tests present similar problems. A number of machines imitate the abrasion due to normal wear and tear but on a much faster time scale. Different types of abrasion testing machines often give different answers for the same material. Thus abrasion tests can be used to compare the abrasion resistance of two similar materials, but they cannot give an absolute measure of durability.

Resistance to biological organisms is a particular problem in timber. If the temperature, moisture content, and air movement are known, the conditions can be copied in the laboratory and the resistance of various timbers to fungi, borers, and termites can be checked.

Movement due to changes of moisture and chemical action causes cracks in ceramic and cement products. Laboratory tests show that most cements and some bricks contract and some bricks expand. In either case the extent of the movement must be known so that adequate, but not excessive, joints can be provided.

Water penetration of windows and curtain walls is harder to predict because there is some doubt about the amount of water and the speed and direction of the air currents that a building might encounter. In one building research laboratory the wind is generated by the propeller of an old aeroplane and the water is simply the greatest amount that several hoses can deliver. A more accurate result can be obtained by placing a model of the building in a wind tunnel, determining the wind pressure and suction that occurs at the various parts of the curtain wall due to the maximum wind velocity expected, and applying these pressures and suctions to a full-size section of a part of the wall, together with an amount of water corresponding to the heaviest rainfall recorded.

Thermal insulation can be determined with much greater certainty. The K-value is obtained by placing the material across an insulated box. One side of the material is heated electrically and the temperature is measured on both sides. The K-value is a measure of the thermal conductivity per unit thickness of a *single material;* for example, particle board. The U-value is a measure of the thermal transmittance of a *structural assembly;* for example, aluminum sheet 3 mm thick, backed by 40 mm of aerated concrete and faced internally with 7 mm of gypsum plaster, including the effect of the air surface on both sides. The principle of the test is the same, but the equipment needs to be larger.

Sound absorption can also be determined relatively simply and accurately with an impedance tube, which consists of a tube with the test sample at one end and a loudspeaker at the other. The incident sound wave is partly reflected and partly absorbed by the sample, and the reflected wave interferes with the incident wave. The pressure distribution is examined with a traveling microphone that is small enough not to disturb the sound field. From the ratio of the pressures at the maximum and minimum positions the sound reflection and sound absorption coefficients can be determined.

Sound insulation requires more elaborate equipment, for the sound must be generated in a well-insulated full-size room. This has a hole in which the test panel is fixed, and the sound transmitted through the panel is measured in a second insulated room.

Tests on durability and thermal insulation have been carried out since the late nineteenth century, tests on sound insulation since the early twentieth century, and rain penetration tests since the 1950s. Various government-sponsored and private bodies have been set up to standardize tests, of which the American Society for Testing Materials (ASTM), founded in Philadelphia in 1898 and still supported by the subscriptions of its members, has been particularly influential. The International Organization for Standardization (ISO) was established in Geneva in 1947.

In 1920 the British government established the Building Research Station (Ref. 9.21), the first organization devoted specifically to building research, and in 1926 laboratories for it were built at Garston in Hertfordshire, where it still is. Its director, Sir Frederick Lea, in 1951 organized the first international congress on building research, which led in 1953 to the establishment in Rotterdam of the International

Council for Building Research, Studies and Documentation (CIB). Most industrialized countries and some developing countries set up building research institutes in the 1940s and 50s. Initially most of these organizations were concerned with tests on building materials, although the emphasis shifted later to environmental problems and more recently to building economics and planning.

CHAPTER TEN

The Industrialization

of Building

I may have said the same thing before, but my explanation, I am sure, will always be different.

OSCAR WILDE

The most interesting, although not the first, prefabricated building of the nineteenth century to employ a modular, standardized design was the Crystal Palace. It was greatly admired in its own time, and it influenced the development of glazed roofs, and much later, of modern architecture. Prefabricated timber buildings have been exported since the seventeenth century, and from about 1830 to 1856 there was a flourishing trade in prefabricated iron houses. Precast concrete was first made in the 1890s.

A major program of prefabricated construction was launched in 1919. It produced many good ideas but few buildings. Prefabrication was more effective after 1945, particularly in the construction of large-panel multistory concrete apartment buildings. In the United States the most successful product of prefabrication is the mobile home.

Because of the restrictive nature of closed systems, which use only the components prefabricated for a particular building type, open systems have been in existence since the 1950s. These are facilitated by modular coordination and by preferred dimensions. An alternative approach to industrialized building is to concentrate on its managerial aspects and on the advantages to be gained from data processing.

10.1   THE CRYSTAL PALACE

The Society for the Encouragement of Arts, Manufactures and Commerce (now the Royal Society of Arts) was founded in 1754. In 1761 it held its first industrial exhibition in London. In 1791 a similar exhibition was held on the Champs-de-Mars in Paris, and exhibitions of locally manufactured articles were held in London, Paris, and the capital cities of some German principalities in later years.

The proposal to hold a great exhibition in London was first mooted by Henry Cole, an Assistant Keeper at the Record Office. In addition to his job at the Record Office, he was editor of the *Journal of Design*, and in that capacity he visited the Paris Exposition of 1849. A friend of Richard Cobden and a firm supporter of Free Trade, he conceived the idea of an international exhibition at which the goods of "all nations" could be shown together: the competition would encourage a higher quality in design. He approached the Society of Arts which had held exhibitions of "art manufacture" in London in 1847, 1848, and 1849.

Prince Albert had accepted the presidency of the Society of Arts in 1847, and he enthusiastically supported the idea as soon as it was put to him. Cole submitted to the Prince a proposal for a Royal Commission, and twenty-four commissioners were appointed in January 1850 under the presidency of Prince Albert. They included the Prime Minister, Lord John Russell; Sir Robert Peel; William Gladstone; Richard Cobden; Sir Charles Barry, the architect of the Houses of Parliament, then under construction; Charles Robert Cockerell, then engaged on the completion of St. George's Hall in Liverpool; Thomas Donaldson, Professor of Construction and Architecture at University College, London; Isambard Kingdom Brunel, the engineer of the Great Western Railway; Robert Stephenson, the engineer of the Britannia Bridge over the Menai Strait; and Sir William Cubitt, an eminent building contractor. In March 1850 the commissioners advertised a competition for a building to be erected on the chosen site in Hyde Park, the plans to be submitted within the space of three weeks; 245 entries were received. The Building Committee reported that they were "penetrated with admiration and respect for these gratuitious and valuable contributions," but they were unable to recommend any of them for implementation. Instead they submitted a scheme of their own.

It has generally been assumed that the official scheme was conceived by Brunel (Ref. 10.31, p. 17), who had by that time designed numerous railway stations, the Clifton Suspension Bridge near Bristol, and the Great Eastern, the largest steamship of the nineteenth century. The proposal envisaged a number of parallel sheds, built mainly of brick, the tallest only 60 ft (18 m) high, surmounted by a sheet-iron dome 150 ft (46 m) high and 200 ft (61 m) in diameter. The largest iron dome then in existence was J. B. Bunning's London Coal Exchange, which spanned 60 ft (18 m).

The scheme caused a public outcry, partly because of its unsatisfactory proportions, partly because it would have required the removal of a few large trees, and partly because it required fifteen million bricks. It would have been difficult to lay all these bricks in the nine months available, and the expense of laying them and removing them again was questioned.

In June 1850 Joseph (later Sir Joseph) Paxton expressed doubts about the official design to a friend, John Ellis MP, and mentioned that he had "a notion in his head." Ellis took him to see Cole, who had become greatly concerned at the public reaction; Cole promised that Paxton's scheme would be considered if he was able to submit detailed drawings within a fortnight.

Paxton's work was already well known. A farmer's son, he had become the head gardener at the Duke of Devonshire's Chatsworth House in 1824. He designed the Great Conservatory (Section 9.3), other greenhouses, fountains, bridges, and the stone cottages in the model village of Edensor near Chatsworth.

Paxton went from Cole's office to Hyde Park to inspect the site and, according to his own account, his mind was made up when he reached home. The Crystal Palace was built substantially in accordance with Paxton's first sketch, an enlarged version of a greenhouse he had built earlier that year at Chatsworth.

The commissioners requested only one change to preserve some trees which had already caused a debate in the House of Commons; thus the central transept was given a curved roof, and this was the only part of the structure for which the standard girders could not be used.

The initial plans were drawn at the Chatsworth estate office (Ref. 10.30, No. 114). Paxton asked Chance Brothers (Section 9.3), the largest glass manufacturers in England, whether they could supply a million square feet of glass, roughly a third of the annual production of the United Kingdom, and Fox and Henderson, a firm of engineering contractors, whether they could produce the iron parts and erect the building in time.

When tenders closed on July 10 Paxton submitted working drawings made by Fox and Henderson, a specification listing all the parts required, and a firm price of £79,800. The tender was accepted, and Fox and Henderson took possession of the site on July 30. They had nine months in which to fabricate and erect a building covering 772,824 ft² (71,800 m²) of ground-floor space and 217,100 ft² (20,200 m²) of gallery space. The central transept was to be 108 ft (33 m) high (Ref. 10.45, p. 23). The buildings used 3300 columns, 2224 girders, 205 miles (330 km) of window sash

The Industrialization of Building

bar, and 300,000 panes of glass. The design was based on a high degree of prefabrication and standardization, planned on standard center-to-center spacing of the columns. The outside diameter of the cast-iron columns was standard throughout, but their thickness varied according to the load they carried. There were only three sizes of truss, and the connections were standard (Fig. 10.1). All window glass was the same size.

Although the concept was Paxton's, the standardized design was due to Charles (later Sir Charles) Fox. It was, in fact, the only way in which a building of this size could have been manufactured and erected in the short time available. A prototype of each component was given a test load to check its strength, and the safety of the structure was thus not in doubt.

The erection procedure was highly mechanized. The girders were lifted into position by horsepower (Fig. 10.2), and a machine was designed for the glazing (Fig. 10.3). The Exhibition opened on time on May 1, 1851. After its conclusion the Crystal Palace was dismantled, and reerected in Sydenham, a southern London suburb, where it remained in use until destroyed by fire in 1936.

Ralph Nicholson Wornum, in an essay printed at the end of the *Exhibition Catalogue*, said that the Exhibition "is of all things best calculated to advance our National Taste, by bringing in close contiguity the various productions of nearly all the nations of the earth in any way distinguished for ornamental manufacturers." (Ref. 10.1, p. 7). In retrospect the Exhibition, on the contrary, encouraged the already evident tendency towards overornamentation.

The building that housed the Exhibition, however, came in the twentieth century to be greatly admired as an early example of elegantly functional architecture (Fig. 10.4). A few contemporary architects also saw it that way, but this did not become the official view. In 1872 Charles Eastlake, at that time the Secretary of the Royal Institute of British Architects, commented:

One architect at least . . . did not hesitate to avow his conviction that Mr. Paxton, guided by the light of "his native sagacity," had achieved a success which proved incontrovertibly how mistaken we had been in endeavouring to copy from ancient examples; that the architecture of the future should be the architecture of common sense; and that if the same principles which had inspired the designer of the Exhibition building had been applied to the Houses of Parliament, to the British Museum, and to the new churches then in course of erection, millions of money would have been saved and a better class of art secured.

Sanguine converts to the new faith began to talk as if glass and iron would form an admirable substitute for bricks and mortar, and wondrous changes were predicted as to the future of our streets and squares. The failure of the professional competition invited by the Royal Commissioners in 1850 was pointed out with triumph, and architects were warned that if they still fondly clung to the traditions of the past they had better abandon their vocations altogether. . . .

It did not take many years to dissipate the dreams of universal philanthropy to which the Exhibition scheme had given rise, and with these dreams to charming visions of a glass-and-iron architecture may also be said to have vanished. If the structural details of the Crystal Palace teach us any lesson, it is that they are strictly limited in application to the purpose for which that building was erected, and that even for such a purpose their adoption is not unattended by drawbacks (Ref. 10.32, pp. 281–282).

10.1

Bolting the cast iron girders to the cast iron columns of the Crystal Palace.

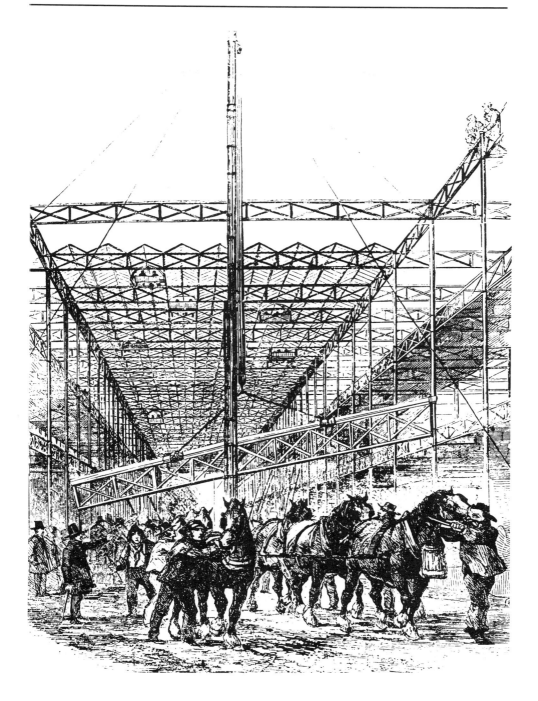

10.2

Raising the girders in the central aisle of the Crystal Palace.

10.3

Wagon used for glazing the Crystal Palace roof, with the cover removed. The sheets of glass measured 49 × 10 in. and were $\frac{1}{13}$ in. thick (1.24 m × 0.25 m × 2 mm). The wagon ran in grooves and was pulled backward by one of the boys as the work progressed. As the wagon was covered, work proceeded regardless of the weather.

Although the Crystal Palace* was not accepted in its own time as "architecture" it was much praised, even by conservative architects, as a marvellous piece of industrial construction, and similar buildings were erected for exhibitions elsewhere. Notable successors were *Palais de l'Industrie* at the Paris Exposition of 1855, which had a central nave of glass spanning 50 m (150 ft), more than twice that of the Crystal Palace, supported on a series of wrought-iron lattice girders; and the *Galerie des Machines,* erected for the Paris Exposition of 1889, which used spectacular three-pin steel portal frames (Section 3.3) to support the glass walls and glass roof. The Crystal Palace also influenced the design of glass-roofed, iron-framed shopping arcades and department stores built in the second half of the nineteenth century (Section 9.3).

The extraordinary speed with which the Crystal Palace was designed and erected and the ease with which the prefabricated pieces fitted together excited general admiration, but the modular standardized design, which in retrospect is its most remarkable feature, received little comment at the time. Because the column size

*The term is used here to denote only the Crystal Palace of London. Several more Crystal Palaces were built subsequently, the most notable being the Crystal Palace of New York (1853) and the Crystal Palace of Chicago (1873). Apart from using glass and iron, however, they differed greatly from Paxton's design.

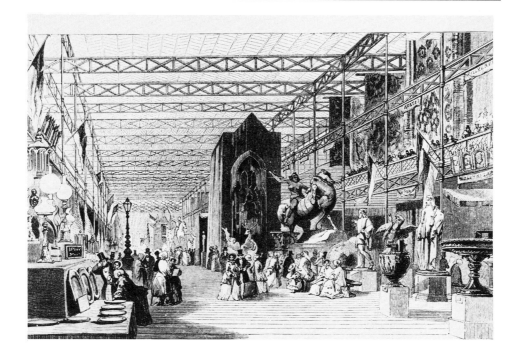

10.4

The Main Avenue East of the Crystal Palace. The unornamented structure contrasted with the "ornamental manufactures." The large statue was described in the catalogue as a zinc casting by M. Geiss of Berlin of "the famous Amazon by Professor Kiss," a bronze statue facing the entrance of the New Museum in Berlin. "We have no doubt that the exhibition of these statues, so admirably calculated for gardens in England, will be followed by a large importation of similar works" (Ref. 10.1, p. 37).

was increased by reducing the internal diameter, both the center-to-center distance and the clear distance of the columns was kept constant, and there was no need for cutting or making special components. All the glass was the same size; when a consignment arrived on the site, it could be used wherever some glass was needed. The fabrication was all done in the factory; on the building site it was necessary only to dig the foundation and join the pieces together.

10.2 PREFABRICATION IN THE NINETEENTH CENTURY

Although the Crystal Palace was the most remarkable prefabricated building of the nineteenth century, it was not the first. There was already in 1851 a flourishing export trade in prefabricated houses, and models of some were exhibited in the Crystal Palace. Settlers proceeding to some country in which building materials or skilled labor were not available frequently brought their houses with them. The first timber houses were taken from England to North America. In 1624 (Ref. 10.2) one Edward Winslow brought *Great House* from England to Cape Ann, Massachusetts. By 1727 timber houses were prefabricated in the North American colonies and shipped from there to the West Indies.

Both cast-iron and wrought-iron prefabricated houses were exported from about 1830 on, the main destinations being Australia, the West Indies, and California. After the development of corrugated sheeting in the 1820s and of galvanizing in the late 1830s, galvanized corrugated iron sheets became common for prefabrication.

Port Philip was discovered in 1802, and permanently settled in 1835. The first Superintendent (later Lieutenant Governor) Charles Joseph La Trobe arrived in 1839. He had a wooden cottage constructed in England and sent to Melbourne in the form of panels where it was erected in 1840; it has been rebuilt in another location in The Domain.

In the 1840s and the early 1850s Hemming of Bristol, Bellhouse of Manchester, and Walker of London exported whole streets of iron buildings to several towns in Victoria, including shops, churches, hotels, and houses. Some were also sent to Adelaide in South Australia, first settled in 1836.

In 1855 Coppin's Olympic Theatre was erected in Melbourne from completely prefabricated iron parts (Ref. 10.3). E. T. Bellhouse, of Manchester, undertook to make the parts and have them on board ship within thirty days of the signing of the contract. The erection in Melbourne took six weeks. The building measured 88 × 44 ft by 24 ft high (27 × 13 by 7 m), and it was demolished many years ago.

Corio Villa (Ref. 10.4, p. 172) in Geelong, Victoria, is still standing. The pieces were made in Glasgow and unloaded in the port of Geelong also in 1855, but not claimed. The port authorities sold them to Alfred Douglas who built the villa. The columns are of ornamented cast iron, the walls are of ½-in. (13-mm) iron boiler plate and corrugated iron, and the roof is of thick corrugated iron.

Gold was discovered in California in 1848, and in 1849 a vast number of prefabricated houses were shipped from the American East Coast, various parts of England, Hong Kong, Hobart (Australia), and Auckland (New Zealand). Five thousand were sent from New York alone (Ref. 10.2). Most were of timber, but iron houses were sent from England and New York. Within a few months there was a great surplus and some of the cargoes were diverted to Hawaii.

Prefabricated iron houses were also sent to Jamaica, St. Lucia (West Indies), Bermuda, Chagres (Panama), Natal (South Africa), and in small numbers to other African colonies (Ref. 10.5).

The Crimean War broke out in 1854, and complaints soon reached England of inadequate accommodation, particularly for the wounded. In 1855 Isambard Kingdom Brunel (Ref. 10.33, p. 16) was invited by the War Office to design a prefabricated field hospital. By the standards of its time it was well equipped, fresh air was drawn in by fans, a small boiler was heated by candles for hot water. Prefabricated barracks were also sent.

When the war ended in 1856, the production of prefabricated iron buildings came to an end and was not resumed until the twentieth century. In Victoria the shortage of building materials was passing, and local foundries were supplying cast iron. The Californian gold rush had come to an end. The invention of the balloon frame in Chicago in 1833 (Section 2.6) had gradually spread to other countries. It obviated the need for complicated joints in timber. A bag of nails, a hammer, a saw, and a pile of sawn timber were now sufficient to make a frame. However, corrugated iron sheets were still exported for the roof, and this export increased.

In North America the demand for prefabricated timber buildings continued. As the West was being settled, building materials had to be brought in and, once the railways were built, it was cheaper to import prefabricated panels for houses. Cincinnati and Chicago became major centers for fabricating the panels (Ref. 10.6). In the 1870s the fabricators began pointing out in their advertisements (Ref. 10.6) the advantages of their panels "for camp grounds, seaside and summer resorts" in addition to their use "for pioneer settlements and the West Indies." The prefabricated house was recognized as being less solid than a home built on the site, but in an increasingly affluent society it was quite adequate for a second house to be used over the weekend or in summer.

We have already mentioned the buildings of iron and glass which were mostly prefabricated (Sections 9.3 and 10.1). In addition, a number of manufacturers made prefabricated facades of cast iron and glass designed to be fixed to otherwise conventional buildings. The most important of these manufacturers was James Bogardus, who described himself as an "architect in iron" in a booklet he published in 1856 (Ref. 10.33, p. 13). Between 1850 and 1880 he produced curtain walls of cast iron and glass, mostly with arched windowheads between neo-Renaissance columns; the office built in 1854 for the publishing house Harper and Brothers in New York is his best-known design. Several other architects in the United States and Britain used the same type of design; there were some good examples in New York, Chicago (Ref. 2.19), St. Louis (Ref. 3.12, p. 202), and Glasgow (Ref. 10.33, p. 33).

10.3 PREFABRICATION IN THE EARLY TWENTIETH CENTURY

The increasing use of precast concrete about 1900 was at least partly due to an inadequate design theory for site-cast concrete. The equations for the strength of reinforced concrete were derived only in 1894 and the water-cement ratio law, in 1918 (Section 2.8). Building regulations varied from country to country and from city to city; some had nothing to say on reinforced concrete and others were excessively stringent.

It is difficult to ascertain the strength of site-cast concrete, except by theory. Precast concrete, like timber or steel, can be tested. Thus a precast concrete component was more acceptable than site-cast concrete to some building authorities, designers, and clients.

The first precast concrete was used for engineering construction in 1893 and for building construction in 1905 (Ref. 10.34). Many of the precast concrete systems developed in the early years of this century were widely used (Fig. 10.5). Tilt-up construction, by which walls are cast horizontally on the concrete floor slab and tilted up after the concrete has hardened, was also developed about 1910.

During World War I all civilian building was suspended. David Lloyd George became Prime Minister of Great Britain at the end of 1915 and formed a Ministry of

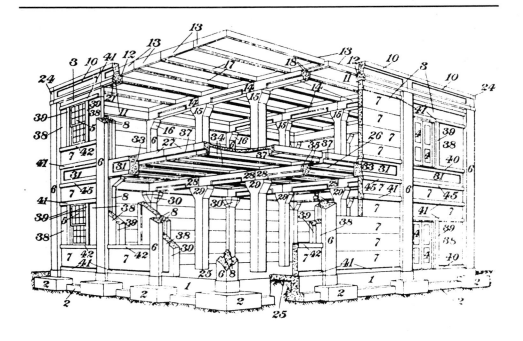

10.5

Completely precast concrete building system, patented in the United States by J. E. Conzelman in 1912. Many buildings of this type were built before 1916 (Ref. 10.34, p. 484).

Reconstruction in August 1917; one of its tasks was to provide "homes fit for heroes to come home to." This was not a mere catch phrase. It was important to keep the army fighting, but Lloyd George was a genuine social reformer. On July 26, 1917, he established a committee under the chairmanship of Sir John Tudor Walters, a member of the House of Commons, to "consider questions of building construction in connection with the dwellings for the working classes." The Tudor Walters Report, published three weeks before the end of the war, recommended a drastically new approach. The nineteenth-century houses, although mostly sound structurally, provided a standard of accommodation no longer acceptable (Section 7.5), and 200,000 needed replacing. In addition, there was a shortage of 400,000 dwellings, and at the same time a shortage of skilled craftsmen in the building trades. Brick was still the main material for houses, but the number of bricklayers had declined from 116,000 in 1901 to 74,000 in 1914; in 1920 only 53,000 remained.

The recommendation that components should be standardized had some effect. Standard sizes for steel windows were adopted in the early 1920s, and the much larger number of manufacturers of wooden windows was obliged to follow suit.

The recommendation that prefabricated houses should be mass produced could not be implemented, even though between April 1919 and April 1920 a total of eighty-eight building systems, which employed timber, steel, concrete, or clay blocks, was approved. R. B. White described some of them (Ref. 10.33, pp. 53–88) and evaluated their contribution (p. 90):

Post-war "prefabrication" had petered out by 1928. At a price not often really competitive with traditional methods it had provided a small but useful additional number of houses during a period of shortages, and this is about all that can be said for it. During this time it had contributed nothing to any other class of buildings, nor had any move been made towards the industrialization of building which remained essentially traditional both in design and execution. The increased number of steel-framed structures for flats, offices and the like in the thirties may have included some features that could be accounted "prefabrication" but they owed little to experiments in the domain of housing.

As the first post-war decade drew to a close and the industrial recession approached, the cost of building dropped. Whereas in 1919 the average cost of 6000 houses of which tenders were examined was £704 excluding land, roads and sewers, and rose still higher in 1920-21, by the end of the decade this figure had been approximately halved without any significant increase of "industrialization" in the building trades; it happened mainly through external economic influences, but partly through severe commercial competition and a general lowering of standards of quality. The cost of building in the mid-thirties was much nearer to its level in 1911.

In the United States and Sweden, where timber was the dominant material for one-family houses, prefabrication was more successful. In both countries partial prefabrication was adopted; that is, components and subassemblies were produced in factories and other members often arrived on the site precut. The number of buildings required to make this sort of prefabrication worthwhile is, however, much smaller than in the building systems attempted in Britain.

The USSR considered prefabrication of buildings a part of its strategy for the Third Five Year Plan 1938–1942, but it was not implemented until after World War II (Section 10.4).

In France a number of prefabricated systems were developed for multistory buildings, one of which, devised by the civil engineer Eugène Mopin, was used in a number of blocks of flats at Drancy, Bagneux, and Vitry-sur-Seine, all near Paris, and Leeds in England. This system employed unusually light hot-rolled steel sections encased in site-poured and vibrated concrete. This framework was clad with precast concrete units. It was claimed that the savings produced made it possible to introduce some unusual amenities, such as the Garchey system of refuse disposal through pipes.

The Leeds City Council began construction of 938 flats at Quarry Hill in 1935, but they were not completed until 1940. The first signs of failure of the cladding were observed in 1954. Water penetration behind the panels had corroded the corner steel angles and supporting brackets. The Garchey system was also giving trouble. Both may have been ahead of their time. The joints between concrete panels had been filled with sand-cement mortar; in the late 1930s mastic sealants that might have prevented the water penetration were invented. After some costly and unsuccessful repairs it was decided in 1964 to demolish the flats progressively by 1978 (Ref. 10.7).

10.4 PREFABRICATION SINCE 1940

The period between 1920 and 1939 was therefore less successful for prefabrication than the years between 1835 and 1855. The European housing problem in 1945 was more difficult than in 1918. There had been no civilian building in Britain since the outbreak of the war in 1939 and in Russia since the invasion in 1941. The destruction by aerial and artillery bombardment was enormous; it was most serious in Germany and Poland, but a great deal of damage had also been done in England, Russia, and in the countries occupied by Germany during the war.

In Britain some houses described in 1918 as unfit for human habitation were still occupied. In France, where there had been relatively little residential construction between the two world wars, the position was worse. In Russia several families often shared a single room.

In Britain this time there was no talk about "homes fit for heroes" because it was taken for granted that housing would be a top priority after the war. The Inter-departmental Committee on House Construction under the chairmanship of Sir George Burt, a building contractor, was appointed in September 1942; that is, before the American and British invasion of North Africa. In 1944 this committee produced a report on house construction (Ref. 10.8), which laid down standards of performance based on scientific research, particularly the work of the Building Research Station (Section 9.10). In addition, it surveyed the "alternative" methods of construction used in the interwar period. In 1946 and 1948 the Burt Committee published two further reports on prefabricated systems proposed by various organizations (Ref. 10.8). In 1943 the British Government built a number of "demonstration houses" at Northolt, Middlesex, to test public reaction and to estimate the probable cost. Six were built by traditional and seven by "alternative methods."

In spite of this careful preparation at a time when the war was still in a critical stage and its end was not in sight, the alternative methods again failed to solve the problem, for which in retrospect at least three reasons can be found.

British families preferred individual homes rather than multistory apartments. One-family houses have been successfully prefabricated in timber in North America and Sweden, but prefabrication is less economical for single houses in other materials. After housing had ceased to be an urgent problem, contracts were decided mainly on price. Traditional construction was cheaper for low buildings, and industrialized construction only competitive for tall buildings.

Average Tender Price for Dwellings Built by Local Authorities and New Towns in Great Britain in 1965, in Pounds Sterling per sq. ft. (Ref. 10.10)

Type of Construction	All Buildings	Five Storeys or More
Industrialized (I)	3:13:2	4:19:3½
Traditional (T)	3:9:2	5:4:4½
I/T	1.06	0.95

Many of the prefabricated houses looked different from those that working-class people were used to. This applied particularly to the aluminum bungalow of the temporary housing program. At the end of the war aluminum, which had been intended for aircraft construction, became available for civilian use and was tried for a variety of new applications (Section 9.6).

The bungalow was completely factory-built, brought to the site in three parts, connected to the services, and bolted to the foundation and one another. The joints were then sealed and the house was ready for occupancy. In one demonstration this was done in an hour, but a more economical time was half a day. The buildings had thermal insulation comparable to that of brick houses, and facilities in the kitchen and bathroom far superior to those in prewar working class houses. The "tinny" appearance of the buildings, widely criticized, would have been less objectionable if the surroundings had been properly landscaped, but many were erected in bare fields in neat rows, each connected to the access road by a straight concrete path. Aluminum proved to be a more popular material for schools, partly because a school is a single large building, not a group of identical small houses, and partly because a school is expected to be bright and clean rather than comfortable.

The main problem, however, was that the number of houses produced for each system, although bigger than in the 1920s, was still too small for economical prefabrication. Moreover, in England continuity of production was uncertain. The Economic Commission for Europe, in a cost study made in 1963 (Ref. 10.11), estimated that the difference between successful large-scale prefabrication and traditional construction was 10 to 20% and this could be reversed if the plant were kept lying idle from time to time.

Thus prefabrication has been more successful in the USSR, where government instrumentalities produced the plan, made the components, erected the buildings, and rented the accommodations. Industrialized building was to form part of the strategy of the Third Five Year Plan (Section 10.3) which had been interrupted by the war. The first fully factory-made buildings were assembled in 1949, but it took a number of years before the project gathered momentum. In the 1960s it produced a great increase in residential construction. Vladimir Promyslov, the chairman of the Moscow City Soviet, that is, the Mayor of Moscow, gave the following figures for the amount of fully prefabricated home building in Moscow, in million square meters ($1m^2 = 10.76$ ft^2) (Ref. 10.35, p. 49).

Year	1956	1957	1958	1959	1960	1961	1962
Area	0.090	0.160	0.444	0.650	0.980	1.766	2.077

Year	1963	1964	1965	1966	1967	1968
Area	2.180	2.300	2.472	2.980	3.300	3.400

Promyslov estimated that prefabrication was 8 to 10% cheaper than traditional brick construction; its main advantage was, however, that it provided much-needed accommodation with a relatively small skilled labor force.

Most of this construction was done with large concrete panels, like all recent prefabricated construction in western Europe. The wall panels are normally a full

10.6

Park Towers, Melbourne, Australia, erected by the Victorian Housing Commission in 1970. This 31-story building is the tallest built entirely from precast concrete panels without a frame. However, it was necessary to prestress the panels vertically which added to the cost. It seems unlikely that prefabricated dwellings will in future be built so high.

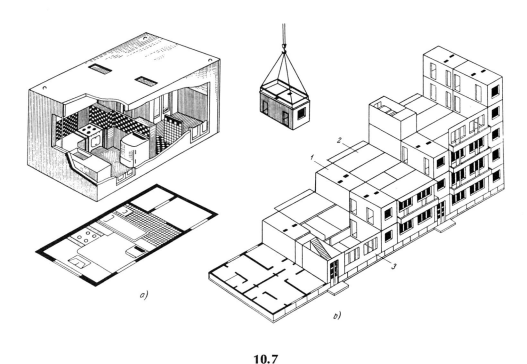

a)

b)

10.7

Erection of a block of flats in the USSR from precast concrete cells, including fully prefabricated kitchen-bathroom units (Ref. 10.12, p. 360).

story high and a full room wide. Smaller concrete units, of the type employed in the 1900s (Fig. 10.5), 1920s, and 40s , have fallen into disuse. Both in western and eastern Europe the economical height of these buildings is about 15 stories, although taller buildings with large concrete panels (Fig. 10.6) have been erected.

The collapse in 1968 of the twenty-four story apartment building at Ronan Point in London caused by a gas explosion that blew out a few loadbearing precast concrete panels (discussed in Section 4.9) brought about a number of revisions in the structural design of such buildings.

In practice, more serious problems develop from the minor damage to precast concrete units that may result during transport and handling, the difficulty of ensuring that the units are perfectly rectangular after curing and of erecting them with evenly spaced and properly sealed joints. In Russia the emphasis has been on the quantity of production to fill an urgent need for housing, and damaged concrete panels and uneven joints that would be visually unacceptable in western Europe or Australia are frequently evident.

Even larger units have been used for the construction of apartments in Russia (Fig. 10.7) but have not become common because of the cost of handling. The tendency in eastern as well as western Europe has been away from *closed systems*, by which components are produced for a specific building type only, toward *open systems*, which use components from a standard catalog.

10.5   MOBILE HOMES

We noted in Section 10.4 that the English aluminum bungalow, and some other prefabricated houses, were fully factory-produced and arrived on the site in as few as three parts. These buildings have not been successful because a great deal of precision was needed to ensure that the parts would fit together on the site, and the cost of transport and erection was high. These negated the savings produced by factory production, which in the 1940s were widely expected to reduce the cost of housing.

White in his summary of the experience gained during the 1940s and 50s wrote:

The theorists of prefabrication could be divided into two classes: those who espoused the teachings epitomized in the Bauhaus (architect-industry partnership), which have to some extent been put into practice, e.g., in the post-war schools in Britain, and those who adopted a more romantic, individual or even roguish approach. The latter have been largely discredited, but we have by no means heard the last of them. One of their favourite arguments, in the present writer's [i.e., White's] opinion a fallacious one, has been based on an attempted comparison between houses and certain industrial products such as motor cars, refrigerators and the like. The car, the "fridge," the gas or electric cooker, the vacuum cleaner, etc., are produced in their millions and are bought, not because they are inherently cheap, but because they present certain advantages over pony-traps, blocks of ice, coal ranges and dust-pans. The point is eventually reached when the new product, although more "expensive" to produce in the first place, presents such real (or imaginary) advantages that it becomes established. From then on, the old product gradually falls into disuse, and so becomes relatively more and more "expensive" as the skills required to produce and operate it vanish from the scene. To a great extent the situation already exists in regard to such standardized building components as windows and doors, where the numbers made result in production costs well below those of comparable articles made to measure. Any accompanying disadvantages of a qualitative or aesthetic nature are widely tolerated for the more immediate and tangible advantages of economy and speed of delivery (Ref. 10.33, p. 301).

The mobile home is, however, an example of full factory prefabrication which has reduced costs. Holiday caravans, or travel trailers, have been manufactured since the 1920s for the purpose of making people independent of hotels without having to sleep in the open or in a tent. In the 1930s, largely as a result of the Depression, these

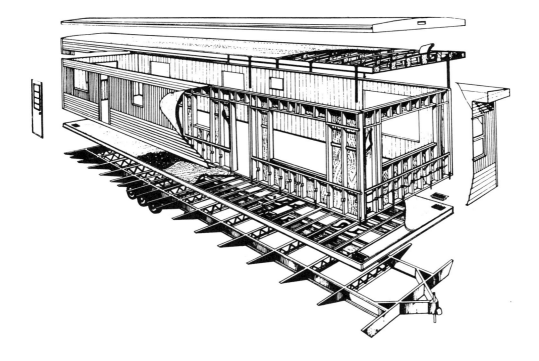

10.8

Frame of Mobile Home. In the United States this is made up to 14 ft (4.3 m) wide and up to 85 ft (25.9 m) long. The mobile home has a kitchen and a bathroom connected to services in a mobile-home park. The wheels are normally removed and the frame placed on permanent blocks.

trailers were used for continual residential accommodation, and permanent caravan or trailer parks sprang up. Traditional trailers could not provide space comparable even to a modest house because the narrow width made it impossible to provide a satisfactory room.

Nevertheless, this use of trailers was accelerated in the United States by the dislocation caused by World War II and by the reconstruction after its cessation. By 1950 45% of trailer sales were for primary housing, and in response to this trend the first trailers with bathrooms came on the market in that year; in 1951 toilets were added (Ref. 10.13, p. 176).

In 1955 American manufacturers began the production of units 10 ft (3.0 m) wide. In 1959 a 12-ft (3.7-m) model was introduced in spite of opposition from some state highway departments. Since then a 14-ft (4.3-m) model has been made and is now permitted in 27 states. Some mobile homes have been built in Australia, but the main production has been in America.

A unit of this width and a length as much as 85 ft (25.9 m) could not, like the traditional trailer, be towed about the country by the family car. It required a special hauler. On arrival at its destination the wheels were removed, the home put on blocks, and connected to services. In the majority of cases it was never moved to another site. The mobile home is therefore essentially a prefabricated house, fully assembled in a factory, and pulled to its destination on its own wheels (Fig. 10.8). With increased production, the cost has fallen and the quality improved. In 1972 the price of a furnished mobile home averaged $8.35/ ft² against an average of $16 for an unfurnished site-built house, exclusive of land. In the same year mobile homes in the United States accounted for 95% of all new one-family houses priced below $15,000 and for 45% of all new houses selling at any price. A mobile home usually has a bathroom, kitchen-dining room, living room, and two bedrooms, all of minimal size, but larger or more numerous rooms can be provided by buying two or three units and linking them permanently.

A majority of those who have bought mobile homes are young couples who do not wish to wait until they have saved enough money for a permanent house and old people who have retired to one of the warmer states. Mobile homes deteriorate over a period of time, unlike permanent houses which may actually increase in value as building costs rise. However, the useful life of a building is determined more and more by the building services, and the concept of an expendable house is becoming more acceptable. The concept is not new. Coignet in 1855 (Section 2.7) thought that one house should shelter only one generation.

10.6 OPEN SYSTEMS AND MODULAR COORDINATION

Some local authorities in the United States and Australia have effectively prohibited mobile homes by classifying them as vehicles and introducing limited-stay provisions. In addition, there is the less tangible objection to the appearance of a "park" of similar small structures, relatively close to one another. It does not seem likely that mobile homes can solve the problem of industrialized building by themselves any more than the closed systems of the 1920s and 40s, in which buildings were assembled entirely from components made especially for a particular system.

The concept of modular design is largely due to Albert Farwell Bemis, an American manufacturer, who began taking an interest in the dimensional coordination of building materials and equipment about 1921. In 1936 he published "Rationalized Design" (Ref. 10.36), in which he proposed the *cubical modular method* "to improve the inefficient methods of assembly of unrelated materials and reduce the cost of building construction by applying industrial production techniques." He considered that 4 in. provided the largest dimensional incremental for "minimising problems resulting from stocking and distributing standardized components, while preserving design flexibility to meet practical and aesthetic requirements." Four inches equals 101.6 mm, and metric countries adopted a 100-mm module. Bemis' concept was that components made by different manufacturers would fit together in the way that any electric bulb fits anybody's light fixture, and any electric plug fits into any electric power point in any country (regrettably this does not necessarily hold true when one crosses a national boundary). The problem of the electric fixture is, however, much simpler because it is linear, whereas the general building problem is cubical.

Bemis died in 1936, and his family established the Modular Service Association. In 1939 the organization combined with the American Standards Association (A.S.A.) to sponsor Project A 62 to "develop a basis for coordinating dimensions of building materials and equipment, to correlate building plans and details with such dimensions, and to recommend sizes and dimensions as standards suitable for dimensional correlation." In 1945 the A.S.A. approved the 4-in. module for dimensional coordination (Ref. 10.37).

In 1953 the European Productivity Agency of the Organization for European Economic Cooperation (OEEC) started a modular study project in which eleven continental countries participated. The Modular Society was formed in Britain in the same year. Its publications, the *Modular Quarterly,* and the *Modular Catalogue,* made the modular concept generally familiar. In 1958 it built the Modular Assembly, a room that utilized a variety of modular components. The fact that this experimental room could not be assembled entirely without cutting (Ref. 10.15) illustrated the complexity of the problem.

The difficulty lies in the need to make joints between the components of a building in three dimensions. Joints must occupy some space, and allowance must be made for the thermal expansion and contraction of all materials and for the moisture movement of some. There are differences between the thicknesses of various finishes (e.g., carpet versus linoleum), and an increment of 4 in. (or 100 mm) is too great for concrete floor slabs and partitions. All these considerations led to the admission of fractions of the module.

Some proponents of modular design consider that not only should any multiple of the module of 4 in. (or 100 mm) in accordance with Bemis' philosophy, and reasonable fractions of the module, such as 1 in. (or 25 mm) be considered modular, but also the sum of the two. This would mean that any multiple of 1 in. or 25 mm is a modular dimension. We would thus find ourselves in the same position as Molière's Tartuffe who was told by his elocution teacher that there are only two forms of speech, poetry and prose, and concluded that he had spoken prose all his life without knowing it. If any multiple of 1 in. is modular, there is no point in modular design. If modular design is to be more economical than nonmodular design, there would have to be a series of preferred dimensions. This argument received much attention during the 1950s (Section 10.8).

In the meantime significant progress was made in the use of nontraditional construction that employed a restricted range of dimensions derived from a basic module. The best known of these in Britain was CLASP (Consortium of Local Authorities' Special Programmes), devised in 1957 by D.E.E. (now Sir Donald) Gibson for a number of local authorities for the construction of schools. By 1963 the Consortium had twenty-three full and associate members so that it was able to place orders for components in considerable quantities and with a reasonable assurance of continuity. The schools were built mainly with prefabricated components, but they did not belong to a single closed system. Orders for the various components were placed with different manufacturers on the basis of competitive bids. After initial teething troubles, the components fitted together without cutting, and the cost was greatly reduced as a result of the competition (Ref. 10.16).

The USSR is trying the opposite approach to open systems in the Moscow Building Catalogue, which lists components that conform to a modular standard but are not manufactured for any specific type of building and are available for use in any prefabricated construction. A. B. Samsanov explained that its objective was "not a building, but the industrially produced building elements themselves from which buildings differing in their spatial composition and architecture are constructed. In other words, the method 'from project to product' will be replaced by the 'from product to project' method." (Ref. 10.17, p. 186).

10.7 INDUSTRIALIZED BUILDING WITHOUT PREFABRICATION

System building has made little progress in the United States. Most attempts in the 1920s and 40s failed. In 1969 the Federal Housing Administration of the Department of Housing and Urban Development launched Operation Breakthrough to provide manufactured housing specifically for the underprivileged. A great many schemes were approved but few houses were built. Nevertheless, American housing is in some respects the most industrialized.

In assessing the cost of American houses, we must allow for the high wages earned by carpenters and other building operatives and for the relatively large size of the dwellings—about four times that considered appropriate in eastern Europe. If the cost is expressed in terms of the number of hours a carpenter (or an architect) has to work to earn the cost of a square meter of an average normal house, it is one of the cheapest in the world. The failure of the American building industry to use prefabrication, except on a limited scale for components, particularly of timber houses, has not therefore affected the industry adversely. (Mobile homes are not produced by the building industry.)

Burnham Kelly, then director of the Albert Farwell Bemis Foundation at the Massachusetts Institute of Technology, said at the International Congress for Housing and Town Planning in Amsterdam in 1950:

A prefabricator does not succeed simply because he has a sound production idea . . . . Indeed it is possible for a company in the United States to bring about reduced capital costs without making use of new materials, production methods, or designs, but by relying only on expert management, on careful bulk purchasing, and on the marketing advantages of a product which is readily available and which may be quickly put in place. . . . Indeed many prefabricators are *less* industrialized than such highly organized site-builders as Levitt of Long Island (Ref. 10.33, p. 311).

Another argument used against system building is the fact that (except for mobile homes) a system-built house is not much cheaper than a house erected to a client's order to accommodate his personal requirements. Although custom-made suits, shoes, and motor cars are at least twice as expensive as those that are mass-produced, there is only a small difference for a one-family house. This was brought out in a study made by the Harvard Graduate School of Business Administration about 1960:

The Industrialization of Building

The production manager (of Sunshine Builders Inc., a Florida builder of single-story houses) explained that by and large "prefabbing" does not pay unless the customer is not permitted to make any changes in the house he buys. If Sunshine adopted such a policy we would lose sales. For instance prefabbed roof trusses, which many construction firms use, are more expensive than "on-site" construction because we have so many models (Ref. 10.18).

American house building is cheap in terms of local wages because American labor is provided with good equipment, ample power tools, and motorized hoisting equipment (for small units), because management is efficient, and because the communications between the supplier of materials and the building site are well organized. All are characteristics of a highly industrialized country. In most underdeveloped countries the cost of building, to a comparable standard, is high in terms of local wages because the country is not industrialized. When a factory for prefabricated housing is installed, as on a number of occasions mainly through foreign aid, it has not reduced the cost of housing unless the cost of the factory was ignored.

Taking a world-wide view, prefabrication has not been more successful in reducing costs since 1945 than it had in the interwar period, 1919 to 1939, except for multistory buildings. However, it provided an alternative method of construction at a time when labor and materials were in short supply. In theory, mass production should reduce the cost, and undoubtedly housing authorities and building research workers will keep the problem under constant review.

In the meantime there are three avenues for reducing the costs of the housing industry. One is to make more standardized components off the site. Since the 1920s windows and doors have to an increasing extent been mass-produced. This type of partial prefabrication could be extended. The most expensive parts of the one-family house are the bathroom and kitchen. As long ago as 1937 Buckminster Fuller proposed to stamp complete bathrooms out of sheet metal (Ref. 10.19, p. 90), but this was too drastic a step to be accepted by a conservative public and there were other problems of repair and maintenance. In the 1960s assemblies for more conventional bathrooms were factory produced (Ref. 10.38, pp. 148–154). Prefabrication of the building services for tall buildings and installation in large units with a minimum of site connections are also becoming more common.

The second method is to make greater use in the less industrialized countries of power machinery for handling materials, cutting holes, fixing pipes, and sealing joints.

The third method is to increase the standardization of components by an international effort, bearing in mind that many are sold across national boundaries, and to confine the dimensions of buildings to a limited number of preferred dimensions.

10.8 PROPORTIONAL RULES AND PREFERRED DIMENSIONS

In 1949 Le Corbusier wrote *Le Modulor* (Ref. 10.39) in which he proposed a system of modular coordination based on the golden section. He was then at the height of his fame. Hence the golden section, which has since the nineteenth century been used as a rule of proportion, became linked with modular design.

The golden section, discovered by Eudoxus in the fourth century B.C., is a construction for the division of a line into two parts, a and b, in which the shorter part (a) is to the longer (b) as the longer (b) is to the whole line (a + b). This gives $a/b = b/(a + b)$, and the solution is $b/a = $ ½ $(1 + \sqrt{5}) = 1.618034$. Euclid included the construction in Book XIII of the *Elements* and used it for drawing the regular pentagon.

De Divina Proportione by Luca Pacioli, a Franciscan friar who was professor of mathematics at Milan, was published in Venice in 1509. In it he derived Euclid's construction and called it *sectio aurea* (the golden section). However, Pacioli's discussion of the proportions of architecture followed Vitruvius (Ref. 10.40). Johann Kepler, the great astronomer who was born sixty-one years after Pacioli's death, also described the golden section and called it one of the two treasures of geometry; but he was concerned with its unusual mathematical not its aesthetic qualities. One of these is that a Fibonacci series based on the golden section gives the same answer as a power series. This series, named after the thirteenth century mathematician Fibonacci (Leonardo of Pisa—see section 5.1), is produced by adding each number to its predecessor in the series. Thus the simplest Fibonacci series is

$$1, 1 + 1 = 2, 2 + 1 = 3, 3 + 2 = 5, 5 + 3 = 8, \text{ etc.}$$

which gives

$$1, 2, 3, 5, 8, 13, 21, \ldots$$

Using Cook's notation (Ref. 10.21) for the golden section ϕ, the Fibonacci series based on ϕ is

$$1, \phi, 1 + \phi, 1 + 2\phi, 2 + 3\phi, 3 + 5\phi \ldots \qquad (10.1)$$

The power series based on ϕ is

$$1, \phi, \phi^2, \phi^3, \phi^4, \phi^5 \ldots \qquad (10.2)$$

Both series produce the same arithmetical sequence:

$$1, 1.618, 2.618, 4.326, 6.854, 11.090 \ldots \qquad (10.3)$$

As we noted, Euclid's construction produced

$$\phi = \text{½} (1 + \sqrt{5})$$

Thus

$$1 + \phi = 1 + \text{½} (1 + \sqrt{5})$$

and

$$\phi^2 = \text{¼} (1 + 2\sqrt{5} + 5) = \text{¼}(4 + 2 + 2\sqrt{5}) = 1 + \text{½}(1 + \sqrt{5})$$

Furthermore,

$$1 + 2\phi = 1 + (1 + \sqrt{5})$$

and

$$\phi^3 = [1 + \text{½}(1 + \sqrt{5})] \cdot [\text{½}(1 + \sqrt{5})] = \text{½}(1 + \sqrt{5}) + [\text{½}(1 + \sqrt{5})]^2$$
$$= \text{½}(1 + \sqrt{5}) + [1 + \text{½}(1 + \sqrt{5})] = 1 + (1 + \sqrt{5})$$

and so on.

This useful and unusual property of the golden section probably accounts for the mysterious qualities attributed to it.

There is no definite information on the rules of proportion, if any, used by the ancient Egyptians, the ancient Greeks, or the Gothic masterbuilders (Ref. 1.1, Section 6.3). Vitruvius (Ref. 10.40) gave as beautiful proportions for rooms three ratios derived from musical harmony, that is, 2:1, 5:3, and 3:2, and in addition the proportion derived from the diagonal of a square $\sqrt{2}$:1 (Ref. 1.1, Section 7.1). Alberti (Ref. 10.41) specifically stated in 1452 that the same numbers that delight the ears also delight the eyes. Palladio (Ref. 10.42) in 1570 restated Vitruvius' proportions and added two further harmonic proportions, namely 1:1 and 4:3. Rudolf Wittkower (Ref. 10.43) has demonstrated that these were the proportions employed in Renaissance architecture.

The concept that the golden section had special aesthetic qualities seems to have originated only in the nineteenth century. In 1854 A. Zeising wrote a book entitled *Neue Lehre von den Proportionen des menschlichen Körpers* (New Theory of the Proportions of the Human Body), published in Leipzig (Ref. 10.44). In 1876 Gustav Theodor Fechner, Professor of Physics at Leipzig University, one of the pioneers of psychophysics and experimental psychology, reported experiments on the proportions of rectangles in his book *Vorschule der Ästhetik*. Three hundred and forty-seven people were asked which of a number of rectangles, ranging from 1:1 to 2.5:1, they liked best; 35% picked 34:21, which equals 1.619:1. Of its two nearest neighbors 21% liked 3:2 best, 20% liked 23:13 = 1.759:1 best, and fewer than 10% opted for any of the others (Ref. 10.20, p. 35). Fechner was a highly respected scientist, and these experiments have been quoted as the scientific basis for the golden section.

In 1914 Sir Theodore Cook, in *The Curves of Life* (Ref. 10.21), gave the letter ϕ (the first letter of the name of the Greek sculptor Pheidias) to the golden section. Cook thought that there was ample evidence to show that the Greeks had used the golden section and used for his analysis the power series (10.2)

$$1, \phi, \phi^2, \phi^3, \phi^4, \phi^5, \ldots$$

This became the Red Scale of *The Modulor*.

In 1921 F. Macody Lund, in *Ad Quadratum* (Ref. 10.22), applied this theory to a large number of Greek temples and Gothic cathedrals.

During the next thirty years more and more evidence was accumulated to show that the ancient Greeks used the golden section mainly by placing geometric constructions over photographs or measured drawings of buildings, vases, and statues.

There were several competitors. Constructions have been made to show that the buildings were, in fact, designed by harmonic proportions, notably the ratio 5:3 = 1.667. The diagonal of the square has been used, and $\sqrt{2} = 1.414$ is also reasonably close to the golden section. The most recent proposal came from Tons Brunés (Ref. 10.23), who gave the name *sacred cut* to the inverse ratio; that is, $10/\sqrt{2} = 7.071$. He showed that it explained the construction of the pyramids, the most famous Egyptian and Greek temples, and the best known Gothic cathedrals.

The constructions reproduced by most proponents are quite accurate, although some used distances between center lines and some used clear distances. In the latter case the fit of the construction depended quite appreciably on the height chosen for the measurement.

The fact that several different rules can be used to explain the same building is, at least, confusing. Pheidias cannot have used the golden section, harmonic proportion, *and* the sacred cut, if indeed he used any of them.

The difference between the golden section (1.618:1) and the harmonic proportion 5:3 (1.667:1) is so small that it is questionable whether many people could tell the difference unless they saw them superimposed on one another; in other words, the argument between those who support the golden section and those who support the harmonic proportion 5:3 on aesthetic grounds seems academic.

It is evident why Alberti favored harmonic proportions. The principal proportions of musical harmony have been known since the fifth century B.C., and it seemed reasonable at the time to assume that the same proportions would please the eye, although there is no evidence to support this view.

The reason for the concept that the golden section has aesthetic qualities is not clear; it has the interesting property that its (additive) Fibonacci series gives the same answer as its (multiplicative) power series (10.1 to 10.3). This might make it useful as the basis for an aesthetic scale without automatically creating beauty in the same way that a musical scale provides the basis for composing music without automatically producing beautiful music.

Its complexity, however, makes it inappropriate for a modular system. Vitruvius had used integral numbers in his analysis of the human figure (Fig. 10.9). Cook in 1914 substituted the powers of the golden section

$$1, \phi, \phi^2, \phi^3, \phi^4, \phi^5, \ldots$$

which equals

$$1, 1.618, 2.618, 4.326, 6.854, 11.090, \ldots$$

Le Corbusier (Ref. 10.39, p. 65) rounded the figures off to form his Red Series (converted in this book from centimeters to millimeters):

$$100, 160, 270, 430, 700, 1130, \ldots \text{mm}$$

and the Blue Series, which is twice the Red Series:

$$200, 330, 530, 860, 1400, 2260, \ldots \text{mm}$$

Whatever the aesthetic merits of the golden section, if any, it is ill-suited for a modular system because it is an incommensurable ratio, and the series thus cannot be expressed in rational numbers.

Having explained why Bemis' module is too simplistic and Le Corbusier's Modulor too complicated, I would be happier if I could now give the solution to the problem. However, no generally accepted system of preferred dimensions has been produced so far.

One difficulty lies in the decimal system, based on the number 10, which divides into only two prime numbers, namely 2 and 5; this limits the range of possible combinations of integral numbers. Ernst Neufert, whose *Bauordnungslehre* has been a standard work in Germany since 1943 (Ref. 10.24), advocated division of the module successively by 2 and its multiplication by 3. Thus we obtain a series $\frac{1}{4}$, $\frac{1}{2}$, 1, and 3. Multiplying by a module of $M = 4$ in., we obtain 1 in., 2 in., 4 in., and 12 in.; this might be interpreted as a vindication of those people who have said for many

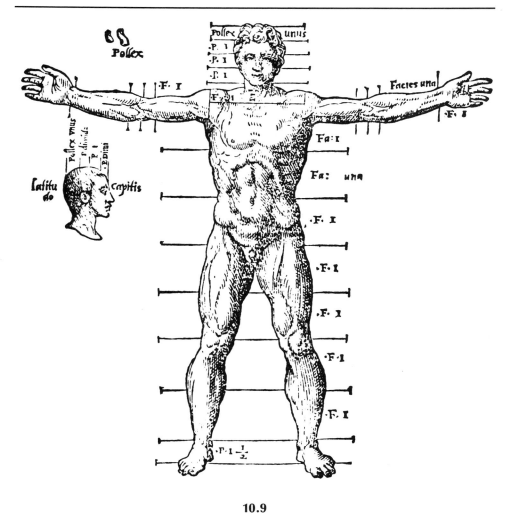

10.9

Proportion of the human figure, as given in Barbaro's edition of Vitruvius' *De Architectura,* published in Venice in 1567. Most of these proportions are given by Vitruvius in Book III, Chapter 1 (Ref. 10, 40, pp. 72–75). The proportions specifically stated by Vitruvius are simple numbers: 10, 8, 6, 4, and 3.

years that the metric system was unnatural and useless. Yet, the metric modular system has produced useful results in most European countries, both east and west.

The module $M = 100$ mm is now almost universally accepted. The multimodule of $3\,M$ and the submodule of $\frac{1}{2}\,M$ are also widely accepted. A multimodule of $6\,M$ and a submodule of $\frac{1}{4}\,M$ are common, except in eastern Europe where $\frac{1}{5}\,M$ is used (Ref.

10.38, pp. 75–81). Design based on these numbers is now normal practice in Denmark (Ref. 10.38).

These rules still leave more dimensions than are appropriate for a system of preferred dimensions. The British ministry responsible for public buildings has published several detailed sets of recommended dimensions (Ref. 10.25) which apply only to particular types of building. Similar statements of preferred dimensions have been produced by other clients.

10.9 INTEGRATION OF THE STRUCTURE AND SERVICES AND DATA PROCESSING

An increasing tendency in Europe is to consider industrialized building as dependent on modular coordination. This is not so in the United States, though the concept originated there. Ezra D. Ehrenkrantz, whose Building Systems Development Inc., has been particularly successful in rationalizing design procedures, said at a summer session on Industrialized Building at M.I.T. in 1969 (Ref. 10.14, pp. 160–161):

The skin of buildings is not nearly as expensive as the services within a building. We have become more concerned with the sophistication of electrical, mechanical and plumbing services. The structure of the enclosure only presents an armature which holds or through which passes the more expensive portions of the total building. As such, dimensional coordination based on old proportional systems alone, or modular coordination based on additive principles of enclosure elements is no longer applicable. This is particularly true with respect to enclosure elements in that we no longer have the common brick as a basis of our everyday construction; in the past, as long as a window or a door was sized to a brick multiple, it would fit. Today we are talking about a variety of planning grids, a variety of modules and largesized products. The basis of their relationship to one another must be much more sophisticated, because we are no longer fitting something of 12 brick widths into a brick wall. Rather, what we are trying to do is match up something of 9 brick widths to another of 12, and this calls for a completely different basis of coordination. We are at a point, today, when coordination of the old type, based on the dimensions of skins, results in buildings designed in a vernacular. This is akin to the way in which a child begins to speak. By saying the same syllable over and over again he uses only a single dimension as the basis of coordination. More sophisticated dimensional disciplines for relating products to one another are not yet available.

If we are going to look at any kind of development which is going to handle more sophisticated building requirements, we have to be able to have large products work together with one another to accomodate small differences that may relate, for example, to joining different column or wall thicknesses, or to a variety of surface requirements. These cannot be handled quite so easily in a simplistic manner. In this respect, I think the challenge is out to us now; if we want to have an approach to construction which offers a considerable number of options for design of our buildings, we must have a system by which we can coordinate those products in the exterior and with the rest of the fabric of the building so that all the products in the building work together. This must be dealt with in a much more sophisticated manner than has been necessary heretofore, and this is the beginning of the third era of system building.

The design of a large modern office building is far more complex, although less difficult, than the design of a Gothic cathedral. The structure of the cathedral had to be proportioned so that it did not fall down. That was much more difficult at the time

10.10

Demolition of three 11-story blocks of the Pruitt-Igoe public housing project in St. Louis, Missouri, in 1972. The complex was designed by Minoru Yamasaki, built in 1954, and regarded as a model for high-quality low-cost housing. The social relations deteriorated to such an extent that the police force was unable to maintain order, and demolition was considered the only possible solution (Ref. 10.29).

(Ref. 1.1, Chapter 6) than the structural design of a modern building is now because the modern engineer has a mathematical theory and empirical data at his disposal (Chapter 4). The Gothic master builder had to understand the style of the building and express the religious purpose of the cathedral, and this also required a great mastery of the art. On the other hand, he was not expected to achieve any specific environmental standards, incorporate building services with the fabric, or even keep to a specified budget or time schedule. If the building could not be completed in one generation, there were more generations to come.

A modern office building has to meet certain standards of thermal comfort, illumination, sound attenuation, and audibility. For this purpose it incorporates a variety of building services, many with conflicting requirements (Chapter 8). The cost and the completion time must be predicted with some measure of accuracy in order to determine whether the project is economical, and the building contractor is expected to keep within the predetermined limits. None of these problems is so difficult as ensuring in the fourteenth century that a Gothic cathedral would not fall down, but the integration of all the design aspects is more complex than the design of the cathedral.

Thus the integration of these various, to some extent conflicting, requirements has become a major design problem. Its complexity and the amount of numerical information involved have led to the use of systems analysis and the processing of data by digital computer (Ref. 10.26). Digital computers have therefore become increasingly important in the design of buildings, quite apart from the design of the structure (Section 5.4).

Data processing is also being applied to social problems. Of those associated with buildings, many have existed for a long time, but the attempt to apply scientific methods to their solution is of recent date. We noted in Section 1.6 that a building that falls down ceases to exist, but a building that has a poor physical environment is likely to remain in service and continue to have the unsatisfactory environment if it is impossible to rectify the fault. The same applies to mistaken social judgments that cannot be corrected. Only in an extreme case (Fig. 10.10) is the building likely to be demolished.

It is self-evident that buildings should be structurally sound and that the materials used in their construction should be durable (Chapters 2 to 6 and 9). Only as society became more affluent was a greater effort spent on improving the physical environment (Chapters 7 and 8). The scientific techniques for investigating social problems, which became available in recent years, have been used so far mainly for welfare and planning. Investigations of the social requirements of the people who are going to live in a particular building are relatively new (Refs. 10.27 and 10.28), but it seems likely that building research and architectural design will to an increasing extent be concerned with social and economic problems.

REFERENCES

1.1 H. J. COWAN: *The Masterbuilders,* Wiley, New York, 1977, 304 pp.

1.2 M. KRANZBERG and C. W. PURSELL (Eds.): *Technology in Western Civilization.* Oxford University Press, New York, 1967. Volume I, 802 pp.

2.1 Report of the ASCE-ACI Joint Committee on Ultimate Strength Design. *Proc. American Society for Civil Engineers,* Vol. 81 (1955), pp. 36–46.

2.2 A. ROBERTSON: *The Strength of Struts.* Selected Engineering Paper No. 28, Institution of Civil Engineers, London, 1925. 55 pp.

2.3 DONALD HOFFMAN: *The Architecture of John Wellborn Root.* Johns Hopkins University Press, Baltimore, 1973. 263 pp.

2.4 JACOB FELD: *Lessons from Failures of Concrete Structures.* Iowa State University Press, Ames, 1964. 179 pp.

2.5 HENRY J. COWAN and IGOR M. LYALIN: *Reinforced and Prestressed Concrete in Torsion.* Arnold, London, 1965. 138 pp.

2.6 T. TREDGOLD: *Practical Essay on the Strength of Cast Iron.* J. Taylor, London, 1824. 305 pp. A standard textbook in the early nineteenth century, still available in many libraries.

2.7 J. F. BAKER: *The Steel Skeleton.* Cambridge University Press, London, 1954 and 1956. Two volumes, 206 and 408 pp.

2.8 S. P. TIMOSHENKO: *History of Strength of Materials.* McGraw-Hill, New York, 1953. 452 pp.

2.9 J. P. M. PANNELL: *An Illustrated History of Civil Engineering.* Thames and Hudson, London, 1964. 376 pp.

2.10 F. A. RANDALL: *History of the Development of Building Construction in Chicago.* University of Illinois Press, Chicago, 1949. 385 pp.

2.11 G. HAEGERMANN, G. HUBERTI, and H. MOLL: *Vom Caementum zum Spannbeton.* (From Caementum to Prestressed Concrete.) Bauverlag, Wiesbaden, 1964. Two volumes, 491 pp.

2.12 P. COLLINS: *Concrete, the Vision of a New Architecture.* Faber, London, 1959. 307 pp.

2.13 W. F. CASSIE: Early reinforced concrete in Newcastle-upon-Tyne. *Structural Engineer,* Vol. 33 (1955), pp. 134–137.

2.14 C. BERGER and V. GUILLERME. *La Construction en Ciment Armé.* Dunod, Paris, 1902. Two volumes, 886 pp. and 69 plates.

2.15 C. F. MARSH: *Reinforced Concrete.* Constable, London, 1904. 545 pp.

2.16 Anniversary issue, *J. American Concrete Institute,* Vol. 35 (1954), pp. 409–524.

2.17 E. MÖRSCH: *Concrete Steel Construction.* (English translation of *Der Eisenbetonbau.*) Engineering News, New York, 1909, 368 pp.

2.18 Preliminary and Second Reports of the Committee on Reinforced Concrete. Institution of Civil Engineers, London 1913. Two volumes, 262 and 187 pp.

2.19 C. W. CONDIT: *The Chicago School of Architecture.* University of Chicago Press, Chicago, 1964. 238 pp.

3.1 PHILIPPE BOUDON: *Lived-in Architecture—Le Corbusier's Pessac Revisited.* Lund Humphries, London, 1972. 200 pp.

3.2   ROLT HAMMOND: *The Forth Bridge and its Builders*. Eyre and Spottiswode, London, 1964. 226 pp.

3.3   W. K. HATT: Notes on the effect of time element in loading reinforced concrete beams. *Proc. Amer. Soc. Testing Materials,* Vol. 7 (1907), pp. 421–433.

3.4   F. R. McMILLAN: Shrinkage and time effects in reinforced concrete. *Bulletin No. 3, Studies in Engineering,* University of Minnesota, 1915, 41 pp.

3.5   L. T. C. ROLT: *Victorian Engineering.* Allen Lane, London, 1970. 300 pp.

3.6   H. J. COWAN: *Architectural Structures.* Second edition. American Elsevier, New York, 1975. 442 pp.

3.7   A. L. A. HIMMELWRIGHT: *The San Francisco Earthquake and Fire, 1906.* The Roebling Construction Company, New York, 1907.

3.8   HENRY J. DEGENKOLB: *Earthquake Forces on Tall Structures.* Bethlehem Steel Corporation, Bethlehem, Pennsylvania, 1970. 24 pp.

3.9   T. Y. LIN: *Design of Prestressed Concrete Structures.* Second edition. Wiley, New York, 1963. 614 pp.

3.10 The test of time—the centenary of Auguste Perret's birth. *Concrete Quarterly (London),* No. 103 (October-December 1974), pp. 8–11.

3.11 LE CORBUSIER (Translated by F. ETCHELLS): *Towards a New Architecture.* Architectural Press, London, 1970. 269 pp. This is a paperback facsimile of the first English edition, published in 1927. The original *Vers Une Architecture* was published in Paris in 1923.

3.12 S. GIEDION: *Space, Time and Architecture.* Fifth edition. Harvard University Press, Cambridge, 1967. 897 pp.

3.13 M. BILL: *Robert Maillart.* Pall Mall Press, London, 1969. 184 pp.

3.14 A. CASTIGLIANO: *The Theory of Equilibrium of Elastic Systems and Its Application.* Dover, New York, 1966. 360 pp. This is a facsimile of the English translation by E. S. Andrews published in 1919. The translation was made from the French version published in Turin in 1879.

4.1   LE CORBUSIER: *The City of Tomorrow.* Architectural Press, London, 1971. 302 pp. (First French edition by Editions Crés, Paris, 1924).

4.2   HENRY-RUSSELL HITCHCOCK: *The Rise of an American Architecture.* Pall Mall Press, London, 1970. 241 pp.

4.3   CARL W. CONDIT: *Chicago 1910–29.* University of Chicago Press, Chicago, 1973. 354 pp.

4.4   H. MANDERLA: Die Berechnung der Sekundärspannungen. (The Calculation of Secondary Stresses.) *Allgemeine Bauzeitung,* Vol. 80 (1880), p. 34.

4.5   OTTO MOHR: *Abhandlungen aus dem Gebiete der Technischen Mechanik.* (A Treatise on Engineering Mechanics.) Second edition. W. Ernst, Berlin, 1914. 567 pp.

4.6   A. BENDIXEN: *Die Methode der Alpha Gleichungen zur Berechnung von Rahmenkonstruktionen.* (The Method of the Alpha Equations for Calculating Structural Frames.) Berlin, 1914. (Quoted in Ref. 2.8 p. 422.)

4.7   W. M. WILSON and G. A. MANEY: *Wind Stresses in Office Buildings.* Bulletin No. 80. Engineering Experiment Station, University of Illinois, Urbana, 1915.

4.8   A. KLEINLOGEL: *Rigid Frame Formulas.* Twelfth edition. Crosby Lockwood, London, 1957. 480 pp. (First German edition by W. Ernst, Berlin 1913).

4.9   A. KLEINLOGEL and A. HASELBACH: *Multibay Frames.* Seventh edition. Crosby Lockwood, London, 1963. 469 pp.

4.10 W. L. SCOTT and W. H. GLANVILLE: *Explanatory Handbook on the Code of Practice for Reinforced Concrete.* Concrete Publications, London, 1934. 143 pp.

4.11 R. L. SANDERSON: *Codes and Code Administration——An Introduction to Building Regulations in the United States.* Building Officials Conference of America, Chicago, 1969. 241 pp.

4.12 H. MARCUS: *Die Theorie elastischer Gewebe und ihre Anwendung auf die Berechnung biegsamer Platten.* (The Theory of Elastic Networks and Its Application to the Flexural Design of Thick Plates.) Julius Springer, Berlin, 1924.

4.13 E. MÖRSCH: *Der Eisenbetonbau.* (Reinforced Concrete Construction.) Sixth edition. Konrad Wittwer, Stuttgart, 1929. Vol. I, Part II, p. 499.

4.14 H. M. WESTERGAARD and W. A. SLATER: Moments and stresses in slabs. *Proc. American Concrete Institute,* Vol. 17 (1921), pp. 415–525.

4.15 K. A. FAULKES: The design of flat slab structures—an historical survey. *Univciv Report No. R-129,* University of New South Wales, Kensington (Australia), 1974. 42 pp.

4.16 T. CRANE and R. NOLAN: *Concrete Building Construction.* Wiley, New York, 1927. 689 pp.

4.17 J. R. NICHOLS: Statical limitations upon steel requirements in reinforced concrete flat slab floors. *Trans. American Society of Civil Engineers,* Vol. 77 (1914), pp. 1670–1736.

4.18 F. GRASHOF: *Theorie der Elastizität und Festigkeit* (Theory of Elasticity and of the Strength of Materials.) Berlin, 1878, p. 358. Quoted by A. NADAI: *Die elastischen Platten* (Elastic Plates). Julius Springer, Berlin 1925 (reprinted 1968). 326 pp.

4.19 HENRY T. EDDY and C. A. P. TURNER: *Concrete-Steel Construction,* Part I—Buildings. Second edition. Privately published, Minneapolis, 1919. 477 pp.

4.20 V. LEWE: *Pilzdecken.* (Mushroom Floors.) Wilhelm Ernst, Berlin, 1926. 182 pp.

4.21 H. D. DEWELL and H. B. HAMILL: Flat slabs and supporting columns and walls designed as indeterminate structural frames. *Proc. American Concrete Institute,* Vol. 34 (1938), pp. 321–343.

4.22 J. DI STASIO: Flat plate rigid frame design of low cost housing projects in Newark and Atlantic City, N.J. *Proc. American Concrete Institute,* Vol. 37 (1941), pp. 309–324.

4.23 F. A. BLAKEY: Towards an Australian structural form—the flat plate. *Architecture in Australia,* Vol. 54 (1965), pp. 115–127.

4.24 G. T. KIRCHHOFF: *Vorlesungen über die mathematische Physik.* (Lectures on Mathematical Physics.) Third edition, G. Teubner, Leipzig, 1883. 449 pp.

4.25 ALFRED PUGSLEY: *The Safety of Structures.* Arnold, London, 1966. 156 pp.

4.26 F. A. RANDALL: Historical notes on structural safety. *Proc. American Concrete Institute,* Vol. 70 (1973), pp. 669–679.

4.27 E. H. SALMON: *Columns.* Frowde, Hodder and Stoughton. London, 1921.

4.28 O. G. JULIAN: Synopsis of the first progress report of the Committee on Factors of Safety. *Proc. American Society of Civil Engineers.* Vol. 83 (July 1957), Separate No. 1316, 22 pp.

4.29 INSTITUTION OF STRUCTURAL ENGINEERS: Report on Structural Safety. *Structural Engineer,* Vol. 34 (May 1955), pp. 141–149.

4.30 E. TORROJA: Superimposed loads and safety factors. *CIB Bulletin* (International Council for Building Research Studies and Documentation, Rotterdam). Vol. 1958, No. 1–2, pp. 1–7.

4.31 J. O. BRYSON and D. GROSS: Techniques for the survey and evaluation of live floor loads and fire loads in modern office buildings. *Building Science Series No. 16.* National Bureau of Standards, Washington, D.C., 1967. 30 pp.

4.32 (a) G. R. MITCHELL and R. W. WOODGATE: Floor loadings in office buildings. (CP 3/71. 30 pp. (b) *Ibid.* Floor loading in retail premises—the results of a survey. CP 25/71. 37 pp. *Building Research Station Current Papers,* London 1971.

4.33 A. I. JOHNSON: Strength, safety, and economical dimensions of structures. *Institutionen för Byggnadsstatik,* Meddelanden Nr. 13. Kungl. Tekniska Högskolan, Stockholm, 1953. Pp. 109–113.

4.34 (a) *Cyclone "Althea"*—Part 1, Buildings. James Cook University of Northern Queensland, Townsville, 1972. 211 pp. (b) GEORGE R. WALKER: *Report on Cyclone "Tracy"—Effect on Buildings, December 1974.* Department of Housing and Construction, Canberra, 1975. Volume I, 66 pp.

4.35 H. van KOTEN: Wind measurements in high buildings in the Netherlands. *Wind Effects on Buildings and Structures.* University of Toronto Press, Toronto, 1968. Volume I, pp. 685–704.

4.36 M. J. LIGHTHILL and A. SILVERLEAF (Eds.): A discussion on architectural aerodynamics. *Philosophical Transactions of the Royal Society of London,* Vol. 269 A (1971), pp. 323–554.

4.37 DAVID B. STEINMAN and SARA RUTH WATSON: *Bridges and their Builders.* Dover, New York, 1957. Pp. 357–364.

4.38 ALFRED KNOTT: *Shelter Design in New Buildings.* Office of Civil Defense, Washington, D.C., 1967. 83 pp.

4.39 PETER S. RHODES: The structural assessment of buildings subjected to bomb damage. *Structural Engineer,* Vol. 52 (1974), pp. 329–339.

4.40 *Report of the Inquiry into the Collapse of Flats at Ronan Point, Canning Town.* H. M. Stationery Office, London, 1968. [Reviewed by S. FIRNKAS in *Civil Engineering (New York),* November 1969, p. 96.]

4.41 The implications of the Report of the Enquiry into the Collapse of Flats at Ronan Point, Canning Town. *Structural Engineer,* Vol. 47 (1969), pp. 265–284.

4.42 *The Building (Fifth Amendment) Regulations 1970.* Statutory Instruments 1970, No. 109. Building and Buildings. H. M. Stationery Office, London, 1970.

4.43 R. H. FERAHIAN: Design against progressive collapse. *Technical Paper No. 332, Division of Building Research,* National Research Council, Ottawa, 1971. 30 pp.

4.44 J. D. RIERA: On the stress analysis of structures subjected to aircraft impact forces. *Nuclear Engineering and Design,* Vol. 8 (1968), pp. 415–426.

4.45 Empire State Building intact after bomber hits 79th floor. *Engineering News Record,* Vol. 135 (1935), pp. 129–130.

4.46 MARKUS REINER: *Deformation and Flow.* H. K. Lewis, London, 1960, 347 pp.

4.47 N. F. MOTT: *Atomic Structure and the Strength of Metals.* Pergamon, London, 1956. 64 pp.

4.48 L. H. van VLACK: *Elements of Materials Science.* Addison-Wesley, Reading, 1965. 445 pp.

4.49 GABOR KAZINCZY: *Kiserletek bafalazott tartokkal.* (Experiments with Clamped Girders.) *Betonszemle,* Hungary, Vol. 2 (1914), pp. 68, 83, and 101.

4.50 *First, Second and Final Reports of the Steel Structures Research Committee.* H. M. Stationery Office, London, 1931, 1934, and 1936.

4.51 C. E. MASSONET and M. A. SAVE: *Plastic Analysis and Design.* Blaisdell, New York, 1965. Volume I, pp. 1–8.

4.52 H. KEMPTON DYSON: What is the use of the modular ratio? *Concrete and Constructional Engineering.* Vol. 17 (1922), pp. 330–336, 408–415, and 486–491.

4.53 C. S. WHITNEY: Design of reinforced concrete members under flexure and combined flexure and direct compression. *Proc. American Concrete Institute,* Vol. 33, pp. 483–498.

4.54 R. H. WOOD: Effective lengths of columns in multi-story buildings. Part 3. *Structural Engineer,* Vol. 52 (1974), pp. 341–346.

4.55 HARDY CROSS: *Selected Papers in Arches, Continuous Frames, Frames and Conduits.* University of Illinois Press, Urbana 1963. 265 pp.

4.56 F. EMPERGER (Ed.): *Handbuch für Stahlbetonbau.* (Handbook of Reinforced Concrete.) Third edition, Volume 1, The elements of the historical development of reinforced concrete, experiments and theory (in German). W. Ernst, Berlin, 1921, 800 pp.

4.57 W. W. ROUSE BALL: *A Short Account of the History of Mathematics.* Dover, New York, 1960. 522 pp. This is a reprint of the fourth edition of 1908; it excludes the twentieth century.

4.58 J. W. DUNHAM et al.: Live loads on floors in buildings. *National Bureau of Standards, Building Materials and Structures Report 133,* Superintendent of Documents, Washington, D.C., 1952. 27 pp.

4.59 G. R. MITCHELL: Loadings on buildings—a review paper. *Building Research Station Current Paper 50–69.* Ministry of Public Buildings and Works, London, 1969. 9 pp.

4.60 W. R. SCHRIEVER and V. A. OSTVANOV: Snow loads—preparation of standards for snow loads on roofs in various countries, with particular reference to the USSR and Canada, in *On Methods of Load Calculation,* CIB Report No. 9, International Council for Building Research Studies and Documentation. Rotterdam, 1967, pp. 13–33.

4.61 J. D. McCREA: Wind Action on Structures. *Bibliography No. 5, Division of Building Research,* Ottawa, 1952. 6 pp.

4.62   *Proceedings of the Third International Conference on Wind Effects on Buildings and Structures.* Saikon, Tokyo, 1972. 1267 pp.

4.63   *Planning and Design of Tall Buildings.* American Society of Civil Engineers, New York, 1973. Five volumes, 5138 pp.

4.64   H. J. COWAN et al.: *Models in Architecture.* Elsevier, London, 1968. 228 pp.

4.65   *Encyclopaedia Britannica.* Ninth edition. Adam and Charles Black, Edinburgh, 1875–1888. Twenty-five volumes.

4.66   ARTHUR MORLEY: *Strength of Materials.* Eighth edition. Longmans, London, 1934. 569 pp.

5.1   R. R. FENICHEL and J. WEIZENBAUM (Eds.): *Computers and Computation.* Freeman, San Francisco, 1971. 282 pp.

5.2   G. E. BEGGS: The accurate mechanical solution of statically indeterminate structures by the use of paper models and special gauges. *J. American Concrete Institute,* Vol. 18 (1922), pp. 58–78.

5.3   O. GOTTSCHALK: Mechanical calculation of elastic systems. *J. Franklin Institute,* Vol. 202 (1926), pp. 61–88.

5.4   W. J. ENEY: New deformeter apparatus. *Engineering News-Record,* Vol. 122 (February 16, 1939), pp. 221.

5.5   *Model Tests of Boulder Dam.* Boulder Canyon Project—Final Reports, Part V—Technical Investigations, Bulletin 3. Bureau of Reclamation, Denver, 1939, pp. 22–23.

5.6   W. B. PREECE and J. D. DAVIES: *Models for Structural Concrete.* C. R. Books, London, 1964. 252 pp.

5.7   E. FUMAGALLI: *Statical and Geomechanical Models.* Springer, Vienna, 1973. 182 pp.

5.8   G. MURPHY, D. L. SHIPPY, and H. L. LUO: *Engineering Analogies.* Iowa State University Press, Ames, 1963. 255 pp.

5.9   A. NADAI: Der Beginn des Fliessvorganges in einem tordierten Stab. (The Beginning of Plasticity in a Bar Subject to Torsion.) *Zeitschrift für angewandte Mathematik und Mechanik.* Vol. 3 (1923), p. 442.

5.10   A. NADAI: *Theory of Flow and Fracture of Solids,* Volume 1. McGraw-Hill, New York, 1950. 512 pp.

5.11   K. K. KEROPYAN and P. M. CHEGOLIN: *Electrical Analogues in Structural Engineering.* Arnold, London, 1967. 274 pp.

5.12   VANNEVAR BUSH: Structural analysis by electric circuit analogues. *J. Franklin Institute,* Vol. 217 (1934), p. 289.

5.13   FREDERICK L. RYDER: Electrical analogs of statically loaded structures. *Proc. American Society of Civil Engineers,* Vol. 79 (December 1953), Separate No. 376. 24 pp.

5.14   R. K. LIVESLEY: The pattern of structural computing: 1946–1966. *Structural Engineer,* Vol. 46 (1968), pp. 169–182.

5.15   G. KRON: Tensorial analysis and equivalent circuits of elastic structures. *J. Franklin Institute,* Vol. 238 (1944), p. 400.

5.16   R. S. FENVES, C. L. MILLER, and E. H. KING: Guide to the development of electronic computer programs. *Proc. American Society of Civil Engineers, J. Structural Division,* Vol. 90, No. ST 6 (December 1964), 57 pp.

5.17   S. J. FENVES, R. D. LOGCHER, S. D. MAUCH and K. F. REINSCHMIDT: *STRESS—A User's Manual.* M.I.T. Press, Cambridge, 1964, 57 pp.

5.18   R. V. SOUTHWELL: *Relaxation Methods in Engineering Science.* Oxford University Press, London, 1940. 252 pp.

5.19   (a) J. H. ARGYRIS: *Energy Theorems and Structural Analysis.* Butterworth, London, 1960. (Reprinted from *Aircraft Engineering.*) (b) J. M. TURNER, R. W. CLOUGH, H. C. MARTIN, and L. J. TOPP: Stiffness and deflection analysis of complex structures. *J. Aeronautical Sciences,* Vol. 23 (1956), p. 805. (Paper presented at the National AIA meeting in New York in 1954.)

5.20   T. R. W. CLOUGH: The finite element in plane stress analysis. *Proc. Second Conference on Electronic Computation.* American Society of Civil Engineers, Pittsburgh Section, Pittsburgh, 1960, pp. 345–378.

5.21 T. M. CHARLTON: *Model Analysis of Plane Structures*. Pergamon, Oxford, 1966, 112 pp.

5.22 CHARLES and RAY EAMES: *A Computer Perspective*. Harvard University Press, Cambridge, Mass., 1973. 174 pp.

5.23 C. B. BOYER: *A History of Mathematics*. Wiley, New York, 1968. 717 pp.

5.24 R. K. LIVESLEY: Analysis of rigid frames by an electronic digital computer. *Engineering,* Vol. 176 (1953), pp. 230 and 277.

6.1 F. STÜSSI: *Das Problem der grossen Spannweite* (The Problem of Large Spans.) Verlag V.S.B., Zurich, 1954. 46 pp.

6.2 BERNARD E. JONES: *Cassell's Reinforced Concrete*. Waverly, London, 1920. 432 pp.

6.3 D.A.L. SAUNDERS: The reinforced concrete dome of the Melbourne Public Library, 1911. *Architectural Science Review,* Vol. 2, pp. 39–46.

6.4 D. HILBERT and S. COHN-VOSSEN: *Geometry and the Imagination*. Chelsea, New York, 1952. Chapter 4, Differential Geometry, pp. 172–271.

6.5 W. S. WLASSOW (V. Z. VLASOV): *Allgemeine Schalentheorie und ihre Anwendung auf die Technik*. Akademie Verlag, (East) Berlin, 1958. 661 pp. There is no English translation at present.

6.6 *Design of Cylindrical Shells*. Manual of Engineering Practice, No. 31, American Society of Civil Engineers, New York, 1952. 175 pp.

6.7 D. RÜDIGER and J. URBAN: *Circular Cylindrical Shells*. Teubner, Leipzig, 1955. 270 pp.

6.8 MIRCEA SOARE: *Application of Finite Difference Equations to Shell Analysis*. Pergamon, Oxford, 1967. 439 pp.

6.9 R. W. CLOUGH and C. P. JOHNSON: Finite element analysis of arbitrary thin shells. *Concrete Thin Shells*. Publication SP-28. American Concrete Institute, Detroit, 1971, pp. 333–364.

6.10 F. AIMOND: Étude statique des voiles minces en paraboloide hyperbolique travaillant sans flexion (Statical theory for hyperbolic paraboloid shells without bending) *Publ. International Association for Bridge and Structural Engineering,* Vol. 4 (1936), pp. 1–112.

6.11 Y. C. YANG: Record concrete shell roof featured in Coliseum. *Civil Engineering* (New York), Vol. 2 (1972).

6.12 M. SOARE: Zur Membran-theorie der Konoidschalen. *Der Bauingenieur,* Vol. 33 (July 1958) pp. 256–265. (Translated as *Membrane Theory of Shells,* library translation of the Cement and Concrete Association, No. 82, London, 1959.)

6.13 JOACHIM BORN: *Hipped-Plate Structures*. Crosby Lockwood, London, 1962, 250 pp.

6.14 G. WINTER and M. PEI: Hipped-plate construction. *Proc. American Concrete Institute,* Vol. 43 (1947), pp. 505–531.

6.15 DAVID YITZHAKI: *Prismatic and Cylindrical Shell Roofs*. Haifa Science, Haifa 1958, 252 pp.

6.16 S. UTKU: ELAS-A general purpose digital computer program for the equilibrium problems of linear structures. *Concrete Thin Shells*. Publication SP-28. American Concrete Institute, Detroit, 1971, pp. 383–417.

6.17 J. W. A. SCHWEDLER: Theorie der Kuppelflächen. (Theory of Dome Surfaces.) *Zeitschrift für Bauwesen* (Berlin), 1866, p. 7.

6.18 B. S. BENJAMIN: *The Analysis of Braced Domes*. Asia Publishing House, London, 1963. 110 pp.

6.19 R. W. SOUTHWELL: Primary stress determination in space frames. *Engineering*. Vol. 109 (February 6, 1920), p. 165.

6.20 L. BASS: Lamella domes in the United States. *Ref.* 6.45, pp. 955–964. The author was one of the structural designers of the Astrodome.

6.21 PAUL H. COY: *Structural Analysis of "Unistrut" Space-Frame Roofs*. University of Michigan Press, Ann Arbor, 1959. Two volumes, 267 and 59 pp.

6.22 F. N. SEVERUD and R. G. CORBOLLETTI: Hung roofs. *Progressive Architecture*. Vol. 37 (March 1956), pp. 99–107.

6.23 SEYMOUR HOWARD: Suspension structures. *Architectural Record.* Vol. 128 (September 1960), pp. 230–240.

6.24 An elegant sports and recreation center. *Architectural Record.* Vol. No. 143 (June 1968), pp. 121–127.

6.25 F. LEONHARDT and W. ANDRÄ: Entwurf eines Leichtbetonhängedaches. (Design of a Lightweight Concrete Suspension Roof). *Bauingenieur,* Vol. 32 (No. 7, 1957), p. 344.

6.26 *Eero Saarinen on his Work.* Yale University Press, New Haven, 1968. 118 pp.

6.27 ULRICH FINSTERWALDER: Vorgespannte Schalenbauten. (Prestressed Shells.) *Vorträge Hauptversammlung 1954.* Deutscher Beton-Verein, Wiesbaden, 1954. pp. 145–161.

6.28 The Philips Pavilion at the 1958 Brussels World Fair. *Philips Technical Review,* Vol. 20 (No. 1, 1958–1959), pp 1–36.

6.29 FREI OTTO: *Das hängende Dach.* (The Hanging Roof.) Ullstein, Berlin, 1954. 160 pp.

6.30 F. LEONHARDT, H. EGGER, and E. HAUG: Der deutsche Pavillon auf der Expo '67 Montreal—eine vorgespannte Seilnetzkonstruktion. (The German Pavilion at the Expo '67 Montreal—a Prestressed Net Structure.) *Der Stahlbau,* Vol. 37 (Nos. 4 and 5, 1968). A reprint of 19 pp.

6.31 *The Work of Frei Otto.* Museum of Modern Art, New York, 1972. 128 pp.

6.32 D. H. GEIGER: U. S. Pavilion at Expo 70 features air-supported roof. *Civil Engineering (New York),* Vol. 40 (March 1970), pp. 48–50.

6.33 G. J. POHL and H. J. COWAN: Multi-storey air-supported building construction. *Build International,* Vol. 5 (March/April 1972), pp. 110–118.

6.34 BUCKMINSTER FULLER: *Ideas and Integrities.* Prentice-Hall, Englewood Cliffs, New Jersey, 1963. Figure between pp. 192 and 193.

6.35 I. TODHUNTER and K. PEARSON: *A History of the Theory of Elasticity.* Dover, New York, 1960. Three volumes, 2244 pp. A reprint of the 1886–1893 edition.

6.36 G. von KLASS: *Weit spannt sich der Bogen* (Great is the Span of the Arch.) Dyckerhoff und Widmann. Munich 1955. 234 pp.

6.37 A. PFLÜGER: *Elementary Statics of Shells.* Dodge, New York, 1961. 121 pp.

6.38 Nicolas Esquillan—*Cinquante Ans à l'Avant-Garde du Genie Civil.* (Fifty years in the Forefront of Civil Engineering.) Syndicat National du Béton Armé et des Techniques Industrialisées. Paris, 1974. 118 pp.

6.39 C. FABER: *Candela: Shell Builder.* Reinhold, New York, 1963. 240 pp.

6.40 J. JOEDICKE: *Schalenbau* (Shell Construction). Karl Kramer Verlag, Stuttgart, 1962. 304 pp.

6.41 A. M. HAAS: *Thin Concrete Shells,* Wiley, New York, 1962 and 1967. Two volumes. 129 and 242 pp.

6.42 Z. S. MAKOWSKI: *Steel Space Structures.* Michael Joseph, London 1965. 214 pp.

6.43 R. W. MARKS: *The Dymaxion World of Buckminster Fuller.* Reinhold, New York 1960, 232 pp.

6.44 K. WACHSMANN: *The Turning Point in Building.* Reinhold, New York, 1961. 239 pp.

6.45 *Proc. First International Conference on Space Structures,* Blackwell, Oxford, 1967. 1233 pp.

6.46 J. NEEDHAM: *Clerks and Craftsmen in China and the West.* Cambridge University Press, London, 1970. 470 pp.

6.47 *Hanging Roofs.* North-Holland, Amsterdam, 1963. 335 pp. Proceedings of the Paris Colloquium held in 1962 by the International Association for Shell Structures.

6.48 D. W. LANCHESTER: *Span.* Butterly and Wood, Manchester, 1939. 100 pp. An invitation lecture delivered to the Manchester Association of Engineers.

6.49 R. N. DENT: *Principles of Pneumatic Architecture.* Architectural Press, London, 1971. 236 pp.

6.50 F. OTTO: *Tensile Structures.* M.I.T. Press, Cambridge, 1967. Volume 1. Pneumatic Structures. 320 pp. Volume 2. Cables, Nets and Membranes. 171 pp.

6.51 *Tension Structures and Space Frames.* Architectural Institute of Japan, Tokyo, 1972. 1042 pp.

6.52 The Structures of Eduardo Torroja—*An Autobiography of Engineering Accomplishment.* Dodge, New York, 1958. 198 pp.

7.1   E. W. MARCHANT (Ed.): *A Complete Guide to Fire and Buildings.* MTP Publishing Co., Lancaster, England, 1972. 268 pp.

7.2   E. GRANDJEAN: *Ergonomics in the Home.* Taylor and Francis, London, 1973. 344 pp.

7.3   J. W. KERR: Historic fire disasters. *Fire Research Abstracts and Reviews,* National Academy of Sciences, Washington, D. C., Volume 13 (1971), pp. 1–16.

7.4   STEPHEN BARLAY: *Fire.* Hamish Hamilton, London, 1972. 293 p.

7.5   *Report of the Summerland Fire Commission,* Government Office, Isle of Man, Douglas, 1974. 82 pp.

7.6   ROBERT CROMIE: *The Great Chicago Fire.* McGraw-Hill, New York, 1958. 282 pp.

7.7   *Encyclopaedia Britannica,* Chicago, 1965. Vol. 9, *Fire Protection.*

7.8   *Notes on Building Construction.* Third edition. Rivingtons, London, 1877, Part II, 256 pp.

7.9   *Fire Test Performance.* Special Technical Publication 464. American Society for Testing Materials, Philadelphia, 1970. 243 pp.

7.10 M. GALBREATH: A survey of exit facilities in high office buildings. *Building Research Note No. 64.* National Research Council of Canada, Ottawa, 1968. 8 pp.

7.11 G. W. SHORTER and K. SUMI: Fire and water—the newest developments. *Fire Fighting in Canada.* April 1962 (reprinted as Fire Research Note No. 2, National Research Council of Canada, Ottawa, 1965) 3 pp.

7.12 Centenary of the automatic sprinkler system. *AFPA Fire Journal (Melbourne),* Vol. 1, No. 1, September 1974, p. 31.

7.13 F. W. FITZPATRICK et al.: *Cyclopaedia of Fire Protection.* Volume II. American School of Correspondence, Chicago, 1912. 460 pp.

7.14 F. W. ROBINS: *The Story of Water Supply.* Oxford University Press, London, 1946. 207 pp.

7.15 W. M. FRAZER: *A History of English Public Health—1834–1939.* Ballière, Tindall and Cox, London, 1950. 498 pp.

7.16 EDWIN CHADWICK: *Report on the Sanitary Condition of the Labouring Population of Great Britain,* published in 1842. Abridged and edited by M. W. FLINN. Edinburgh University Press, Edinburgh, 1965. 443 pp.

7.17 J. H. CLAPHAM: *An Economic History of Modern Britain.* Vol. 2: *Free Trade and Steel, 1850– 1886.* Cambridge University Press, London, 1963 (First published in 1932).

7.18 C. ROETTER: *Fire is their Enemy.* Angus and Robertson, Sydney, 1962. 184 pp.

7.19 N. FITZSIMONS (Ed.): *The Reminiscences of John B. Jervis—Engineer of Old Croton.* Syracuse University Press, Syracuse, 1971. 196 pp.

7.20 T. T. LIE: *Fire and Buildings.* Applied Science Publishers, London, 1972. 276 pp.

7.21 S. B. HAMILTON: *A Short History of the Structural Fire Protection of Buildings.* H. M. Stationery Office, London, 1958. 73 pp.

7.22 *Proceedings of the Symposium on the Fire Protection of High Rise Buildings.* Chicago Committee on High Rise Buildings and Illinois Institute of Technology, Chicago, 1972. 303 pp.

8.1   H. M. WINGLER: *The Bauhaus.* M.I.T. Press, Cambridge, 1969, 653 pp.

8.2   REYNER BANHAM: *Age of the Masters.* Architectural Press, London, 1975. 170 pp.

8.3   J. M. RICHARDS and N. PEVSNER (Eds.): *The Anti-Rationalists.* Architectural Press, London, 1973. 210 pp.

8.4   LEONARD HILL: *Report on Ventilation and the Effect of Open Air and Wind on the Respiratory Metabolism.* Reports of the Local Government Board of Public Health, New Series, No. 100. London, 1914.

8.5   D. C. WHITE: Energy, the economy and the environment. *Technology Review,* Vol. 70 (October-November 1971), pp. 18–31.

8.6   A. F. DUFTON: The equivalent temperature of a room and its measurement. *Building Research Technical Paper No. 13.* H. M. Stationery Office, London, 1932. 8 pp.

8.7   F. C. HOUGHTEN and C. P. YAGLOU: Determination of the comfort zone. *Trans. Amer. Soc. Heating and Ventilating Engineers,* Vol. 29 (1923), p. 361.

8.8   F. P. ELLIS: Thermal comfort in warm and humid atmospheres—observations on groups and individuals in Singapore. *J. Hygiene,* Vol. 51 (September 1953), pp. 386–404.

8.9   P. O. FANGER: *Thermal Comfort.* Danish Technical Press, Copenhagen, 1970. 244 pp.

8.10   T. C. ANGUS and J. R. BROWN: Thermal comfort in the lecture room. *J. Inst. Heating and Ventilating Engineers,* Vol. 25 (1957), pp. 175–182.

8.11   R. K. MacPHERSON in Chapter 2, *Physiological Background of Air Conditioning,* in N. R. SHERIDAN (Ed.): *Air Conditioning.* Queensland University Press, Brisbane, 1963. pp. 19–39.

8.12   AMOS RAPOPORT: *House Form and Culture.* Prentice-Hall, Englewood Cliffs, New Jersey, 1969. 150 pp.

8.13   M PAWLEY: *Mies van der Rohe.* Thames and Hudson, London, 1970, 134 pp.

8.14   *Frank Lloyd Wright–The Early Work.* Horizon, New York, 1968. 144 pp.

8.15   A. F. DUFTON and H. E. BECKETT: The heliodon—an instrument for demonstrating the apparent motion of the sun. *J. Scientific Instruments,* Vol. 9 (1932), pp. 251–256.

8.16   *American Steam and Hot Water Heating Practice, Being a Selective Reprint of Descriptive Articles from the Engineering Record.* Engineering Record. London, 1895. 317 pp. (The Engineering Record was published in New York).

8.17   MARGARET INGELS: *William Haviland Carrier, Father of Air Conditioning.* Country Life Press, Garden City, New York, 1952. 176 pp.

8.18   FRANK LLOYD WRIGHT: *The Natural House.* Pitman, London, 1971. 223 pp.

8.19   LE CORBUSIER: *When the Cathedrals Were White.* McGraw-Hill, New York, 1947. 217 pp. (Published in French in 1937; first English edition published by Harcourt, Brace and Jovanovich.)

8.20   *Planning and Design of Tall Buildings,* Volume Ia. Environmental Systems, pp. 1–301. American Society of Civil Engineers, New York, 1972.

8.21   CARRIER AIR CONDITIONING COMPANY: *Handbook of Air-Conditioning Design.* McGraw-Hill, New York, 1965. 803 pp.

8.22   (a) *ASHRAE Handbook of Fundamentals.* American Society of Heating, Refrigeration and Air Conditioning Engineers, New York, 1969. 544 pp. (b) *Ibid.,* New York, 1972. 688 pp.

8.23   L. J. OVERTON: *Heating and Ventilating.* Sutherland, Manchester, 1944. 340 pp.

8.24   *Rivingtons' Notes on Building Construction.* Longmans, London, 1915. Part I. 306 pp.

8.25   PERCY DUNSHEATH: *A History of Electrical Engineering,* Faber, London, 1962. 368 pp.

8.26   J. MELLANBY: *The History of Electrical Wiring.* Macdonald, London, 1957. 244 pp.

8.27   FARRINGTON DANIELS: *Direct Use of the Sun's Energy.* Yale University Press, New Haven, 1964. 374 pp.

8.28   S. V. SZOKOLAY: *Solar Energy and Building.* Architectural Press, London, 1975. 148 pp.

8.29   A. THAU: Architectural and town planning aspects of domestic solar water heaters. *Architectural Science Review,* Vol. 16 (1973), pp. 89–104.

8.30   *The First One Hundred Years.* Otis Elevator Company, New York, 1953. 44 pp.

8.31   GEORGE R. STRAKOSCH: *Vertical Transportation.* Wiley, New York, 1967. 365 pp.

8.32   Spiral lifts for the Eiffel Tower. *The Engineer,* Vol. 66 (August 3, 1888), pp. 100–102.

8.33   J. M. TOUGH and C. A. O'FLAHERTY: *Passenger Conveyors.* Ian Allan, London, 1971. 176 pp.

8.34   S. WELLS: Edison's electrical light. *IES Lighting Review* (Sydney), Vol. 37 (August 1975), pp. 97–99.

8.35   W. HARRISON and E. A. ANDERSON: Effect of reflection from surroundings in an experimental room. *Trans. Illuminating Engineering Society,* Vol. 11 (February 1916), pp. 67–91.

8.36   J. W. GRIFFITH: *Predicting Daylight as Interior Illumination.* Libby-Owens-Ford Glass Company, Toledo, 1958. 27 pp.

8.37   BUILDING RESEARCH STATION: *BRS Daylight Protractors.* H. M. Stationery Office, London, 1968. 25 pp.

8.38 R. G. HOPKINSON, P. PETHERBRIDGE, and J. LONGMORE: *Daylighting*. Heinemann, London, 1966.

8.39 P. MOON and D. E. SPENCER: Illumination from a nonuniform sky. *Illuminating Engineering,* Vol. 37 (December 1942), pp. 707–726.

8.40 *Code of Practice for Daylighting of Buildings*. Indian Standards Institution, Delhi, 1969. 34 pp.

8.41 H. C. WESTON: *Sight, Light and Work*. H. K. Lewis, London, 1962. 283 pp.

8.42 R. G. HOPKINSON and J. B. COLLINS: *The Ergonomics of Lighting*. Macdonald, London, 1970. 272 pp.

8.43 FRIEDRICH BRUCKMAYER: *Handbuch der Schalltechnik im Hochbau*. (Handbook of Architectural Acoustics.) Deuticke, Vienna, 1962. 808 pp.

8.44 P. H. PARKIN and K. MORGAN: "Assisted resonance" in the Royal Festival Hall. *J. Sound and Vibration,* Vol. 2 (1965). pp. 74–85.

8.45 P. H. PARKIN, H. J. PURKIS, and W. E. SCHOLES: Field measurements of sound insulation between dwellings. *National Building Studies, Research Paper No. 33.* H. M. Stationery Office, London, 1960. 571 pp.

8.46 CLIFFORD R. BRAGDON: *Noise Pollution*. University of Pennsylvania Press, Philadelphia, 1970. 280 pp.

8.47 LEO L. BERANEK (Ed.): *Noise Reduction*. McGraw-Hill, New York, 1960. 752 pp.

8.48 C. C. ZWICKER, C. W. KOSTEN, and J. van den EIYK: Absorption of sound by porous materials. *Physica* Vol. 8 (1941), pp. 149–160, 469–476, 1094–1101, and 1102–1106.

8.49 G. P. MITALAS: Cooling load caused by lights. *Trans. Canadian Society for Mechanical Engineering,* Vol. 29 (1973–1974), pp. 169–174. Reprinted as Research Paper No. 639, Division of Building Research, NRC, Ottawa.

8.50 M. R. SCHROEDER et al.: Acoustical measurements in Philharmonic Hall (New York). *J. Acoustical Society of America,* Vol. 40 (1966), pp. 434–440.

8.51 D. BLAGDEN: Experiments and observations in a heated room. *Philosophical Transactions of the Royal Society of London,* Vol. 65 (1775), p. 111.

8.52 T. BEDFORD: *Basic Principles of Ventilation and Heating*. Second edition. H. K. Lewis, London, 1964. 438 pp.

8.53 S. F. MARKHAM: *Climate and the Energy of Nations*. Oxford University Press, London, 1947. 240 pp.

8.54 R. R. ADLER: *Vertical Transportation for Buildings*. Elsevier, New York, 1970. 228 pp.

8.55 W. T. O'DEA: *A Short History of Lighting*. H. M. Stationery Office, London, 1958. 40 pp.

8.56 J. BACKUS: *The Acoustical Foundations of Music*. Murray, London, 1970. 312 pp.

8.57 L. L. BERANEK: *Music, Acoustics, and Architecture*. Wiley, New York, 1962. 586 pp.

8.58 W. C. SABINE: *Collected Papers on Acoustics*. Dover, New York, 1964. 279 pp.

8.59 R. BANHAM: *The Architecture of the Well-tempered Environment*. Architectural Press, London, 1969. 295 pp.

8.60 L. C. HACKER: William Strutt of Derby. *J. Derbyshire Archaeological and Natural History Society.* Vol. 80 (1960). pp. 49–70.

9.1 *SOM—Architecture of Skidmore, Owings, and Merrill, 1950–62*. Architectural Press, London, 1962. 232 pp.

9.2 R. J. SCHAFFER: The weathering of natural building stones. *Special Report No. 18.* Building Research Station, Garston, England, 1931. 149 pp.

9.3 S. B. HAMILTON: The history of hollow bricks. *Trans. British Ceramic Society,* Vol. 58 (February 1959), pp. 41–62.

9.4 T. RITCHIE: Notes on the history of hollow masonry walls. *The Association of Preservation Technology (APT) Bulletin,* Vol. 5, No. 4 (1973), pp. 40–49. Reprinted as Technical Paper No. 415, Division of Building Research, NRC, Ottawa, Canada.

9.5 *Rivingtons Notes on Building Construction*. Third edition. Rivingtons, London, 1887. Part 2, 246 pp.

9.6 D. FOSTER: Reinforced brickwork and grouted cavity construction. *Consulting Engineer,* Vol. 32 (July 1968), pp. 46–50.

9.7 J. S. GERO and H. J. COWAN: *Design of Building Frames*. Applied Science, London, 1976. 482 pp.

9.8 *Masonry Structures* in *Proc. of the Conf. on Planning and Design of Tall Buildings*. American Society of Civil Engineers, New York, 1972. Vol. III, pp. 955–1144.

9.9 *Masonry Building Systems. Ibid.,* Vol. Ia, pp. 567–589.

9.10 C. L. V. MEEKS: *Italian Architecture 1750–1914*. Yale University Press, New Haven, 1966. 546 pp.

9.11 DEPARTMENT OF THE ENVIRONMENT: *Thermal Insulation of Buildings*. H. M. Stationery Office, London 1971. 157 pp.

9.12 BERNARD E. JONES (Ed.): *Cassell's Reinforced Concrete*. Waverley, London, 1913. 432 pp.

9.13 J. G. WILSON: *Concrete Facing Slabs*. Cement and Concrete Association, London, 1955. 23 pp.

9.14 J. H. CALLENDER et al.: *Curtain Walls of Stainless Steel*. School of Architecture, Princeton University, Princeton, 1955. 187 pp.

9.15 WILLIAM DUDLEY HUNT: *The Contemporary Curtain Walls*. Dodge, New York, 1958. 454 pp.

9.16 IRVING SKEIST (Ed.): *Plastics in Building*. Reinhold, New York, 1966. 466 pp.

9.17 W. A. CHUGG: *Glulam—The Manufacture of Glued Laminated Structures*. Ernest Benn, London, 1964. 423 pp.

9.18 E. R. BALLANTYNE: Energy costs of dwellings. Preprints of the *Fifth Australian Building Research Congress, Melbourne 1975*. Session 1B. 7 pp. Reprinted as Reprint 704, Division of Building Research, CSIRO, Melbourne.

9.19 B. A. HASELTINE: Comparison of energy requirement for building materials and structures. *Structural Engineer,* Vol. 53 (September 1975), pp. 357–365.

9.20 S. B. HAMILTON, H. BAGENAL, and R. B. WHITE: A qualitative study of some buildings in the London area. *National Building Studies, Special Report 33*. H. M. Stationery Office, London, 1964. 165 pp.

9.21 F. M. LEA: *Science and Building—A History of the Building Research Station*. H. M. Stationery Office, London, 1971. 203 pp.

9.22 T. HEATH and T. MOORE: Sydney's arcades. *Architecture in Australia,* Vol. 52 (June 1963), pp. 85–90.

9.23 R. McGRATH et al.: *Glass in Architecture and Decoration*. Architectural Press, London, 1961. 712 pp.

9.24 E. I. BRIMELOW: *Aluminium in Buildings*. Macdonald, London, 1957. 378 pp.

9.25 E. G. COUZENS and V. E. YARSLEY: *Plastics*. Penguin, Harmondsworth, 1968. 386 pp. (The first and second editions were published in 1941 and 1956.)

9.26 A. E. HURST: *Painting and Decorating*. Griffin, London, 1963. 482 pp.

9.27 BANISTER FLETCHER: *A History of Architecture by the Comparative Method*. Seventeenth edition. Athlone, London, 1961. 1366 pp. (The eighteenth edition was published in 1975.)

10.1 *The Crystal Palace Exhibition—Illustrated Catalogue*. Dover, New York, 1970. 328 pp. A reprint of a special issue of *The Art—Journal Illustrated Catalogue, London 1851*.

10.2 RITA ROBISON: Prefabs: an old technique. *Architectural and Engineering News,* Vol. 9 (June 1967), pp. 65–69.

10.3 E. GRAEME ROBERTSON: The Australian verandah. *Architectural Review,* Vol. 112 (April 1960), pp. 238–245.

10.4 DAVID SAUNDERS (Ed.): *Historic Buildings of Victoria*. Jacaranda, Brisbane, 1966. 278 pp.

10.5 GILBERT HERBERT: Corrugated iron and prefabrication. *Working Paper No. 3, Centre for Urban and Regional Study, Israel Institute of Technology,* Haifa, 1971. 20 pp.

10.6 MARGARETTA JEAN DARNALL: Innovations in American prefabricated housing; 1860–1890. *J. Society Architectural Historians,* Vol. 31 (March 1972), pp. 51–55.

10.7   A. E. J. MORRIS: Life ends at forty. *Building,* Vol. 227 (November 1974), pp. 131–133.

10.8   *House Construction.* Post-War Building Study No. 1. H. M. Stationery Office, London, 1944. 152 pp.

10.9   (a) *House Construction—Second Report.* Post-War Building Study No. 23. H. M. Stationery Office, London, 1946. 68 pp. (b) *House Construction—Third Report.* Post-War Building Study No. 25. H. M. Stationery Office, London, 1948. 73 pp.

10.10   MINISTRY OF HOUSING AND LOCAL GOVERNMENT: *Housing Statistics, Great Britain, No. 1, 1966.* H. M. Stationery Office, London, 1966. 57 pp.

10.11   *Cost, Repetition, Maintenance; Related Aspects of Building Prices.* United Nations, Geneva, 1963.

10.12   O. LEDDERBOGGE et al.: *Montage von Beton und Stahlbetonfertigteilen* (Erection of Prefabricated Concrete Components). V.E.B. Verlag für Bauwesen, Berlin, 1964. 440 pp.

10.13   A. D. BERNHARDT: The Mobile-Home Industry: A Case Study in Industrialization. Published in *Industrialized Building Systems for Housing.* M.I.T. Press, Cambridge, 1971. pp. 172–215.

10.14   EZRA D. EHRENKRANTZ: The Third Era of System Building. *Ibid.,* pp. 160–171.

10.15   First and Second Public Forum on the Modular Assembly. *Modular Quarterly,* No. 27 (Winter 1958–1959), pp. 14 and 25.

10.16   *CLASP 1963 — Report of Sixth Year's Work.* Printed by the Nottinghamshire County Council, Nottingham, 1963. 17 pp.

10.17   A. B. SAMSANOV: Moscow Building Catalogue. *Second CIB Symposium on Tall Buildings.* VÚVA Research Institute for Building and Architecture, Prague, 1975. pp. 185–197.

10.18   A. R. DOOLEY et al.: *Casebooks in Production Management.* Wiley, New York, 1964. *Sunshine Builders Inc.,* pp. 164–184.

10.19   ROBERT W. MARKS: *The Dymaxion World of Buckminster Fuller.* Reinhold, New York, 1960. 232 pp.

10.20   M. BORISSAVLIEVITCH: *The Golden Number.* Philosophical Library, New York, 1958. 91 pp.

10.21   THEODORE ANDREA COOK: *The Curves of Life.* Constable, London, 1914. 479 pp.

10.22   FREDRIK MACODY LUND: *Ad Quadratum.* Batsford, London, 1921. Two volumes, 385 pp.

10.23   TONS BRUNÉS: *The Secrets of Ancient Geometry.* Rhodos, Copenhagen, 1967. Two volumes, 331 + 252 pp.

10.24   E. NEUFERT: *Bauordnungslehre.* (Handbook for Rational Building.) Bauverlag, Wiesbaden, 1965. 336 pp.

10.25   *Dimensional Co-ordination for Building.* D.C. 10. Ministry of Public Building and Works, London, 1969. 84 pp.

10.26   C. E. EASTMAN (Ed.): *Spatial Synthesis in Computer-Aided Building Design.* Applied Science Publishers, London, 1975. 333 pp.

10.27   U. HERLYN: *Wohnen im Hochhaus.* (Living in Tall Buildings.) Krämer, Stuttgart, 1970. 275 pp.

10.28   DEPARTMENT OF THE ENVIRONMENT (formerly MINISTRY OF HOUSING AND LOCAL GOVERNMENT): *Design Bulletins,* 17, 19, 20, and 21 (1970); 22 and 23 (1971); 25 (1972); and 27 (1973). H. M. Stationery Office, London.

10.29   City Life: St. Louis Project razing points up public housing woes. *New York Times,* December 16, 1973, p. 72.

10.30   J. PHYSICK and M. DARBY: *Marble Halls—Drawings and Models for Victorian Secular Buildings.* Victoria and Albert Museum, London, 1973. 220 pp.

10.31   C. HOBHOUSE: *1851 and the Crystal Palace.* Murray, London, 1937. 181 pp.

10.32   CHARLES L. EASTLAKE: *A History of the Gothic Revival.* Leicester University Press, Leicester 1970. 209 and 372 pp. Facsimile of the first edition published by Longmans Green, London, 1872.

10.33   R. B. WHITE: Prefabrication—A history of its development in Great Britain. *National Building Studies, Special Report No. 36.* H. M. Stationery Office, London, 1965. 354 pp.

10.34   J. L. PETERSEN: History and development of precast concrete in the United States, in *Ref. 2.16*, pp. 477–500.

10.35   V. F. PROMYSLOV: *Moscow in Construction*. MIR Publishers, Moscow, 1967. 365 pp.

10.36   A. F. BEMIS: *The Evolving House*. Volume 3: Rational Design. Technology Press, Cambridge, 1936. 625 pp.

10.37   *Modular Practice*. Wiley, New York, 1962. 198 pp.

10.38   H. NISSEN (translated by P. KATBORG): *Industrialized Building and Modular Design*. Cement and Concrete Association, London, 1972. 443 pp. This is translated from a standard Danish text published in 1966.

10.39   LE CORBUSIER (translated by P. de FRANCIA and A. BOSTOCK): *The Modulor*. Faber, London, 1954. 243 pp. Published in French in 1950.

10.40   MARCUS VITRUVIUS POLLIO (translated by M. MORGAN): *The Ten Books of Architecture*. Dover, New York, 1960. 331 pp. This is a modern translation based on the original first-century text, as far as that can be ascertained, and deleting modern additions.

10.41   LEONE BATTISTA ALBERTI (translated by J. LEONI): *Ten Books on Architecture*. Alec Tiranti, London, 1955. 256 pp. This is a facsimile of the first English edition of 1755. Alberti reputedly presented the *Ten Books* to Pope Nicholas V in 1452. It was first printed in Latin in 1485 and in Italian in Venice in 1546. Leoni, a Venetian architect, used the Italian version.

10.42   ANDREA PALLADIO: *The Four Books of Architecture*. Dover, New York, 1965. 110 pp. and 94 plates. This is a facsimile of the English edition published by Isaac Ware in London in 1738. It was first published in Italian in 1570.

10.43   R. WITTKOWER: *Architectural Principles in the Age of Humanism*. Third edition. Alec Tiranti, London, 1962. 173 pp. and 48 plates.

10.44   P. H. SCHOFIELD: *The Theory of Proportion in Architecture*. Cambridge University Press, London, 1958. 156 pp.

10.45   P. BEAVER: *The Crystal Palace*. Evelyn, London, 1970. 151 pp.

GLOSSARY

This glossary of technical terms used in the preceding chapters is intended to assist the general reader. Words in italics denote a cross-reference.

aqueduct   A channel or conduit for the conveyance of water, especially the part carried above ground.

arch   A structure which supports loads across a horizontal opening mainly by compression. See also *corbeled masonry arch,* and *true masonry arch.*

beam   A structural member that supports a load across a horizontal opening by bending.

bending moment   *Moment* in a member of a structure caused by the loads acting on the structure.

beton   Concrete.

brittle material   A material that breaks with little or no *plastic deformation.* Stone, brick, cast iron, glass, and concrete are brittle materials.

buckling   Failure of a compression member by deflection at right angles to the load. The material is not necessarily damaged by buckling.

built-in   Rigidly restrained at the ends to prevent rotation.

buttress   A projecting structure built against a wall to resist a horizontal force.

camber   A slight upward curvature given to a beam, girder, or truss to compensate for its anticipated deflection.

came   Lead strip of H-section used to hold pieces of glass in a window.

cantilever   A projecting beam, slab, truss, or column *built-in* at one end and free at the other.

cast iron   Iron with a total carbon content between 1.8 and 4.5%. It is hard and strong in compression, but also a *brittle material* and weak in tension.

catenary   Curve assumed by a cable hanging under its own weight; that is, the cable is purely in tension. A similar structure turned upside down forms a catenary *arch* which is purely in compression.

channel section   A metal section shaped [.

chord   A horizontal member in a *truss.*

coefficient of variation   Statistical measure of the degree of dispersion of experimental data.

concrete   Originally an artificial rock usually made of pieces of stone or gravel, sand, and a binding material, such as lime or cement; now almost invariably made with cement.

continuous beam   A beam that is continuous over intermediate supports and thus *statically indeterminate,* as opposed to a *simply supported beam.*

corbel   A stone or brick, laid horizontally and projecting from the surface of a wall; it is in fact a short cantilever.

corbeled masonry arch   An arch formed by a series of corbels which gradually close the opening. See also *true masonry arch.*

creep   Deformation that occurs over a period of time without an increase in the load.

cross section   Section at right angles to the span.

crown   The highest point of an *arch* or *dome.* The stone at the crown of a masonry arch is called the *keystone.*

cupola   A *dome.*

curtain wall   A thin external wall hung from a *skeleton frame,* as opposed to a *loadbearing wall.* The frame supports the roof and the floors.

cylindrical shell   A thin *cylindrical vault.*

cylindrical vault   A vault formed by a portion of a hollow cylinder, most commonly half a circular cylinder; that is, any cross section of the vault is a semicircle.

dead load   The weight of the building, as distinct from the *live load* carried by the structure.

decibel   Unit for measuring sound levels.

deflection   Deformation due to bending.

dial gauge   A mechanical device for measuring *deflection* which employs a train of gears to magnify the deflection.

diaphragm   A relatively thin, usually rectangular plate used to stiffen a structure. It does so by acting as a *shear panel.*

dome   A vault of double curvature formed by the rotation of a curve around a vertical axis. The most common type of dome is spherical, formed by the rotation of a part of a circle. See also *hemispherical dome* and *shallow spherical dome.*

drum   A vertical wall, in the shape of a hollow cylinder, supporting a dome.

dry-bulb temperature   The temperature of the air. See also *wet-bulb temperature.*

elastic deformation   Deformation fully recovered when the load is removed.

elevator   Synonym for lift.

entasis   An almost imperceptible swelling given to Greek and later Classic columns to correct the optical illusion of concavity which would result if the sides were actually straight.

extensometer   A device for measuring *strain.*

factor of safety   The ratio of the stress at failure to the *maximum permissible stress,* or working stress. See also *load factor.*

fixing moment   *Statically indeterminate* moment at the end of a structural member due to the restraint of the support or an adjacent member.

flat plate   In reinforced concrete construction a concrete slab supported directly on the columns without enlarged column heads.

flat slab In reinforced concrete construction a concrete slab supported directly on the columns by enlarged column heads.

flexure Bending.

formwork Temporary structure used during construction, particularly for supporting wet concrete to which it gives its form.

frequency The number of cycles of a periodic phenomenon which occur in a given time interval. It is usually measured in *hertz*.

fully restrained See *built in*.

geodesic line The shortest possible line that can be drawn from one point of a curved surface to another.

giga Prefix meaning a thousand million; a gigapascal is thus 1000 megapascal.

girder A large *beam*. A lattice girder is a *truss*.

golden section A geometric construction for dividing a line into two unequal parts *a* and *b*, such that $b/a = \frac{1}{2}(1 + \sqrt{5}) = 1.618$.

header A brick (or block or stone) laid across the wall to bond the *stretcher* bricks together.

hemispherical dome A *dome* formed by the rotation of a semicircle. Its support *reactions* are vertical only. See also *shallow spherical dome*.

hertz Cycles per second.

high tensile steel Steel that is stronger than normal structural steel; that is, it has a *yield stress* or proof stress greater than 400 MPa (60 ksi).

hinge A joint allowing free rotation.

Hooke's law *Stress* is directly proportional to *strain*.

hoop force Internal horizontal force in a dome.

hoop tension The tension that occurs in the lower portion of a *hemispherical dome*, or in the *tension ring* of a *shallow spherical dome*.

horizontal reaction A reaction that acts horizontally or the horizontal component of a support *reaction* inclined to the vertical. It must be resisted by a *tie, a tension ring,* or *buttresses*.

hydraulic mortar A mortar that is not washed out by water.

hypar Hyperbolic paraboloid.

hyperbolic paraboloid A saddleshaped surface generated by a straight line moving over two other straight lines inclined to one another.

hypocaust An underfloor heating system used by the ancient Romans.

hypotenuse The longest side of a right-angled triangle; it is opposite to the right angle.

incommensurable ratio A ratio that cannot be expressed as a fraction of two *rational numbers*.

irrational number A real number that is not *rational*; for example, π, $\sqrt{2}$, and $\sqrt{5}$.

jack arch Short-span arch supporting a floor between closely spaced beams.

keystone The stone at the top of a masonry arch.

kilo Prefix meaning a thousand.

lantern Small open or glazed structure crowning a roof, particularly a dome.

lever arm   See *moment arm*.

lift   Synonym for elevator.

lintel   A short-span beam, usually over a door or window opening.

live load   A load due to people or contents of a building, which may or may not be acting, as opposed to a *dead load*.

loadbearing wall   A wall strong enough to support the roof and the floors above it, as opposed to a *curtain wall*.

load factor   A factor by which the *ultimate load* or collapse load should exceed the *service load*.

longitudinal   Parallel to the span.

maximum permissible stress   The highest stress permitted by the building regulations under the action of the working load or *service load*. Also called a working stress.

mega   Prefix meaning a million.

membrane structure   Structure free from bending.

meridional force   Internal force acting along the meridians of a *dome* at right angles to the *hoop forces*.

mild steel   Low-carbon *steel* of moderate strength and high ductility.

milli   Prefix meaning one thousandth.

modulus of elasticity   Measure of elastic deformation, defined as the *stress* which would produce a unit *strain*.

moment   A force multiplied by the distance at which it acts.

moment arm   Distance between the resultant tensile and compressive flexural forces in a cross section (Fig. 3.12). Also called lever arm.

moment distribution   A method for designing *statically indeterminate* structures (Section 4.2).

moment of resistance   Internal moment in a beam, which for equilibrium must equal the bending moment acting on the beam.

monolithic   Cast in one piece and therefore continuous.

mortise-and-tenon joint   Traditional timber joint formed by a rectangular slot (mortise) into which a tongue (tenon) from another piece fits.

nave   The body of a church or cathedral, usually separated from the aisles by lines of pillars

neutral axis   Line at which the flexural stresses change from tension to compression.

newton   The unit of force and weight in the SI metric system, abbreviated N. One newton is the force which, applied to a mass of 1 kilogram, produces an acceleration of 1 meter per second per second.

pascal   The unit of stress in the SI metric system, abbreviated Pa. It is defined as 1 *newton* per square meter.

party wall   A wall forming part of two buildings.

pin joint   A joint allowing free rotation, whether or not formed by a pin.

plain concrete   Concrete without reinforcement.

plastic deformation   Deformation not recovered when the load is removed.

plastic hinge   A *pin joint* formed by the *plastic deformation* of the structural material when the load is increased above the *service load.*

plastic theory   Theory for determining the *ultimate load* of a structure, based on its collapse due to the formation of *plastic hinges.*

plenum duct   A duct containing air under a pressure slightly above that of its surroundings.

point of contraflexure   A point at which curvature changes sign; that is, it changes from convex to concave or vice versa.

precast concrete   Concrete placed in position after casting instead of being cast in place.

prestressed concrete   Concrete that is precompressed in the zone in which tensile stresses occur under load, usually by tensioning strands of *high-tensile steel.*

purlin   A horizontal beam in a roof at right angles to the roof trusses or *rafters.*

Pythagorean triangle   A right-angled triangle whose sides are in the proportion 3 : 4 : 5. The enclosed angles are 90°, 53°7', and 36°53'.

rafter   A sloping piece of timber extending from the wall plate to the ridge or the sloping upper member of a roof truss.

rational number   An integer (whole number) or a fraction. See also *incommensurable ratio and irrational number.*

reaction   Force exerted by the ground or by another structural member in opposition to the loads.

redundant member   A structural member in excess of those required for a *statically determinate* structure.

resistance moment   See *moment of resistance.*

reinforced concrete   Concrete reinforced with iron or steel to improve its resistance to tension.

reinforcement   Bars or mesh of iron or steel used in *reinforced concrete.*

restrained beam   A beam with restraints on the rotation at its ends, as opposed to a *simply supported* beam. A *built-in* beam is rigidly restrained.

rigid frame   A frame in which all or some of the joints are rigid so that it becomes *statically indeterminate.*

rigid joint   A joint that allows no rotation of the members joined in relation to one another; for example, a right-angle joint remains a right angle under load.

rigidly restrained   See *built-in.*

rise   The height of the *crown* of an arch or dome above the *springings.*

service load   The actual *dead load* and *live load* carried by the structure; also called the working load. To obtain the *ultimate load* the various service loads are multiplied by appropriate *load factors.*

shallow spherical dome   A dome formed by the rotation of a circular arc that is less than a semicircle. Its support *reactions* are inclined. If the angle subtended by the circular arc at its center of curvature is less than 104°, the *hoop forces* are entirely compressive. See also *horizontal reaction.*

shear force The resultant force tending to shear or cut through the member of a structure caused by the loads acting on the structure.

shear panel A panel or slab that resists shear forces in its own plane due to wind, earthquake forces, or explosions and thus stiffens the structure against deformation by such forces.

shear wall A vertical *shear panel*.

shell Thin curved structural surface.

simply supported beam A beam supported in a manner that permits free rotation. It is *statically determinate*.

skeleton frame A frame that acts like a skeleton, carrying the walls, the floors, and the roof.

slenderness ratio A measure of the tendency of compression members toward *buckling*.

slope-deflection analysis A method for designing *statically indeterminate* structures (Section 4.2).

span The distance between the supports of a structure.

spherical dome See *hemispherical dome* and *shallow spherical dome*.

springings Supports of an arch or dome.

statically determinate Soluble by statics alone. *determine facture*

statically indeterminate Insoluble by statics alone because there are more members, rigid joints, or reactions than there are statical equations.

statics The branch of physics that deals with forces in equilibrium.

steel An alloy of iron with carbon content between 0.1 and 1.7%. Iron with a lower carbon content is *wrought iron,* and iron with a higher carbon content is *cast iron*.

straight-line theory Any theory based on a linear relationship between two variables; in structural design a theory based on *Hooke's law*.

strain Deformation per unit length.

stress Force per unit area.

stress reversal Change from compressive to tensile stress or vice versa.

stress-strain diagram The diagram obtained by plotting the stresses in a test piece against the strains. It is used to assess the structural suitability of materials.

stretcher A brick (or block or stone) laid with its length parallel to the wall. Stretchers are often interspersed with *headers* to achieve a proper bond.

strut A compression member, the opposite of a *tie*.

temperature See *dry-bulb temperature* and *wet-bulb temperature*.

tension ring A ring that absorbs the horizontal components of the inclined reactions of a *shallow spherical dome*.

terracotta Burned clay units of a shape more complex than a brick or a tile.

theorem of three moments Theorem used in the design of *continuous beams*.

thermoplastic Becoming soft when heated and hard when cooled.

thermosetting Becoming rigid on heating, usually due to a chemical reaction.

three-pin arch   An arch with three *pin joints;* it is *statically determinate.*

thrust   Compressive force or reaction.

tie   A tension member, the opposite of a *strut.*

transept   The transverse part of a cruciform building at right angles to the *nave.*

true masonry arch   An arch whose *voussoirs* are arranged so that the joints are at right angles to the line of *thrust,* as opposed to a *corbeled masonry arch.*

truss   An assembly of straight tension and compression members that performs the same function as a deep *beam.*

ultimate load   The highest load a structure can sustain.

variation   See *coefficient of variation.*

vault   See *cylindrical vault* and *dome.*

vertical reaction   A reaction that acts vertically or the vertical component of an inclined reaction. See also *horizontal reaction.*

voussoir   Wedge-shaped block of masonry forming part of an *arch* or *dome.*

water-cement ratio   The ratio of water to cement in a concrete mix.

wet-bulb temperature   The temperature recorded by a thermometer whose bulb is wrapped in a damp wick dipping into water. The relative humidity of the atmosphere can be determined from a comparison of the wet-bulb temperature and the *dry-bulb temperature.*

working stress   The *maximum permissible stress.*

working load   The *service load.*

wrought iron   Iron with a carbon content lower than *steel* and consequently weaker. It was used as a structural material in the eighteenth and nineteenth centuries. It is not so strong in compression as *cast iron,* but it is more ductile and has a higher tensile strength.

yield stress   The stress at which substantial *plastic deformation* first occurs.

Young's Modulus   The *modulus of elasticity* in tension and in compression.

zenith   The point in the sky immediately above the observer; that is, at altitude 90°.

A NOTE ON UNITS OF MEASUREMENT

Numerical data are given in SI metric and American units (formerly used also in British Commonwealth countries). In the text the basic unit is given first, with the conversion in brackets; that is, the dimensions of a building erected in America or in England in the nineteenth century are stated in feet (with conversion to meters) and the dimensions of a building erected in France or Germany at the same time are given in meters (with the conversion in feet).

The following conversion table lists units used more than once in this book.

LENGTH

kilometers (km), meters (m), millimeters (mm) : miles, feet (ft), inches (in.)

1 km = 1000 m, 1 m = 1000 mm	1 mile = 5280 ft, 1 ft = 12 in.
1 km = 0.621,371 miles	1 mile = 1.690,84 km
1 m = 3.281 ft = 3 ft 3½ in.	1 ft = 0.304,80 m
1 mm = 0.039,87 in.	1 in = 25.400 mm

AREA

square meters (m²) : square feet (ft²), square inches (in²)

1 m² = 1,000,000 mm²	1 ft² = 144 in.²
1 m² = 10.764 ft²	1 ft² = 0.092,903 m²

VELOCITY

meters per second (m/sec) : miles per hour (mph), feet per second (ft/sec), feet per minute (ft/min)

$$1 \text{ mph}   = 1.467 \text{ ft/sec}$$
$$1 \text{ ft/sec} = 60 \text{ ft/min}$$

1 m/sec = 2.237 mph   1 mph   = 0.447,04 m/sec
        = 3.281 ft/sec   1 ft/sec = 0.304,80 m/sec
        = 196.860 ft/min   1 ft/min = 0.005,08 m/sec

DENSITY

kilograms per cubic meter (kg/m³) : pounds per cubic foot (lb/ft³)

1 kg/m³ = 0.062,428 lb/ft³   1 lb/ft³ = 16.0185 kg/m³

STRESS

gigapascals (GPa), megapascals (MPa), kilopascals (kPa), pascals (Pa) : pounds per square foot (psf or lb per sq ft), kilopounds per square inch (ksi), pounds per square inch (psi or lb per sq in)

1 GPa = 1000 MPa, 1 MPa = 1000 kPa, 1 kPa = 1000 Pa
1 MPa = 1 newton per square millimeter (N/mm²), 1 Pa = 1 newton per square meter (N/m²)

                          1 ksi   = 1000 psi, 1 psi = 144 psf

1 GPa = 145.038 ksi   1 ksi = 6.894,76 MPa
1 MPa = 145.038 psi   1 psi = 6.894,76 kPa
1 kPa   = 0.145,038 psi   1 psf = 47.880,3 Pa
1 kPa   = 20.885 psf

TEMPERATURE

degree Celsius (°C): degree Fahrenheit (°F)
$°C = \frac{5}{9} (°F - 32)$ $°F = (\frac{9}{5} °C) + 32$

PEOPLE, PLACES, AND STRUCTURES

by Dr. Valerie Havyatt

(Titles of books, societies, commercial companies and institutions appear in the General Index.)

Page numbers in *italics* refer to illustrations.

CNIT Exhibition Hall, Paris, 161, *163*, 189, 190
Coalbrook Dale Bridge, 148
Coal Exchange, London, 304
Cockrell, Charles Robert, 303
Cocoanut Grove, Boston, fire at, 195
Coignet, Edmond, 38
Coignet, François, 31, 32, 35, 36, 320
Cole, Henry, 303, 304
Coliseum, Charlotte, North Carolina, 172, 336
Collins, P., 331
Colosseum, Rome, 72, 184
Colossus Bridge, Fairmont, Pennsylvania, 146
Concertgebouw, 261
Condit, Carl W., 74, 331, 332
Conway River Bridge, Wales, 22, *22*
Conzelman, J. E., 312
Cook, Theodore, 324, 325, 326, 342
Coolidge, D., 257
Cooper-Hewitt, Peter, 257
Coppin's Olympic Theatre, Melbourne, 310
Corio Villa, Geelong, Australia, 310
Cornelius, W., 181
Cort, Henry, 18
Costa, Lúcio, 227
Cottancin, 54
Coulomb, Charles Augustin de, 4
Cowan, Henry J., 331, 332, 335, 337, 341
Cramer, Stuart W., 241
Cristophe, Paul, 38
Cross, Hardy, 76, 127, 334
Crystal Palace Exhibition, Chicago (1873), 249, 308
Crystal Palace Exhibition, London (1851), 24, 148, 218, 244, 275, 276, 303, *306, 307, 308, 309,* 341, 342, 343
Crystal Palace Exhibition, New York (1853), 249, 308
Ctesiphon, Iraq, arch at, 151
Cubitt, William, 303
Culmann, Karl, 13, 172
Cummings, Alexander, 210

Dallas, Atlantic Refining Company Building, 252
Dalton, John, 21
Darby, Abraham III, 4
D'Asnières, Paris, bridge at, 50
Da Vinci, Leonardo, 222
Davy, Humphry, 123, 244, 287
De Ferranti, Sebastian, 244
Deitrick, William H., 180
Demarest Building, New York, 250
Derbyshire General Infirmary, England, 240
De Saussure, Horace Bénédict, 222
Deutscher Werkbund Exhibition (1914), 219
Di Mauro, Ernesto, 276
Dischinger, Franz, 152, 153
Di Stasio, Joseph, 87, 333

Doehring, C. W., 59
Dome of Discovery, London, 173, *174*, 290
Donaldson, Thomas, 303
Dresden, Germany, fire, 194
Dufton, A. F., 222, 228, 338, 339
Dulles Airport Terminal, Washington, D.C., 183
Dunham, J. W., 94, 95, 334
Duomo, Florence, 7, 161, 163, 189, 190
Dutert, Ferdinand, 54
Dyckerhoff and Widmann, *engineers*, 31, 151, 152, 154, 184
Dyson, H. Kempton, 108, 334

Eagle Aviary, Bristol, England, *294*
Eastlake, Charles, 305, 342
Eckert, J. Prosper, 125
Eddystone lighthouse, 61
Edinburgh, fire, 194
Edison, Thomas Alva, 244, 256, 339
Ehrenkrantz, Ezra D., 328, 342
Eiffel, Gustav, 277
Eiffel Tower, 58, 148, 161, 254, 339
Empire State Building, New York, 27, 75, 100, 109, 111, 117, 126, 147, 252, 285, 334
Eney, William, 126
Engels, Friedrich, 208
Equitable Life Building, New York, 249
Eros, Statue of, London, 288
Esders clothing factory, Paris, 46
Esquillan, Nicholas, 163, 337
Euclid, 17, 324
Eudoxus, 324
Euler, Leonard, 18
Ewing, Alfred, 209, 211, 240
Experiment Farm Cottage, Parramatta, Australia, *228*
Expo, 70, 187
E-Z factory, Chicago, 46

Fagus works, Alfeld, Germany, 219, 279
Fahrenheit, Gabriel Daniel, 222
Fairbairn, William, 21, 22, 23, 31, 32, 198
Fair Store, Chicago, *29*
Fanger, P. O., 223, 339
Faraday, Michael, 244
Farnsworth house, Fox River, Illinois, 219, 225
Fechner, Gustav Theodor, 325
Fentiman, A. E., 175
Festspielhaus, Bayreuth, 262
Fibonacci, 122, 324
Fielder, Henry, 18
Florence, Duomo of, 7, 161, 163, 189, 190
    San Miniato al Monte, 271
Fogg Museum of Art, Boston, 262
Forth Bridge, Edinburgh, 51, *52, 53,* 62, 147, 332
Fourcault, Emile, 275

Rand McNally Building, Chicago, 27
Rankine, W. J. M., 20, 80, 90, 91
Ransome, Ernest Leslie, 35, 36
Rapoport, Amos, 225, 339
Rayleigh, *Lord,* 262
Red Fort, Delhi, *226*
Reid, D. Boswell, 222, 239
Reissner, H., 157
Reliance Building, Chicago, 27
Rhodes, Peter S., 98, 334
Richardson, Henry Hobson, 74
Rio de Janeiro, Ministry of Education, 227
Ritter, A., 12
Ritter, W., 38
Robertson, Andrew, 21, 331
Robertson, Leslie E., 111
Rockefeller Center, New York, 242, 290
Roebling, John, 146
Röhm, Otto, 291
Rohles, F. H., 223
Rome, Colosseum, 72, 184
  Palazetto dello Sport, 155, *155*
  Pantheon, 151, 189, 190
  S. Pietro, Basilica of, 189, 288
Rome (Ancient), bricks, 270
  ceramic blocks, 274
  fire, 193, 194
  fire brigades, 201
  fireproofing, 196
  interior environment, 217
  metals, use of, 287
  regulations, 90
  tall buildings, 72, 116
  theaters, 184
  toilets, 208
  traffic noise, 263
  trusses, 11
Ronan Point flats, London, 99, 317, 334
Roof Structures Inc., *engineers,* 173
Root, John Wellborn, 331
Rouen Cathedral, France, 72
Royal Botanical Gardens, Kew, 276
Royal Exchange, London, 276
Royal Exchange Insurance Company, 202
Royal Festival Hall, London, 263
Royal Liver Building, Liverpool, 36, 71
Royal Opera House, Covent Garden, 260
Royal Pavilion, Brighton, 24
Royal Victoria Hospital, Belfast, 241
25B Rue Franklin, Paris, 279
Ryder, Frederick, 137, 335

Saarinen, Eero, 183, 337
Saarinen, Eliel, 74
Sabine, Wallace Clement, 4, 262, 264, 340
Sachs, Edwin, 199
St. Agnan, France, schoolhouse at, 31

St. Andrew's Concert Hall, Glasgow, 261
St. George's Hall, Liverpool, 260, 303
St. Laurent-du-Pont, France, fire at, 195
St. Louis, Air Terminal, 161, *162*
  Climatron, *175*
  Pruitt-Igoe project, *329,* 342
  St. Louis and St. Mary Priory, *167*
St. Pancras Station, London, 53, *54,* 148
St. Paul Building, New York, 27
St. Paul's Cathedral, London, 189, 190, 244
St. Petersburg (now Leningrad), 207
Saint-Venant, Barré de, 101, 134
St. Vitus' Cathedral, Prague, 280
Salmon, E. H., 91, 333
Salvation Army, *see* Cité de Refuge
Samsanov, A. B., 322, 342
San Antonio, Milam Building, 242
San Francisco, earthquake and fire, 64, 97, 193,
    195, 332
  Golden Gate Bridge, 147
  Hallidie Building, 279
  Leland Stanford Junior Museum, 35
  Opera House, 262
San Miniato al Monte, Florence, 271
S. Pietro, Basilica of, Rome, 189, 288
S. Sophia, Istanbul, 189
Sarger, Rene, 181
Schaffhausen, bridge at, 145
Scheerbart, Paul, 219
Schelling, E., 184
Schenectady, General Electric Company turbine
    building, 285
Scheutz, George, 123
Schleyer, F. K., 179
Schmidt, Max E., 254
Schulz, Eugen, 151
Schwarzwaldhalle, Karlsruhe, 184, 190
Schwedler, J. W. A., 53, 148, 171, 172, 336
Schyler, Montgomery, 72
Scott, Gilbert, 54
Scott, Giles, 290
Seagram Building, New York, 220, *286*
Sears Tower, Chicago, 75, 111, *113,* 117
Seven Wonders of the Ancient World, 188
Severud, Fred N., 180, 181, 183, 336
Shaftesbury, 7th Earl of, (formerly Anthony
    Ashley Cooper, later Lord Ashley), 206,
    234
Shaw, Norman, 4
Sheffield, 71, 269
Shreve, Lamb and Harmon, *architects,* 75
Siev, Avinadav, 179
Silsbee, J. L., 254
Simon, John, 205
Singer Building, New York, 27
Skidmore, Owings and Merrill, *architects,* 182,
    340

People, Places, and Structures

GENERAL INDEX

by Dr. Valerie Havyatt

Entries in italics are foreign words, names of institutions or titles of books.

Page numbers in italics refer to illustrations.

diagrams, *17, 48, 64, 105*
  see also Resistance moment
Bending theory, 16
  of shells, 157, 159
Bessemer steel, 18, 26
Béton agglomeré, 31, 32, 35
Le Béton Armé et ses Applications (Cristophe),
  38
*Beton in combination with iron as a building
  material* (Ward), 34
Der Betoneisenbau (Mörsch), 38, 333
Béton pisé, 32
Béton translucide, 279
Birdair, 186
Blast furnace slag, expanded, 280, 281
Boilers for heating, 235, *237*
Borehamwood Fire Testing Station, 199
Bow's notation, 13, *14*
Brass, *286*, 287
Bricks, 270, 271, 272, *273*, 295, 312,
  340
Brickwork, reinforced, 274
Bridges, 145, 333
  comparative spans, 147
  long-span, 145
  prestressed concrete, *57, 58, 59*, 60
  pre-18th century, 145, 148
  railway, 11, 22, *22*, 31, 51, *52, 53,* 61, 90,
    91
  suspension, 146, *146*, 147, 178
  tubular, *22*
Brise soleil, 227
British Fire Prevention Committee, 199
British Standards Institution, 93
British Steel Corporation, 177
(British) Steel Structures Research Committee,
    21, 94, 104
Bronze, *286*, 287
Buckling, 128, 345
  of arches, 180
  of beams, 21
  of columns, 18, 91, 106
  of frames, 104, *133*
  of shells, 158, 164
Building Construction (Rivington), 198, 338
Building economics, 297, 314, 322, 330, 342
Building Research Station (UK), 298, 314
Building services, 140, 217, 222, 234, 241, 249,
    255, 328
Building Systems Development Inc., 328
Building types, *see* Apartment buildings; Cine-
    mas; Concert halls; Department stores;
    Exhibition buildings; Factories; Green-
    house; Hospitals; Libraries; Office build-
    ings; Opera houses; Religious buildings;
    Residential buildings; Sporting arenas;
    and Theaters

Cables, 60, 178, 179, *179*
Calcium silicate bricks, 274
Calculating devices, 121, *124*
Cantilever, 16, 52, 345
Caravan, 318
Carbon arc, 244, 256
Carbon filament, 256
Carnegie, Phipps and Company, 18, 26
Carpets, 213, 264, 296
Cast iron, 12, 18, 22, 26, 31, 197, 306, *306,
    307, 309,* 310, 311, 345
Castigliano method, 66, *66*, 75
Cavity walls, 272, 340, 341
CEB, 93
Celluloid, 291
Cement, natural, 31
  portland, 31, 295
  Roman, 35
Ceramic blocks, 274
Ceramics, 270, 275
Chicago School, 25, 74, 117, 331
Chimneys, 234, *236*
Cholera, 206
Churches, 7, 46, 148, 149, 161, *167*, 189, 244,
    271, 279
CIB, 299
Cinemas, 242, 262
City of Tomorrow (Le Corbusier), 72, 117, 332
CLASP system, 321, 342
Climate, design for, 221
Climate and the Energy of Nations (Markham),
    224, 340
Climatology, 221
Closed systems, *312, 316, 317,* 318
Coal gas, 256
Codes, *see* Regulations
Coiled filament, 257
Collapse, 38, 61, 98, 99, 317
Collapse mechanism, 56, 57, *106,* 334
Collodion, 291
Column, buckling, 18, 91, 106
  capital, *83*, 84, 87, 89
  cast iron, 18, 26, 306
  concrete, 20, 46, 50, 84, 87
  design, 18, *19*
  fire protection, 199
  head, *83*, 84, 87, 89
  long, 19
  short, 18
  steel, 21
Comfort criteria, 221, 240, 260, 329, 339
Communication systems, 254
Compression flange, buckling of, 21
Compression members, 12. *See also* Column
Computer, 255, 335
  analog, 122, 137, *137*
  digital, 123, 138

Computer applications, 67, 128, 131, 137, 138, 171, 177, 185, 252, 330
  autocodes, 139
  data processing, 330
  graphics, 139, 185, 231
  programs, 139, 159
  time sharing, 139
Concert halls, 183, 240, 261, 262, 263
Concrete, 45, 295, 345
  aerated, 281
  codes, 20, 38, 40, 81, 82, 85, 87, 88, 93, 108, 109
  cover, 40
  cracking, 45, 109
  creep, 59, 61, 89, 108, 346
  houses, 31, 32, 33, 34, 46, *312*, 313, 314
  lime, 31
  masonry, 274
  mixer, 31
  plain, 31
  precast, 46, 98, 274, 282, *284*, 303, 311, *312*, 313, 341, 343, 349
  prestressed, 59, 331, 332, 337, 349
  properties of, 38, 59, 94
  reinforced, 31, 36, *37*, 41, 47, 331. *See also* Reinforcement
  structures, 11, 31, 113, 199
  surface finish, 35, 45, 282, *284*, 318
  see also Béton agglomeré; Beton in combination with iron as a building material; Béton pisé; *and* Béton translucide
Conduits, 245, 290
Conoids, 153, 165, *168*, 181
Conservatories, 276, *277*
Continostat, 126
Continuous structures, 37, 42, 47, *50*, *51*, *55*, *56*, *62*, 75, *79*, 104
Contraflexure, point of, *62*, 63, *63*, 349
Cooling, 217, 242
Copper, 287
Corning Steuben Inc., 279. *See also* Owens-Corning, Inc.
Corps des Ingénieurs des Ponts et Chaussées, 4
Corrosion, 284, 285
Corrosion-resistant steel, 285
Cover for concrete, 40
Cracking of concrete, 45, 109
Creep, in concrete, 59, 61, 89, 108, 346
  in steel, 197, 346
Cross-sectional area, 13
Crown glass, 275
Crystal structure, 101, 103, *103*
Curtain walls, 220, 231, *233*, 243, 248, 270, 275, 279, 282, 285, 341, 346
  aluminum, 289, 290
  copper, 287
  precast concrete, 282

stainless steel, 285
  see also Glass, buildings
Curved structures, *53*, *54*, *55*, *57*, 148, 163, 171, 178, 185, 190
Curves of Life (Cook), 325, 342
Cylinder glass, 275
Cylindrical shells, 152, 153, *153*, *156*, *158*, *159*, *162*, 346

Dalle de verre, 280
Damping of vibrations, 96
Dampproof, courses, 272, 290
Daylight factor, 258
Daylighting, 249, 258, *259*, 279, 339, 340
Dead load, 61, 90, 91, 346
Deflection, 51, 58, 61, 66, 80, 89, 108, 170, 177, 184, 346
Deformation, *51*, *59*, *80*, 101, 102, *103*, 104, 108, 179, 334
Deformeter, 125, 126, 127, 335
Department of Housing and Urban Development (USA), 322
Department stores, 197, 249, 258, 277
Deutsche Luxfer Prismen-Gesellschaft, 279
Dewpoint, 241
Digital computers, 123, 138
Dimensional change, 298. *See also* Creep; Shrinkage
Dimensional similarity theory, 127
Discourse on Earthquakes (Hooke), 97
Disease, 6, 205, 206
Domes, 346
  classical, 154, 269
  concrete, 148, 148, *149*, 152, 153, *154*, *155*, *157*, 171, 187, *188*
  geodesic, 173, *174*, *175*, 189, 347
  glass, 276
  great-circle, 173
  lamella, 172, 336
  masonry, 154, 269
  metal, 171, 304
  Schwedler, 172
  shallow, 154, *155*, *157*, 349
  spans, 189, 190
  square, *157*
  triangulated, 171, *172*
Dorman Long and Company, 18
Double shells, *163*, 189
Drain, 211, *212*, 290
Drain pipes, 212, 290
Durability, 40, 296, 341
Dynamite, 269

Earthquake, 96
  Lisbon, 97
  Long Beach, 274
  Los Angeles, 98

Foamed concrete, 281
Foamed plastics, 281
Folded plates, 169, *169*
Foot candle, 256
Forces, 60, *60,* 66
   due to earthquakes, 63, 96, 332
   due to explosions, 98, *99*
   due to wind, 22, 23, 61, 91, 95, 184, 187,
      274, 333, 334, 335
   in plane frames, 13
   in space frames, 171
   thermal, 22
Formwork, 31, 46, 88, 163, 169, 171, 187,
   283
FORTRAN, 139
Foundations, 25, 27, 51
Frames, 23, 109, 138, 139, 140, 341
   analogues for, 136
   balloon, 25, 311
   iron, 23, *24, 25*
   models of, *133*
   plane, 13
   reinforced concrete, 35, 36, 46, 47, 50, *79,*
      98
   rigid, 27, *30,* 36, 50, *62,* 63, 64, 75, 98, 104,
      106, *106,* 136, 274
   skeleton, 23, *29,* 62, *64,* 100, 108
   space, 171, *172, 176, 178,* 293, 336
   square, *56*
   steel, 28, *29, 30,* 61, 71, *73,* 75, 98, 104
   timber, 25
   Vierendeel, *56*

Galvanizing, 281, 285, 310
Gas Light and Coke Company, London, 256
Gauge, strain, 127, 130
Gaussian distribution curve, 93, *94*
General Electric Company, Schenectady, 257
General Theory of Shells (Vlasov), 158, 336
Geodesic domes, 173, *174, 175,* 189, 347
Geometry, 121, 125
Gerber beam, 51, *52,* 84
Girders, 247. *See also* Beams
Glare, 260
Glass, 275, 295
   buildings, 218, *218,* 219, 220, 276, *277,* 304,
      308, 309
   cathedral, 380
   double glazing, 220
   fibers, 280
   heat-reflecting, 220, 232
   stained, 280
   use of, 218, 220, 232, 258, 277, *278,* 279,
      280, 341
   see also Curtain walls
Glassmaking, 275
Glasswool, 28

Glazed ceramics, 275
Glazed tiles, 275
Glazing, 276, *308*
Golden section, 323, 324
Gothic architecture, 72, 92, 98, 328
Grand Junction Water Company, 205
Graphic statics, 13, 125
Grashof-Rankine method, 81
Great Fire, Chicago, 25, 194, 338
   London, 193, 194, 201
   Rome, 193, 194
Greenhouse, 218, *218,* 276, *277,* 304
Greenhouse effect, 231
Grillage foundations, 26

Hangar, at London Airport, *178*
   at Orly, Paris, *150,* 151
   at Orvieto, Italy, 127, *170*
Hanging of floors from the roof, 114, *115, 116*
Hardboard, 292
Heat, radiant, 281
Heating, 217, 234, *237, 238,* 296
Height of buildings, 26, 36, 71, 109, 111, 116
Heliodon, 228, *230*
Hooke's law, 37, 100, 347
Hospitals, 240, 241, 280, 310
Housing, 7, 207, 272, 322, *329*
   concrete, 31, 32, 33, 34, 46, *312,* 313, 315,
      316, 317
   prefabricated, 19th century, 310
      early 20th century, 46, 313
      since, 1940, 314, *317,* 320
Howe's truss, 11
HUD, 322
Humidity, 222, 223, 241
Hygiene, 6, 205, 206
Hygrometers, 222, 241
Hypars, *see* Hyperbolic paraboloids
Hyperbolic paraboloids, 129, 153, 164, *164, 165,*
      166, 167, 181, 184, 347
Hyperboloids, 164
Hypocaust, Roman, 239, 247

IBM, 125
Ice, 241
ICI, 291
Illumination, *see* Lighting
Illustrations of the theory and practice of venti-
      lation (Reid), 239
Impact loads, 98, 334
Imperial Chemical Industries Limited (ICI), 291
Indian Standards Institution, 260
Industrialized building, 303, 314, 318, 320, 322,
   323
Ingeniator, 4
Institution of Civil Engineers, 4, 40
Insulation, sound, 217, 264, 281, 298, 340

thermal, 217, 274, 280, 281, 296, 298, 341
Interior environment, 217
International Business Machines (IBM), 125
International Council for Building Research Studies and Documentation (CIB), 299
International Fire Prevention Congress, London (1903), 199
International Organization for Standardization (ISO), 298
Inventions, 35, 122, 125, 137, 139, 187, 203, 241, 244, 245, 249, 252, 254, 256, 257, 285, 288, 291. *See also* Patents
Iron, cast, *see* Cast iron
  frame, 23, 25
  products, 285
  properties, 18, 38, 59, 90, 100, 197, 284, 331
  structures, 4, *12,* 24, 53, 62, 148, 304, *306, 307,* 310
  wrought, *see* Wrought iron
  see also Steel
I-sections, 18
ISO, 298

Jacking, 58, *58*
Joint Committee on Ultimate Strength Design, 20, 40, 331. *See also* American Concrete Institute
Joints, MERO, 173
  mortise-and-tenon, 25, 348
  Nodus, 177, *177*
  pin, 15, 63, *63,* 348
  timber, 25
  triodetic, 175
  Unistrut, 175
Joists, *see* Beams

Katathermometer, 222
Kitchens, 213
K-value, 298

Lambeth Water Company, 205, 206
Laminated plastics, 291
Lamps, electric, 244, 256, 257
Laws, *see* Regulations, building
Lead, 287, *288*
Least work, principle of, 65
Lehigh University, Pennsylvania, 106
Lehrbuch der Statik (Möbius), 15
Lever principle, 3, 11
Libbey-Owens,Glass Company, Inc., 275. *See also* Owens-Corning, Inc.
Libraries, 24, *149,* 150, *271,* 277, 290, 336
Lifts, *see* Elevators
Lighting, artificial, 217, 244, 245, 249, 255, 296, 340
  electric, 244, 256, 257

gas, 256
  natural, 249, 258, *259,* 279, 339, 340
Lightning conductor, 244
Light wells, 258
Limit states design, 108
Linear shells, 163, 165, 168
Live load, 61, 83, 91, 94, 333, 348
Load, 61, 90, 91, 93, 94, 179
  dead, 61, 90, 91, 346
  fire, 200, 333
  impact, 98, 334
  live, 61, 83, 91, 94, 333, 348
  service, 108
  snow, 95, 184, 187, 334
  wind, 22, 23, 61, 91, 95, 184, 187, 274, 333, 334, 335
Load balancing, 61, 90, 184
Load factor, 93, 348
Logarithms, 122
London Main Drainage Scheme, 207
London Transport Board, 201
Long shells, *156,* 157, *158, 159*
Long span structures, 53, 145, 148, 189
Loudspeaker, 262, 263, 264
Lumber, *see* Timber
Luxfer Prismen Gesellschaft, Deutsche, 279

Magnesium, 295
Maintenance cost, 296
Manchester Literary and Philosophical Society, 21
Manchester and Salford Water Company, 206
Matrai system, *41*
Matrix algebra, 138
Maxwell-Mohr method, 65, *65,* 75, 86
Measurement, units of, 353
Measuring instruments, 125, 127, 222, 259, 263
Mechanism, 16. *See also* Collapse mechanism
Melamine formaldehyde, 291, 293
Membrane structures, 151, *156,* 185, 348
  theory of, 151, 152, 165, *168,* 336
MERO joint, 173
Metal, 284, 295. *See also* Aluminum; Brass; Bronze; Copper; Galvanizing; Iron; Magnesium; Muntz metal; *and* Steel
Metal physics, 101
Method of resolution at joints, 12, 172
Method of sections, 12, *13*
Metric units, 353
Microconcrete, 128
Microphone, 262, 263
Microphone, 262, 263
Middle third rule, 92
Mix proportioning, 40
Mobile homes, 303, 318, *319,* 320, 322, 342
Mode of Communication of Cholera, On the (Snow), 206

'Top hat' structures, 112, *114*
Torsion analogue, 134, *135*, 136
Towers, 58, 148, 161, 254, 274, 339. *See also* Tall buildings
Traffic noise, 263
Trailer, 318
Transportation systems, 72, 117, 235, 249, 253, 255
Tredgold formula, 19
Triodetic joint, 175
Tropical architecture, 224, 227, 229, 260
Truss, Howe, 11
    Pratt, 12, *12*, *56*, *57*
    Roman, 11
    Vierendeel, *56*
    Warren, 12, *15*
Trussed Steel Company, 150
Trusses, 11, *12*, *13*, *14*, *15*, *16*, *56*, 58, 351
    cast iron, *306*, *307*
    plane, *12*, *13*
    space, *176*, *178*
    steel, 11
    timber, 11
    wrought iron, 11
Tube, fluorescent, 257
    pneumatic, 254
    speaking, 254
    structural, 22, 173
Tube concept, 109, *112*, *113*
Tungsten filament, 257
Turnbuckles, 58, 59

UK, *see* British Fire Prevention Committee; British Standards Institution; British Steel Corporation; *and* (British) Steel Structures Research Committee
Ultimate strength, concrete, 94
    wrought iron and steel, 90, 102
Ultimate strength design, 38, 92, 100, 108, 128, 331
Underwriters' Laboratories Inc., 199
Unistrut system, 175, 336
Urea formaldehyde, 291, 293
Urethane foam, 281
USA, *see* ACI; American Institute of Steel Construction; ASCE; ASHRAE; American Society of Heating and Ventilating Engineers; *and* ASTM
US Department of Commerce Building Code Committee, 94
USSR, prefabrication in, 314, 315, 317

U-value, 298

Valve closets, 210, *210*
Veneered timber, 293
Ventilation, 217, 223, 239, 249, 264, 338
Verandahs, 226, 228, *228*, 341
Vermiculite, 199, 280
Vernacular construction, 225
Vers une Architecture (Le Corbusier), 45, 279, 332
Vertical transportation, 23, 25, 116, 117, 201, 249, *250*, *251*, 339, 340
Vibration, 96
Victorian Housing Commission, *316*
Vierendeel truss, *56*
Visco-elastic dampers, 96
Vitreous enamel, 213, 275, 285, 295

Walls, brick, 272, 274
    cavity, 272, 340, 341
    concrete, 31, 100, 282, 315, *316*
    curtain, *see* Curtain walls
    loadbearing, 26, 274
War damage, 98, *99*, 158, 334
Ward-Leonard multivoltage control, 252
Warren truss, 12, *15*
Washdown closet, 210
Washout closet, 210, *211*
Water-cement ratio, 40, 351
Water closets, 207, 208, 209, 210, 211, *211*
Water heaters, solar, 248
Water penetration, 298
Water supply, 6, 205, *208*, 211, 338
Weatherometer, 297
Welsbach mantle, 245, 256
Wind bracing, 23, 24, 25, 27, 62, 75
Wind loads, 22, 23, 61, 91, 95, 184, 187, 274, 333, 334, 335
Windows, 219, 225, 226, *231*, *232*, *233*, *235*, 248, 259, 276, 312
Wind tunnel investigations, 96
Wood, *see* Timber
Wool, mineral, 280
Working loads, 90, 91, 92, 351
Wrought iron, 12, 18, 22, 24, 26, 32, 36, 37, 90, 100, 197, *198*, 310, 351

Yield point, 39, 101, *102*, *103*, 351
Young's modulus (modulus of elasticity), 38, 47, 59, 103, 348

Zinc plating, 281, 285, 310